INTERNATIONAL SCHOOL OF FUSION REACTOR TECHNOLOGY
"Ettore Majorana" Centre for Scientific Culture

TOKAMAK REACTORS FOR BREAKEVEN

A CRITICAL STUDY OF THE NEAR-TERM FUSION REACTOR PROGRAM

Erice-Trapani (Sicily), September 21 – October 1, 1976

Edited by: Dr. H. Knoepfel
Centro Gas Ionizzati, Frascati (Roma), Italy

Published for the
COMMISSION OF THE EUROPEAN COMMUNITIES
by
PERGAMON PRESS

OXFORD–NEW YORK–TORONTO–SYDNEY–PARIS

U.K.	Pergamon Press Ltd., Headington Hill Hall, Oxford OX3 0BW, England
U.S.A.	Pergamon Press Inc., Maxwell House, Fairview Park, Elmsford, New York, 10523, U.S.A.
CANADA	Pergamon of Canada Ltd., 75 The East Mall, Toronto, Ontario, Canada
AUSTRALIA	Pergamon Press (Aust) Pty. Ltd., 19a Boundary Street, Rushcutters Bay, N.S.W. 2011, Australia
FRANCE	Pergamon Press SARL, 24 rue des Ecoles, 75240 Paris, Cedex 05, France
WEST GERMANY	Pergamon Press GmbH, 6242 Kronberg-Taunus, Pferdstrasse 1, West Germany

Published for the Commission of the European Communities, Directorate General Scientific and Technical Information and Information Management, Luxembourg

EUR 5735 e

In order to make this volume available as economically and rapidly as possible the authors' typescripts have been reproduced in their original form. This method unfortunately has its typographical limitations but it is hoped that they in no way distract the reader.

First Edition 1978

British Library Cataloguing in Publication Data

International School of Fusion Reactor Technology, Erice, 1976
    Tokamak reactors for breakeven.
    1. Tokamaks – Congresses
    I. Title   II. *Knoepfel*, Heinz   III. Commission of the European Communities
    621.48′4   TK9204   77–30255

    ISBN 0–08–022034–7

Printed in Great Britain by Unwin Brothers Limited, Old Woking, Surrey, England

ISBN 0 08 022034 7

CONTENTS

TECHNOLOGY (cont.)

STRATEGY

INTRODUCTION

In my experience, sometimes a scientific paper is born by just writing the title on a white, empty sheet of paper, and then subsequently working in slowly all the data, ideas, and what have you, that are scattered on laboratory log books, computer print-outs, memos and similar writing places. Sometimes, on the other hand, one mingles about the title up to the very end, just shortly before the manuscript is actually sent on its way to the editor.

This Course clearly started in the first way. There was a bold short title *"Tokamak Reactors for Breakeven"* with only a small margin for doubts or for immagination left open by the attribute "breakeven". Some people thought that the title was too bold and misleading, and they wanted a question mark (or two, in the Spanish way) added. But, have you ever seen a school program with one or two question marks? The solution out of this title dilemma was the subtitle: *"A critical study of the near-term fusion reactor program"*.

In retrospect, the course evolved along the lines I wished it would when the program of the lectures was set. I sincerely hope that it also has lived up to the expectations of the participants.

In an unusually uninhibited atmosphere, where discussions dominated the scene, all the possible aspects concerning the future Tokamak line of research were analized and debated. I thought it useful (and a typical expression of the atmosphere at the Course) to include with the ordinary papers in these Proceedings on the following pages also some lighter, but not necessarily less meaningful material presented at the Course.

Some of the interesting particularities and also contradictions in this area of applied research became evident, especially during the final discussion sessions. The discussions on *"Plasma Modeling"* and *"Technology and Systems"*, although with some strongly diverging opinions, followed more or less orderly patterns and could, therefore, be resumed by the respective chairmen.

The discussion on *"International Collaboration"* was much more controversial; in fact, some of the participants thought it was not good, others considered it a useful exercise to "cool down tensions".

1

The discussion excerpts published in these Proceedings clearly show, however, that the fusion community is realistically aware of the many important implications of the future fusion research programs.

I think it was clear to all that fusion has now reached a level of political attention and funding (particularly in the United States) that makes an aggressive and goal oriented management necessary. Only gradually the community is becoming aware of the fact that fusion research and development, if it finally may be carried out to success, will most likely surpass in future consequences and in technological and financial dimensions, for example, the NASA Apollo program.

But on the other hand, there is more to aim at, namely a scientific and intellectual motivation for this activity that will make it possible, among others, to survive inevitable pauses or drawbacks during its progress. After all, there is in fusion oriented plasma research and development a challenge of new frontiers in physics, of new numerical and theoretical working methods, of sophisticated diagnostic systems and of advanced technologies that in its totality is hardly matched by any other field of human endeavor towards the new and the better.

The organization of the scientific program benefitted from the many useful suggestions and benevolent criticisms of some of the members of the Scientific Committee, namely Messrs. D. Breton, R. Carruthers and K.H. Schmitter, and of the Director of the Fusion Reactor Technology School, Mr. B. Brunelli.

This Course would not have been possible without the very active collaboration of Mrs. Mariuccia Fiorini, who has managed, before and during the meeting, with enthusiasm and in an independent way nearly all the organizational problems. It is also a pleasure to acknowledge the charming and effective help received during our stay in Erice by Miss Anna Maria Pacifici and the useful assistance provided by the permanent secretariat of the "E. Majorana Centre", namely Miss Pinola Savalli and Mr. S.A. Gabriele.

Finally, I want to express my sincere appreciation to Mrs. Stalpaert of the Directorate General of Scientific and Technical Information and Information Management of the Commission of the European Communities for her competence and care in organizing the publication of these proceedings.

Heinz Knoepfel

DISCUSSION

DISCUSSION SESSION - "PLASMA MODELING"

Chairman: R.J. Bickerton, Culham

Two discussion sessions were held at which the present
state of the art was covered under the headings of Containment, Limiting β, Impurities and Heating, then followed by
a discussion of the future strategy. The points made included
the following:

CONTAINMENT

Mechanisms of anomalous loss in existing machines are
now partially understood, recent measurements show drift waves at
about the right level; but such fluctuations have been seen
before in multipoles and eventually proved to be uncorrela-
ted with loss (Furth).

The multi-regime computer code gives good agreement with
ATC and Pulsator (Duechs) but fails for extreme cases such as
Alcator. The theory behind the multi-regime model is primiti-
ve and particularly for the trapped ion mode should not be
taken seriously (Coppi).

Alcator results can be explained on the basis of current
driven drift waves (Coppi) as can recent results from the
Culham Levitron and Torsatron machines (Bickerton).

The temperature scalings of confinement is unknowable
at present because of the coupling between confinement and
ohmic heating. A major advance in understanding is to be
expected when experiments are due in which auxiliary heating
(e.g. neutral injection) dominates the ohmic power input
(Furth). Confinement in new larger machines might be quite
different from the present ones due to the formation of
complex magnetic topology (Rebut).

LIMITING β

The function of the plasma is merely to illuminate the
first wall of a fusion reactor with an energy flux of the
order 10 MW/m^2; to do this in small size, and cheaply the

plasma β must be higher than 10% (Carruthers).

The proposal to increase the β by shaping the cross-section would be difficult because of control problems if the elongation was large; general opinion was that a valuable gain in β by factor ~ 2 using mild shaping b/a ~ 1.5 was the best that could be expected.

The proposal by Furth to stabilize the resistive m = 2 mode by increasing the current density at the q = 2 surface was thought to be theoretically correct (Coppi) but technically difficult. At higher temperatures, resistive modes should be more stable due to ion-ion collisions (Coppi). Thus it should be possible to reduce q at the wall before disruption in hotter systems.

The proposal to increase β by the so-called *flux conserving tokamak* scheme was heard for the first time by many of the participants and was discussed in some detail. Points included: Maintaining the steady state might require non-physical particle sources and plasma resistivity distribution (Bickerton); there was no limit to the β available; values up to 40% had been computed; the poloidal beta scaled like (toroidal beta)$^{1/3}$ (McAlees); there would be a limiting β corresponding to the local particle diffusion velocities reaching the speed of magnetic field diffusion (Pfirsch); anomalous particle losses observed in present experiments require a large effective perpendicular resistivity; this might reduce the lifetime of this transient system (Rebut); the MHD stability against ballooning modes concentrated on the outside of the torus needed investigation (Furth); during the build-up from low β the toroidal current had to increase and the toroidal flux be conserved (Clarke).

IMPURITIES

There is an urgent need for a complete classical theory of impurity diffusion covering all the collisional regimes and including many species (Duechs). There was general agreement that the present experimental evidence is confused, on occasions impurities are apparently ejected by instabilities, or held outside by temperature gradients. There is no real understanding, but this confused situation is more optimistic than was the simple gloom at the 1974 Culham Fusion Reactor meeting. Main hopes rest on divertor experiments, on careful choice of limiter materials and on the reduced surface to volume ratio of large machines.

HEATING

Neutral injection heating has been an outstanding success story; several machines have now run with the auxiliary power input approaching the ohmic level. Ions continue to behave classically (Furth, Clarke, Sheffield). The electron temperature shows a puzzling reluctance to increase; this could be an ominous sign, but the experiments to date are marginal and could be explained, e.g. on TFR, by arguing most of the electron heating would take place (theoretically) near the wall (Sheffield).

The role of RF heating was not discussed.

FUTURE STRATEGIES

The general question - We don't understand present machines, so why build larger ones ? - was answered by noting the non-linear nature of the key phenomena and the necessity therefore of direct test of the confinement of near thermonuclear plasmas in apparatus of near reactor size (e.g., JET, TFTR, etc.).

A proposal by the Chairman, that the physics problems were so severe and basic that top priority should be given to their solution with the absolute minimum of new technology, was not well-received. The general view was that vigorous programmes on superconducting coils, radiation damage, etc., should be supported even though there is a finite chance (judged small by the majority) that the Tokamak might not prove satisfactorily in the long-term. The main outstanding physics questions of containment, plasma beta, impurities and heating would be answered one way or another by 1980 (Clarke). Consequently now was the time to design the next -stage, a serious tritium burner. The Chairman was in a minority (of one?) in thinking that these questions would barely be answered by 1985. Such *doubting Thomas's* were encouraged to stop rocking the boat, but to sit down and to start rowing.

7

DISCUSSION SESSION - "TECHNOLOGY AND SYSTEMS"

Chairman: J.B. Adams, Geneva

The Chairman opened the discussion by presenting a simplified diagram relating Systems and Technologies to current experiments and to the ultimate Fusion Reactors (see Figure). By Systems is understood the different components of an experiment or a reactor. At the broadest level the Systems include:

- The Plasma System: the system of particles with certain characteristics both individual and collective.

- The Confinement System: in the present context, the magnetic configuration together with its electrical supplies and cathodes.

- The Energy Conversion System: the system to extract energy, and transport and convert it finally to electrical energy.

- The Fuel System: including the generation of new fuel, and the filling and emptying processes of the device.

- The Safety System: the system to ensure safe operation of the device and to protect the local population from any unforseen malfunctions of the device.

- etc.

Each system can be broken down into subsystems: for example, the Confinement System can be broken down into Magnetic Coils, Power Supplies, Vacuum Chambers and Pumps, etc., and the systems and subsystems require for their construction many specific technologies. Finally, in an experiment or a reactor, the systems and subsystems interact together, and at the design stage these interactions impose compromises on permissible solutions and ultimately affect the operations performance of the device. Clearly, many of the systems and subsystem existing in present day experiments will also be present in future reactors, but the reactors will require additional systems not needed in experiments.

The Chairman pointed out that there seemed to be two rather different approaches to the development of systems and their required technologies. Either they could evolve with the experiments and new ones be added as the experiments evolve slowly into the ultimate reactors, or systems specific to the future reactors could be studied now in parallel with current experiments. The first method was probably the most economic in that no unecessary systems or technologies would be generated, since they would only be sufficient for the experiments in operation or being designed. On the other hand, the second method had the advantage that Reactor System Studies could help guide the next generation of experiments and point to future problems.

In the discussion it was pointed out that the first method - which coul be called the evolutionary approach - was only valid if present day experiments were really aiming at future reactors. Several people had claimed during the Sessions that current experiments were not actually aimed at reactor conditions, and many were studying phenomena which would never occur in future reactors. In these circumstances, present day experiments were very misleading as a guide for the development of systems technologies.

In favour of the second method - which could be called the planned approach - it was claimed that Reactor Systems Studies could be used to involve Electricity Utility Organizations and manufacturing industry at an early stage in the development of fusion energy. This brought new ideas and different people into the work and not only were such studies valuable in foreseeing future problems, which could then be tackled experimentally and theoretically, but also this often revealed that some of the current worries about fusion reactors were unjustified. Furthermore, the studies has highlighted certain technical problems, such as materials problems and plasma-wall interaction problems, which would take a long time to resolve and hence should be started in parallel with the present day confinement experiments in order to reduce the time to a fusion reactor.

It became clear in the discussions that the volume of Reactor Studies and reactor technological development work was much higher in the USA them in Europe. There were extensive plans for Materials Testing Facilities in the USA and much involvement of Public Utilities and industry in Reactor Systems Studies. The Chairman asked whether this situation was the result of a conscious decision by the European Laboratories or whether it had happened by default.

Subsequent discussion showed that the American effort was part of an overall plan to reach a fusion reactor by a certain date under well defined assumptions. These plans are summarized in documents ERDA 76/110/0 and ERDA 76/110/1. No such overall planning was reported for western Europe and this was regretted by some speakers who felt that the present rather light coordination of the National Programmes in Europe was insufficient for such an important development effort, which might prove vital for the energy supplies of Europe in the future.

This criticism was further elaborated in the subsequent discussion when some speakers pointed out the need for a single European center for Fusion Development where not only the larger experiments, such as JET, could be built but also teams of physicists and engineers could be built up to construct large apparatus, and carry out Reactor Systems studies and undertake other technological developments. It was claimed that the diffusion of the European effort at the present time led to sub-critical teams in the laboratories.

However, other speakers pointed out that technological studies can easily grow out of control and absorb a great deal of money at the expense of the experiments at presents under way and being planned in the National Laboratories.

11

It was pointed out that even in America long·term studies,
which include Materials Studies and Reactor System studies,
were only consuming a few percent of the annual fusion budgets.
However, the figures of the annual budgets for the American
programme were not only larger than European ones in the current year but planned to be doubled next year and increased substantially thereafter, whereas the European budgets were planned to
remain at their present level for the next few years (see for
example ERDA - 76/110/1 page 29 for the American Budget fore-
casts). In these circumstances it is hardly surprising that
the money left over from the experimental programmes for long-
-term technological developments in Europe is much less.
Nevertheless, the problem of sub-critical technical teams in
Europe remains.

Several speakers raised questions about specific technologies, either for future experiments of for ultimate reactors. One such technologies was super-conducting coils. Some
experimenters said they were not necessary even for new large
experiments such as JET, and would only complicate the experiments unecessarily. Although a reactor would need super-conducting coils,there was time to develop these in the future.
However, no plan to do so was brought forward in the discussions.

The general impression left after these discussions was
that the American effort was highly organized and very well
planned by ERDA in collaboration with the large American Laboratories and the nuclear fusion community in the USA.
Perhaps, as a result, it was much better supported financially than the European effort,which in comparison was much less
organized and lacked an overall planning geared to clear target
dates which were accepted by the European fusion community
and the National Laboratories.

Although this contrasting situation in America and
Europe may not much effect the scientific results obtained
from the present day experiments which, judging by the contributions to scientific conferences by American and European

teams, are not so unequal, it was the opinion of some speakers that the present balance could not be held in the future when experiments became larger, required larger teams of experienced constructors and when the problems of reactor design became pressing. Other speakers, on the other hand, felt that it was premature to launch out technological studies addressed to reactors, and some doubted the wisdom of starting larger scale experiments in Europe at this time preferring to continue with experiments of the present sizes.

It could be concluded,therefore,that the discussions on Systems and Technologies were rather inconclusive, but many points were brought up which merit further thought by the European Laboratories involved in this work.

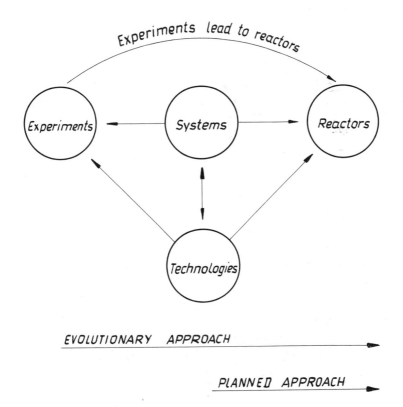

EVOLUTIONARY APPROACH

PLANNED APPROACH

DISCUSSION SESSION - "INTERNATIONAL COLLABORATION"

Chairman: *C. Salvetti, Rome*

<u>SALVETTI</u>:... Introducing this discussion meeting my first question is: do we really need fusion? And, if we need it, why do we need it?

... The question of the motivation, I think, is particularly important in Europe, where big efforts are being made in different directions, both on national levels and on the European level. For instance, breeders are just one direction in which several European countries are strongly committed. Since Europe has much more limited financial resources than US, this question of motivation, according to me, is rather important.

Another point is to see the degree of involvement of scientists, of engineers, technologists and, why not, of industry in fusion.

I would leave aside for the time being the utilities, since they have to face short and medium term problems.

... Passing now to the problem of European collaboration: Do you think that going on with the JET project (and assuming a positive decision) will be enough, or should we do more? And, if we should do more, what should we do? And when? These are very important questions.

... I think that in the future the "Consulting Committee on Fusion" (CCF) may play a considerable role in promoting fusion in Europe and in strengthening European collabo-

---

Editors Note:

Due to the complexity of this double discussion session, no resumé could be prepared by the Chairman, as in the previous two sessions. Instead, a selection of the most relevant discussion points is presented, which has been made under the sole responsibility of the Editor on the basis of taped records. Although the contributions have been shortened substantially, the original phrasing has been maintained; particular care was taken not to alter, whenever possible, the overall content and meaning of the discussion.

15

ration in this field. For the moment being,CCF has been taken only by the problems connected with the JET project.

... Going over to the broader international collaboration, I think that it would be worthwhile to see what can be done in the field of a collaboration between Europe and the United States.

... Well, as I told before, I gave only an example of items that we could maybe discuss and you are absolutely free to add anything you want in this field of international collaboration.

KNOEPFEL: ... We have heard during this week that JET may eventually cost 250 million dollars. We know that TFTR has been projected for 250 million dollars, but if you are in the States, you hear any figure upwards from that.

Now, the European Program in the next years is about half the American one. And my question is really the following: Is the talk about JET realistic in this scenario? I mean, has it a space financially speaking? Or, alternatively, how realistic is it to expect that Europe increases its financial commitment to fusion?

GRIEGER: It's a very difficult question you pose. First I think, up to now, people stick to the official amount of money, at least for the current five-year program and, as far as I understand, there is no provision made at present to exceed this figure.

KNOEPFEL: I was not so concerned about figures, but about a subject which has been, I think, one of the central arguments and concerns here, namely that fusion lives with a certain intellectual mobility and if JET would be drying up all the money (and it should probably within this scenario) we would just kill this mobility which goes along with it.

REBUT: I think that I have to answer to this question. It seems to me that 135 MUC is quite a realistic figure for JET, including the definition of the unit of account, which is 50 Belgian francs at the value of March '75. We are asking five years to put JET in what is called *basic operation* and the 135 MUC correspond to this basic performance, which does not include the tritium phase, nor part of the remote handling (the specific remote tools, which are going with the tritium) and does include only a part of the neutral injection. There is something of the order of 35 MUC, which are for the unforeseen; I don't think that in fact we'll exceed this value if there is a possibility of following the decay of the money.

COPPI: ... We have not solved the fusion problems yet and, frankly, I have many doubts that the large-volume devices will be the answer to a fusion reactor. But it's a personal point of view. Nevertheless, I think JET is valuable in the sense that you set up motor generators, you can set up, for example, a computing centre and a centre for people to go to and discuss and meet.

Now, JET will not be the final experiment. So, the atmosphere with which this centre should be set up is with this long-range point of view. So, before concentrating on a machine, which is an accident (it will be changed in any case), one should really see that there will be the necessary intellectual environment there.

SALVETTI: Just a comment by reflection: this argument of the center and its intellectual environment can be that of the egg and the chicken: who came first?

CARRUTHERS: Well, people have been turning here to second order questions and nobody has looked at the first one. And, in the absence of an answer to that, the rest

17

are almost irrelevant. Until we have decided whether we really need fusion, and if so why, the questions about a wider
involvement and further cooperations are almost irrelevant.
So, how do we decide whether we really need fusion? ....

... Fusion is going to be an energy resource and I am
not at all sure how we establish a need for it without looking at it in an energy picture.

Somehow you have to see whether it fits into the scenario with fossil fuels running out, thermal reactors, fast
breeders . You see, it is difficult to think of a world for
fusion in the abstract. And once you start folding it in into
this other pattern, one finds it difficult to avoid the need
for some demonstration of credibility around the turn of the
century.

... But, let me comment on some of the others points.
I do not fully share Dr. Grieger's enthusiasm for the IEA,
I'm afraid. I find it tends to intrude on other already
working channels of cooperation, where there were direct
arrangements; there were things, which could be done through
the IAEA and in many ways it is forcing cooperation on activities where the programs of the cooperating bodies have not
been clearly established.....

ADAMS: ... Let me comment about the competition between fusion reactors and the breeders. The time score between these
two are very different: I would say we're at least 20 or 50
years away from the breeder with fusion reactors at the moment. And it is difficult to have a competition with something that is 50 years away from you.

But, the two have the following similarity: both aim
to be more or less independent of fuel supply.

... The only thing that seems to me where the breeder
may fail, assuming there is no major technological snag when
one gets to the next stage, is that, in some way or other,
it becomes socially unacceptable, and this could come from
various reasons. It could come from the present problems
about plutonium or waste products. It could also come from

18

irrational fears of the population, and things like that.

... I think society at large, the world at large, should at least afford one other source that is not too widely uncompetitive with the breeder. And this would seem, to me at least, the argument for going ahead with the fusion reactor, with the development work leading towards it.

ZALESKI: Well, I think I agree almost completely with what was said before on breeders....

I think you have to strive for a solution, which is better from an environmental point of view and not too bad from the economic point of view....

TOSCHI: I hear quite often talking about CERN, as an example of collaboration, and some people dreaming that a fusion center would come into operation one day like CERN. I think this is a dream, because there are basic differences. CERN is an excellent facility; JET is an experiment. The facility available at CERN is several orders of magnitude better, larger than any national facility. It is obvious that CERN attracts good scientists from every country.

... It seems to me that the real interest in JET is potentially in the laboratory, which is candidate to house JET, because it will become mostly the experiment of that laboratory.

But all the other associations seem very much preoccupied to have their own program with which they could live independently of JET, or at least being complementary to JET.

We have experienced this sort of attitude in forming the JET team. I mean, out of the 750 professionals that are around in Europe, we could hardly collect 45 people.

... So, it seems to me that hoping for a collaboration in fusion through JET, as it is at CERN, is a pure dream, which I think we should forget about.

And I have another question now to the breeders. I wonder if, looking back, one would consider the failure of Euratom in this field as a regrettable thing or something that would have happened in any case....

ZALESKI:... Euratom failed to make a fully coordinated program; there are many reasons for this. One of these is what I mentioned before, that when you are talking about important commercial programs, you have to have some kind of leadership of one nation and Euratom was not really a nation; they have not a sovereignty ready to do it. So,they had not any power to do it and they embarked in some kind of thing that they were not able to achieve.

 ... And now you are asking if it is a pity. I think, from the general point of view, yes.

 ... I would prefer that Europe would be stronger and better able to make a coordinated program...

GRIEGER:Cooperation inside the community is difficult.

 ... It is very difficult for Brussels to put pressure on the associations, but in principle it is possible....

ADAMS:... The CERN collaboration, compared with the reactor collaboration, is obviously a very easy situation; there was no commercial interest involved; there was indeed quite a strong political interest at that time, but the political interest was to see whether European collaboration would work; I mean, it was a very positive interest. ...

We needed a large device, too big for any member State to build in order to start it and launch it,and we launched it at a time which was very favourable in Europe; only about 5 years or 8 years after the end of the war. The situation now in Europe is very different for launching new things, new collaborations.

For all these reasons, because in the fusion field the commercial interest at the moment is relatively low, it seems to me rather essential to get in fairly soon and do it.

 ... You cannot just decide "we will have a common lab, we will have a collaboration", unless it is pinned on some project, which will focus people's attention; it should be something which the individual countries decide that they cannot or do not want to afford individually. And maybe

20

JET is that.

But, coming to the JET case, and pursuing the argument
that you were just bringing up, you can go on logically to
show you should not start JET at all. If, as you say, only
the house laboratory of JET is really interested in JET, then
the majority of labs in Europe are not, and hence it does not
matter whether it is done in the house laboratory or they all
watch Princeton build the TFTR and get the results from there.
So, why do it in Europe at all? ...

KNOEPFEL: ... There is at the moment a formal collaboration
between the United States and Soviet Union, there are some
bilateral collaborations between European countries and the Soviet Union, but there is no formal collaboration between Europe and the United States. My question is: Should this situation be improved, and what benefits in the long range could
one expect for the partners?

DEAN: ... I think that we really have right here, and we
always have had, such good personal contacts and communications directly between each of the European laboratories and
each of the US laboratories, and such good communications between the heads and scientists of these laboratories that I do
not personally see it as a great detriment to the progress
of fusion if we do not have a formal collaboration.

KULCINSKI: Dr. Dean, do you see any possibility of a joint
machine or a joint facility between the US and Europe? I'm
thinking in particular of some materials testing facilities,
or tritium facilities, or other facilities in auxiliary technology areas. There has been some concern in the community
that we do not have enough money to build both physics devices and technology devices in the individual countries, but
maybe if we pool our efforts, some of these might be built
on a collaborative effort. I haven't seen any proposals
coming in a concrete form, but it seems to me that, if you

go to an expensive neutron-producing facility, you are talking about large sums of money. It could be beneficial to do this together, and then you might have to have some formal agreement.

DEAN: I think that's a useful thing to pursue myself and I think that is something that should be loooked at. I don't see any reason why we can' do that kind of thing ....

But let me ask,for example, if an offer would be made to Europe to pay half of the cost of TFTR in return for half the operating time: Would there be an interest in this kind of collaboration? I am not proposing it, but it is conceivable that someone could be interested in doing something like that....

$$=°=°=°=°=°=$$

ADAMS: Well, let me start by trying to say a few words about the discussion of yesterday. I think the Director's remark that we seem to be singularly unprepared to discuss some of these difficult problems on subjects like JET, or how we develop technology and systems in Europe, is correct.

It seems that in the United States they are more prepared, more organized if you like, to discuss this sort of matters.

... The feeling I got after this discussion was that the present situation in Europe, where we have this structure of independent labs held together through the Brussels Organization, may be is not really strong enough to face up the building of a device, which is too big for any of the national labs to build singularly. And I think that the problem that many people spoke privately to me after the meeting is about a Centre,which would not only be a scientific centre, but a centre where the engineering effort could be brought together into a strong enough force to build things of the size of JET or whatever follows it.

... One point I'd like to raise, not really à propos of JET in particular, is the experience I had when we were trying to get approval for the 400 GeV machine we are just finishing at CERN,which might be useful to you in thinking about JET..

... Between the original design study of 1964 and what was finally approved there were many big changes made, and these changes were made for two reasons. One was to try to fit them to the current political conditions. In '64 the economy of Europe was very different from what it was in '69 or in '70. I don't need to say in what direction it was going; you all know. And furthermore there were a number of political problems concerning the site and things of this sort, so that the design study had to be, if you like, modified to make it acceptable at the political level. This was one sort of thing.

Secondly, there was the community of physicists in Europe, the high energy physicists, which is much more numerous than the plasma physics community in Europe; these physicists had to be convinced that the design was acceptable to them.

Now this is always a very difficult business, because high energy physicists, like plasma physicists, are scientists with very divergent views as to what they want, and most of them have local interests. Some of them raise their sights to a European level and see a general interest in Europe; these are not usually the majority. A lot of work had to be done in discussions in all countries, at different levels and on the European level, to discuss this project before we could safely put it to governments in the knowledge that there wouldn't be some community of physicists in Britain, in France, or I don't know where, who would say to their representatives and their ministers: "*Oh, we don't agree with this project! We think one should do something else*".

...In the end the whole community of these physicists, with all their divergent views, were agreed solidly that what they wanted was this machine as first priority, independent of what happened at home....

... The last point I would make is in connection with what CERN has done in Europe....

... I think CERN not only provides protons through holes in the wall; it also provides a scientific centre and also the sort of technological centre in Europe, where one can advance machine design, bubble chamber design, detector design, everything; where we have teams of engineers, applied physicists, technicians available to tackle any of these problems, and in sufficient number that we can tackle jobs very much bigger than the JET size.

And in addition, it's the scientific centre, and in addition it is the collaborative centre for Europe with all the other continental areas in the world....

... I'd like to leave you to think now in your own terms, in the fusion field, the way you are, what your needs are. Perhaps this kind of background might help you in the discussion.

BRUNELLI:... My opinion on the relevance of experiments to assess anything about the reactor behaviour is the following: The present and next generation of Tokamaks will be capable of finding the scaling laws useful at least for the strategy of the start-up of the reactor.

This is an important point, especially if Tokamaks will be pulsed reactors. Furthermore, I think that the Tokamak based on the high toroidal magnetic field, or based on the flux conserving concept (as exposed by the Oak Ridge people), certainly will be useful in establishing if it is possible to reach a reasonable value of the power density in the reactor. ...

KNOEPFEL: ... I am trying to take up the provocative introduction of Dr. Adams and my question is: Can we scientifically really pretend that we have unanimity ,or 100% approval, for a project like JET? And, instead of answering that, I would like to ask, for instance the A- merican colleagues, what was the final approval of TFTR by the scientific community in the States?

DEAN: I think we never took a vote, which is a very dangerous thing to do actually, and we did not keep any statistics...

... We went about it in a similar way that Dr. Adams descri- bed, namely: we had a series of meetings, a series of groups to look at the review. These were open sessions, documents were passed around, debates were held, the advocates stood up before large groups of peo- ple and presented their views, they answered questions; independent groups of theoreticians and others reviewed the design and made com- ments on it and we gave everyone an opportunity to express their doubts and fears and tried to answer the questions....

... The final process, of course, after we did all the scien- tificl debate, was that it went to our Fusion Power Coordinating Com-

24

mittee for review and endorsement...

... So, that was sort of the process and I believe ever since there hasn't been any significant opposition to the building of the machine, even though the detailed design of the machine is even at his time still changing.

GRIEGER: I cannot avoid the impression that, at least for the initial phases, it was very much similar with the JET machine; there was quite a lot of discussion in Europe, as far as I see it. What to my opinion is missing,is just the final point; the final point to say really everybody now is going to support this thing and to compromise on this thing....

SALVETTI: ... I think that, as far as the work progresses in the field of fusion, the problems which come out are rather similar to those encountered in the fission reactors, although it is true that if we took an experiment like JET we can take out several similarities with what has been done at CERN. But if I look to the after JET, and if we look in perspective to the future reactors, I think that you have not only to inform the scientific community, but we have to involve health physicists, health authorities, engineers and so on....

... This is something to be kept in mind, because to stress too much the fact that only the scientific community has to accept the fusion program and to support it may create a kind of wrong idea about the development of these things and also, in addition to that, you risk to make a kind of ivory tower out of it.....

BRETON: ... It seems to me that there is a great difference between nuclear physics, high energy physics and fusion.

In nuclear physics each experimentalist uses the machine, but he has his own ideas and can make innovations. In fusion you need more unanimity, the number of experiments is rather well defined; you see what I mean? I think it is very very important to obtain a greater unanimity on the fusion machines than what was the case for accelerators.

ADAMS: If this is true for an accelerator, it is not true for the big detectors at CERN, where you have exactly the same kind of community, very mixed, they are all European teams, they even include Americans and Russians usually, and they have to fix the big detector design and they go through the same mechanisms that you are reporting here.

BERTOLINI: I would like to make some comments about the problem of cooperation in Europe as compared to the United States.

... I think the comparison between Europe and United States is inappropriate for two main reasons. There is no national identity in Europe and, I think, this is one obvious problem. The second reason, I think, is that the experience of Europe and the European industries in new technologies in the past 30 years has been such not to support the hope that European technology would compete in selling goods, and, of course, we have obvious examples: the first jet aircrafts for public transportation, the supersonic Concorde, nuclear energy, etc....

... Now, if you want to succeed in having a competition for example with the Americans, or with others, in the field of fusion energy, we should be confident that we can go all-the-way through the four possibilities: scientific, technological, economic and marketing.

BODANSKY: In this discussion, comparing the European program with the American program, I missed one point, and this is the limited budget. Dr. Dean developed his different logics and he showed us that for these different logics we have to pay a certain price.

I would like to raise a question, whether the European think that the budget, which Prof. Salvetti said yesterday is strictly limited to the level which we have now, is enough to come in a logic which means a certain progress.

SALVETTI: ... Well, look, this is a very difficult answer. Maybe we are still in logic 1 and we are approaching logic 2 now, if JET goes on. We have not a logic 3 for the moment being, as far as I know....

GRIEGER: ... The fusion community has to sit together and to say what they really want to do and what is, in their opinion, the time scale and what support they need, and so on.

This is important, because otherwise it is very difficult to talk to political people. They would like to know what they get for the money.

BEHRISCH:... I think the money is a major subject we have to look at in Europe, especially in comparison with the American program. And I think that we won't really have an increase in money, and so the situation for JET is much more difficult than for the TFTR, because the situation in the US would have been also much more difficult if the money wouldn't have increased so much.

... I think there is a general agreement that we want JET. Of course, there are some objections. And I think there is also some agreement that we want a European centre, or there is a need for a European centre, especially as I see the situation in Germany. But the constraint is really the cost.....

REBUT: ... It seems to me that the problem is not so much the money.

... Take one example: the superconductivity program, which has been discussed for two years, without anybody making a very strong effort to have this program carried on.

It seems to me that it is some aggressivity, which is missing to this community. And generally speaking, I know that to have money is very difficult... but if you are sufficiently convinced that your program is right, you will get the money at the end. I am quite sure of this.

CLARKE: ... Considering and listening to these discussions about what is right and what is wrong with the European program, was very interesting...

... But you are not going to get anywhere without aggressive determination in this particular program. This is not a scientific program in the sense that laboratories have traditionally carried on in plasma physics research. If we are ever going to get to a fusion reactor, you've got to make up your mind to do it and get on with it.

This leadership came about in the US through the one man: his name is Bob Hirsch. There was tremendous resistance to the thing

27

that he tried to do with the fusion program in the US....

... But he believed in it and his belief convinced the people that had the money to supply the necessary resources for it. And I think that is what is required.

This question of leadership has two aspects to it: there is a level of political leadership, you can say; you cannot get anywhere without that, without the money. But then there is a question of leadership on a technical level, that is getting back down to the laboratory level; you've got to have an aggressive leadership at the technical level to cut through these problems and make decisions....

ADAMS: ... In addition to aggressivity, I think, also perhaps self-confidence is necessary.

TOSCHI: What I found very interesting in this discussion are the voices in favour of a common centre . In other discussions we had, at the Committee of Directors for instance, we considered that as one of the major fears: Everybody was very much worried about having a new centre for fusion.

Well, can you imagine having a new fusion centre competing with the present national centres....

KNOEPFEL: Well, thank you Dr. Adams and Prof. Salvetti for having guided these difficult discussions. If there are claims or protests on these discussions, I take all the blame and maybe tomorrow or next week I'm fired.

CENTRE FOR SCIENTIFIC CULTURE

CIREE, CILISY, Sept. 1901[*]

AEROPLANES FOR BREAKEVEN

<u>LONG TERM AIM</u>

Economic transport of passengers across Atlantic
(cheaper than ships)

<u>SHORT TERM AIM</u>

To satisfy "take-off" criterion

$$v \tau > L$$

v = velocity of aircraft

τ = time off ground

L = length of aircraft

[*]   PARTICIPANTS (*anagramme*)

Ben Rockit	Sulki Nick	del ffishe
Knee Flop	Chip FRS	le Rack
See Clam	Chet Mill	le Sin
Titch Mers	Choblonk	da Sam
Rush Carter	Rubi Nell	de Such
Hi Zen	Picop	La Bumpo
Charlie Kek	Brute	La Snow
Chris Heb	Ned A	Boner Tili

FIG. X1 - LAWSON'S CONFERENCE 1901

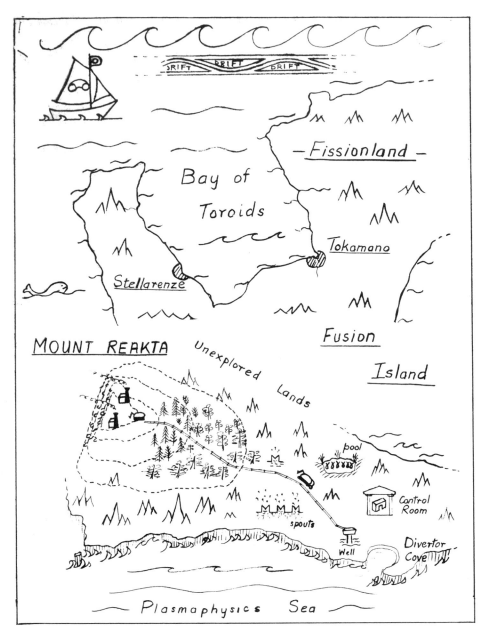

FIG, X2 - MITCHELL'S FUSION ISLAND

According to the reference, the Diet of the Nine Countries of Fissionland sent several expeditions to explore Fusion Land - across the South Plasmaphysics Sea. The records are

30

incomplete......

     After sailing for many years, they sighted - in the distance - Mount Reakta on Fusion Island and knew that the hardest part of their journey was ahead. They made landfall along the First Wall Cliffs - where they saw the terrible Plasmaphysics Waves dashing with thunderous noise, again and again on the Interaction Rocks...... When they had slept, some set about finding water, for it was a dry and inhospitable land...... Then the Explorers were sent off travelling over the Next Steppes, climbing all the time up Mt. Reakta ...... They kept well away from the open steep side of the mountain which was covered with loose blisters, voids and other swellings from the high radiation damage away from the screening woods. There the written story ends......

     "However I had another piece of luck:on the wall of the temple of Venus (2) here in Erice, I found two stones upon which by some curious trick of the light at this place I saw engravings of old maps - reproduced above".

FOR WHICH OF YOU INTENDING TO BUILD A TOWER, SITTETH NOT DOWN FIRST AND COUNTETH THE COST, WHETHER HE HAVE SUFFICIENT TO FINISH IT?

LEST HAPLY, AFTER HE HATH LAID THE FOUNDATION, AND IS NOT ABLE TO FINISH IT, ALL THAT BEHOLD IT BEGIN TO MOCK HIM, SAYING, THIS MAN BEGAN TO BUILD AND WAS NOT ABLE TO FINISH.

St. Luke XIII, VV. 28-30.

FIG. X3 - CARRUTHERS' ADMONITION

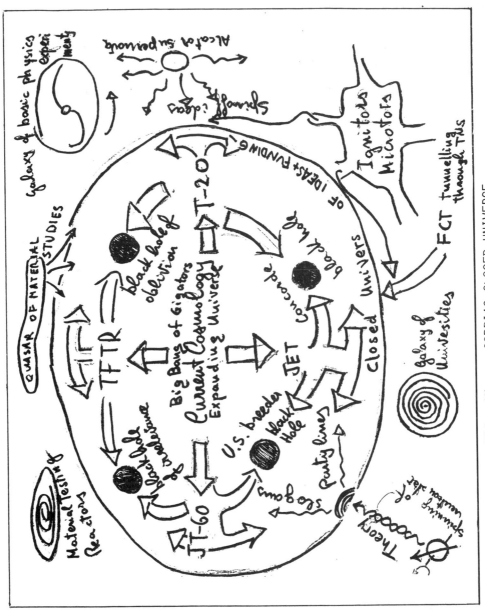

FIG. X4 — COPPI'S CLOSED UNIVERSE

32

FIG. X5 - KULCINSKI'S FUSION WORLD

33

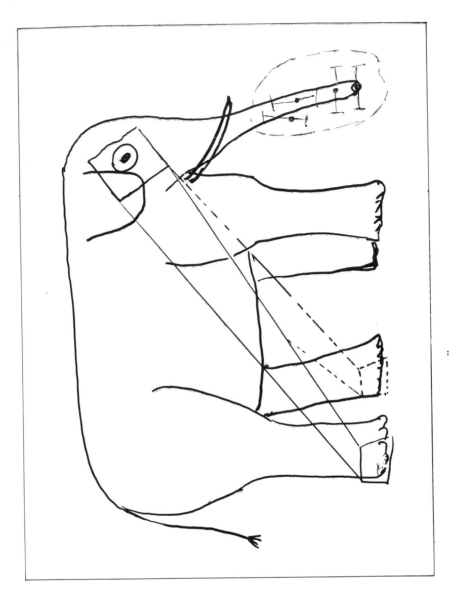

FIG. X6 – DÜCHS'S UNDETERMINED ELEPHANT

PHYSICS

PLASMA-WALL INTERACTIONS

Rainer Behrisch

Max-Planck-Institut für Plasmaphysik, EURATOM Association

8046 Garching bei München, Germany

ABSTRACT

The plasma wall interactions for two extreme cases, the 'vacuum model'
and the 'cold gas blanket' are outlined. As a first step for understanding the plasma wall interactions the elementary interaction processes at the first wall are identified. These are energetic ion and
neutral particle trapping and release, ion and neutral backscattering,
ion sputtering, desorption by ions, photons and electrons and evaporation.
These processes have only recently been started to be investigated in
the parameter range of interest for fusion research. The few measured
data and their extrapolation into regions not yet investigated are
reviewed.

INTRODUCTION

Plasma wall interactions are still one of the most critical but least
understood processes in achieving high density high temperature plasmas.
In todays experimtnts the problem is mostly regarded only from the standpoint of impurity introduction from the first wall, but recycling

37

and refuelling of the plasma particles are even more important areas as about 100 times more particles are involved. In computer simulations the plasma wall interactions are the boundary conditions for the solution of the differential equation. However, due to lack of sufficient data the boundary conditions are generally taken in such a way to reproduce experimental results. Only recently attemps have been made to introduce more realistic boundary conditions /1,2/. In future fusion reactors the plasma-wall interaction is expected to cause additionally an erosion of the first wall which may severely limit the lifetime of the discharge vessel /3,4,5/. Thus it will be essential to quantitatively understand the plasma wall interaction and its determining role for the achievable plasma parameters /6-15/. In the following I will first try to define the problem. Then I will give a summary about the relevant atomic data investigated up to now and finally I will discuss their possible implications for the plasma. A complete picture of the plasma-wall interaction is still missing.

DEFINITION OF THE PROBLEM

In todays tokamak experiments and in future tokamak type fusion reactors the magnetic confinement of the plasma is terminated at the plasma boundary. This is given by the outermost closed magnetic surface touching a limiter or by a magnetic separatrix combined with a divertor. As the magnetic confinement is only weak the plasma is continuously leaking out at this boundary /7/. Neutrals, neutrons and the electromagnetic radiation are not confined at all. The first wall i.e. the solid surface of the discharge chamber facing the plasma is generally at a distance of several centimeters away from the plasma boundary. The first wall is mostly not

smooth but built partly out of bellows and several percent of the surface
are holes. Only at the limiter or the divertor the plasma touches a solid
surface leading to direct plasma solid interactions. This is shown schematically in Fig.1.

Due to the space between the plasma boundary and the first wall the interaction processes of the plasma with the first wall are not straightforward
to define. They depend largely on the density and energy of the ions,
neutrals, photons and electrons which built up in the volume between the
first wall and the plasma boundary. This may be different for different
machines, but up to now it is hardly investigated experimentally. In computer simulations the plasma is terminated at the plasma boundary and the
possible processes in the space up to the first wall are generally only
regarded in order to justify the boundary conditions used /1,16,17/. In the
theoretical treatments of the layer between the plasma boundary and the
first wall on the other hand, fixed plasma parameters and wall conditions
are assumed and thus the influence of the first wall on the plasma cannot
be obtained /18-22/.

Two extreme cases may be regarded for getting a model for the plasma
wall interaction: The 'vacuum model' and the 'high density cold gas blanket'.

THE VACUUM MODEL

Here it is assumed that the density of ions, neutrals and electrons in
the volume between the plasma boundary and the solid surface is so low that
no collisions can take place. Electrons and ions leaving the plasma boundary will predominantly hit the limiter or divertor surfaces with their
full energy. Neutrals injected or released from the first wall will reach

39

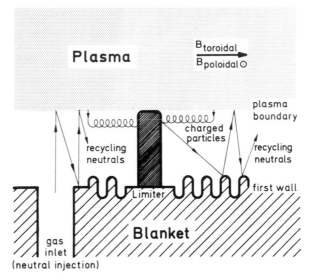

Plasmaboundary near the limiter (toroidal)

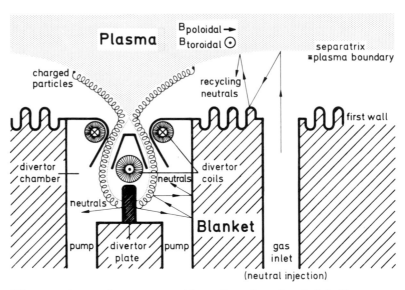

Plasmaboundary near the divertor (poloidal)

FIG.1. *Schematic of the plasma boundary — first wall area for the case of no divertor and with a divertor.*

the plasma boundary with the same energy which they had in leaving the first wall and vice versa. This means that there is no temperature difference for the particles at the plasma boundary and at the first wall. If we further can neglect the flight time between the plasma boundary and the first wall, the first solid wall may as well be shifted right to the plasma boundary and the plasma wall interaction is the same as at the limiter. However the geometrical structure of the first wall generally does not fit the plasma boundary.

In the first attempts to understand the plasma wall interaction mostly the approximations of this model are used /1,2,8-15,23-25/.It is very useful to put up particle balance equations which provide zero order estimates for the particle currents bombarding the first wall.

For steady state operation such particle balance equations give the following mean deuterium and tritium (D,T), helium (He) and neutron (n) currents /13,14,15/ at the first wall.

The deuterium and tritium current:

$$\phi_{DT} = \frac{V}{0} \frac{n_{DT}}{\tau_{DT}} \frac{\eta_{DT}(1-f) + (1-\eta_{DT})A_1}{(1-r_{DT}A)(1-f)+(1-r_{DT})A_1} \tag{1}$$

The helium current:

$$\phi_{He} = \frac{V}{0} \frac{\eta_{He}}{1-r_{He}\eta_{He}} \frac{f}{2} \cdot F_1 \tag{2}$$

The neutron current:

$$\phi_n = \frac{V}{0} \frac{f}{2} \alpha F_1 \tag{3}$$

41

where V is the plasma volume and O the surface area of the first wall,

n_{DT} is the density in the plasma and τ_{DT} the mean particle lifetime of D

and T ions in the plasma. η_{DT} is the fraction of ions passing the separa-

trix which will not reach the divertor (for the case of no divertor $\eta_{DT}=1$)

and f is the fractional burnup with respect to F_1, the D,T throughput[+].

'A_1' is the probability that a charge exchange fast neutral is released

at the plasma boundary for a neutral particle fuelled into the plasma,

while A is the same probability for neutrals reflected or reemitted from

the first wall. 'r'$_{DT}$ is the reemission plus reflection probability for

particles coming to the first-wall from the plasma. In steady state we will

have finally $r_{DT} \simeq 1$. The subscript 'He' and 'n' means the same quantities

for helium and neutrons and apply if fusion reactions take place. 'α' is

the neutron flux enhancement factor at the first wall due to scattering of

the neutrons in the blanket. Neutron flux calculations give values for α

between 5 and 10 /5,6,26/. F_1 is the total fuelling flux of D,T gas.

In a reactor $f \cdot F_1$ will be burnt to He and neutrons while the rest ends up

in the divertor or may be extracted by other means. For constant particle

density in the plasma the total refuelling flux is given by:

$$F_1 = \frac{n_{DT}}{\tau_{DT}} \; \frac{1 - Ar_{DT} - \eta_{DT} \, r_{DT}(1-A)}{(1-Ar_{DT})(1-f)-(1-r_{DT})A_1} \qquad (4)$$

For $r_{DT} = 1$ this gives

$$F_1 = \frac{n_{DT}}{\tau_{DT}} \; \frac{1 - \eta_{DT}}{1 - f} \qquad (4a)$$

(+) In a previous publication of B.B.Kadomtsev and the author /13/, f has
    been defined in respect to the D,T density n_{DT} in the plasma. This
    gave slightly different formulae.

The D,T current, ϕ_{DT}, to the first wall consists of two parts. The first part constitutes of those ions diffusing out of the plasma and for the case of a divertor do not reach the divertor plates and the neutral current induced by them by the recycling process. The second part is the recycling neutral current due to the refuelling flux F_1. If no divertor is used or if the divertor is not very effective the first term will be large, and the second term will be small (screening divertor) while it is just opposite if a divertor is very effective (unload divertor). This means that ϕ_{DT} will always be of the order of $V n_{DT}/0\tau_{DT}$ except that fuelling can be made in a way that A_1 becomes very small as hopefully by the injection of very fast neutrals or large fast clusters on pellets.

For todays plasma experiments as well as for the parameters used in fusion reactor design studies this gives for the D,T current to the first wall values of $\phi_{DT} \simeq 10^{15}$ to $10^{16}/cm^2 sec$, i.e. of the order of 100 μA to $1 mA/cm^2$. The He flux to the first wall will be smaller by one to two orders of magnitude as probably $f \lesssim 10\%$ and $\eta \sim 10\%$. The energy of the particle fluxes will range from a few eV up to several keV with a maximum close to the plasma temperature near the separatrix. There will be a component of higher energy particles near the energy of the injected neutral beams as well as a tail in the distribution of the α-particles ranging up to 3.5 MeV. All particles will have a broad distribution in angles of incidence.

THE COLD GAS BLANKET

In this model the other extreme for the plasma wall interaction is assumed. This means that a high particle density can be built up in the region between the plasma boundary and the first wall, which may be higher than the

plasma density. Ion and neutrals entering this volume suffer many collision and locally thermal equilibrium is obtained. There will be a large
temperature gradient between the plasma boundary and the cold plasma or
neutral gas touching the first wall. This temperature gradient will adjust
according to the energy and particle flux leaving the plasma boundary
as well as the energy accommodation at the first wall. A transition between cold plasma and neutral gas may occur if sufficient neutralization
by radiation and/or tree body collisions takes place. If such a high
density neutral gas and/or plasma layer will build up the plasma boundary may
no longer be determined by the limiter or the separatrix. It adjusts
according to the energy and particle fluxes at the cold plasma-neutral gas
transition.

Several calculations have been performed for this model/6,27,28/ however,
an experimental verification seems to be difficult up to now. The 'vacuum
model' is generally envisaged to be achieved in the divertor experiments
while in experiments with no divertor the cold gas blanket would be most
favourable. In the later case the particle flux to the first wall will be
largely increased compared to equation (1) however, the ions and neutrals
will presumably have an energy below the threshold for sputtering and
damage production at the first wall.

In todays tokamak experimtnts the density in the area between the plasma
boundary and the first wall is not known, but it may be assumed to be of
the same order of magnitude as the plasma density /16/. Further the plasma
boundary is not sharp. Plasma diffuses out of the plasma boundary defined
by the limiter and neutrals from the first wall enter deeply into the
plasma. Thus both models can be applied only with care. For example in a

computer simulation of the plasma obtained in high density discharges in PULSATOR /29/ it is necessary to assume that the neutral gas fed into the discharge vessel at the first wall with thermal energy (< 0.1 eV) enters the plasma boundary defined by the limiter at an energy of ~ 50 eV in order to reproduce the experimentally obtained temperature and density profiles /30/. The dissociation and heating of the gas up to the temperature of ~ 50 eV must have occured in the region between the first wall and the plasma boundary. This occurs presumably predominantly by charge exchange collisions with the ions diffusing out of the plasma.

DATA NECESSARY FOR UNDERSTANDING THE PLASMA WALL INTERACTION

In order to understand these plasma wall interactions quantitatively a large amount of atomic data are needed. These are firstly the elementary processes of the particles at the first wall, and secondly the cross sections for the low energy atomic collisions processes in front of the first wall. In this talk I will deal only with the elementary processes at the first wall due to the bombardment with very low energy hydrogen and helium atoms.

These are:

1) Ion and neutral backscattering, (yields, energy- and angular distribution and charge state of the backscattered particles)

2) Ion and neutral trapping at very high doses, diffusion and reemission of the trapped gas in the solid.

3) Ion and neutral sputtering, (yields, energy- and angular distribution, charge distribution of the sputtered material).

45

4) Other erosion mechanisms by ions and neutrals as blistering and chemical reactions.

5) Desorption by ions, neutrals, electrons and photons. This is of special importance during the start phase of a plasma experimtnt, when the walls are not yet atomically clean but covered with sorbed layers.

6) Evaporation and disintegration due to overheating, nonuniform in space and time.

7) Absorption and reflection of electromagnetic radiation, especially the synchrotron radiation at rough surfaces.

In the following I will mainly deal with the latest results on the points 1 to 4, which are shown schematically in Fig.2. There are only very few

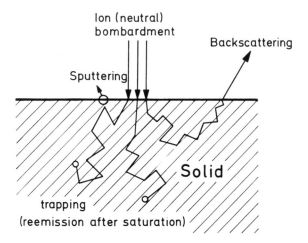

FIG.2. *Trajectories for energetic ions and neutrals impinging on a solid.*

data for point 5 /31,32/ while point 6 has been treated up to now mostly
only theoretically /9/ and nearly no experimental results are available
in the open literature. Point 7 becomes important for the energy balance
of the plasma only in very high temperature plasmas (T $\gtrsim$ 50 keV).

ION AND NEUTRAL BACKSCATTERING

Energetic light ions and neutrals impinging on a solid mostly penetrate
the surface and are slowed down in collision with the atoms of the solid
(Fig.2). The trajectories of the ions are determined by the collisions with
the nuclei while the energy is mostly lost to the electrons /22/. Depending
on the type of ions and their energy as well as the target material some
trajectories will be bent back and thus part of the ions will leave the
surface, i.e. become backscattered. Fast neutrals and ions will behave
in the same way as they all penetrate the solid as ions.

Total backscattering yields and backscattered energy as well as energy-
and angular distributions of the backscattered atoms have been calculated
analytically /34,35/ and by computer simulation programs which follow the
individual ion trajectories /36 to 38/. Experimental determination of
backscattering yields and distributions are difficult, as in the energy
range of interest most of the backscattered particles are neutral. The
most successful way of ionising the neutrals is a gas filled stripping cell,
as used also in neutral particle measurements at plasma experiments /40/.
Such a cell could be calibrated to energies as low as 130 eV /41/.

The results for total backscattering yields and backscattered energy of
hydrogen ions on different materials are shown in Fig.3 /34 to 37,42 to 44/.

47

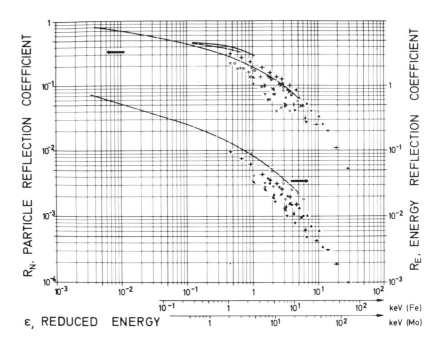

FIG. 3. *Particle and energy reflection coefficients as determined by different authors: o H → SS, ☐ D → SS, ▽ H → Nb (annealed) ▿ H → Nb (bef. ann.), ◇ D → Nb (bef. ann.) (ref. 42), ● H → SS, ▼ H → Nb, ▲ H → Cu, ● H → Al, ◆ H → Mo, ✦ H → Ag, ◣ H → Ta, ✚ H → Au (ref. 43), + H → Zr, ✕ H → Ti (ref. 44),* ——— *ref. 35,* ——— *ref. 36, 37* ————— *ref. 34*

The values are plotted as a function of a dimensionless universal energy ε given by /45/

$$\varepsilon = \frac{M_2}{M_1 + M_2} \cdot \frac{a}{Z_1 Z_2 e_0^2} E$$

where M_1, Z_1 and M_2, Z_2 are the masses and charge numbers of the incident ions and the target atoms, e_0 is the elementary charge ($e_0^2 = 14.39$ eV $\overset{o}{A}$); 'a' is the Thomas Fermi screening length given by:

$a = 0.486 \ (Z_1^{2/3} + Z_2^{2/3})^{-1/2}$ and E is the incident energy in eV. In the two lower scales of Fig. 3 also absolute energies for stainless steel (Fe) and Mo are introduced. Especially for stainless steel the agreement between calculated and measured values is very good, which gives some confidence in the calculated values at low energies where no measurements are available, but which are of special interest in fusion research. The energy distribution of the backscattered atoms for normal incidence

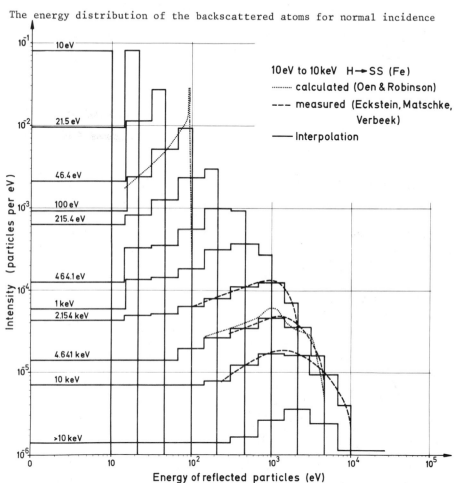

FIG.4. *Energy distribution and intensity of hydrogen atom backscattered from stainless steel. The incident energy is given on the left side above the curves. The measured values are determined at an angle of emergence of 45 deg./42/ while in the calculated values all angles of emergence are included /37/. The histograms are a smooth interpolations between the measured and calculated values.*

49

of hydrogen on stainless steel has been only measured for energies above
2 keV /42/ while for 5 keV and 100 eV a computed spectra for a copper
target have been published /27/. For plasma simulation codes the energy
range between 10 eV to 10 keV is of interest, but only histogramms with
relatively broad energy ranges for the energy distribution of the backscattered particles are needed. Such curves obtained by a smooth interpola-
tion between the known spectra are plotted in Fig.4 together with the
exact spectra. The energy ranges are chosen accordingly to those used in the
Düchs code /16,17/. The intensities give absolute values as backscattering
coefficients are taken into account. For incident energies above 2 keV
there is a maximum in the backscattered intensity at 1 to 1.5 kV, while
at lower incident energies the maximum is close to the incident energy.

As mentioned already most of the backscattered atoms are neutrals and
only less than ~ 5% have a positive or negative charge /42/.

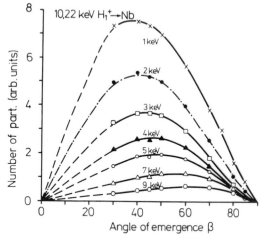

The only example of a
measured angular distri-
bution of the atoms back-
scattered into different
energy ranges for normal
incidence is shown in Fig.5
/46/. While those atoms
backscattered with low
energy show nearly a

FIG.5. *Angular distribution of protons backscattered from polycrystalline*
Nb target into different energy ranges /46/.

cosine distribution, the high energy particles show some predominant

backscatterering at glancing angles (80° to the normal). The distributions

may be different at lower energies but neither calculated nor measured data

are available.

ION AND NEUTRAL TRAPPING AND REEMISSION

Those ions and fast neutrals not directly backscattered come to rest in the

solid. After slowing down they generally occupy interstitial positions and

may diffuse further. Depending on the solubility, the diffusibility, and

the barrier at the surface they may either diffuse into the bulk of the

solid, they may partly leave the surface, or they may be trapped in the

implanted layer, generally at damage sides /47,48/.

In the case of a high solubility as for hydrogen in titanium and zirconium

all ions coming to rest in the solid are trapped if the temperature is

high enough for good diffusion, but low enough so that thermal desorption

is negligible /44,48, 49/. This is shown in Fig.6 for different bombarding

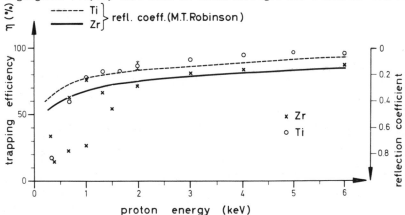

FIG.6. *Trapping efficiency for 0.3 to 6 keV protons bombarding Zr and Ti surfaces (ref. 51).*

energies at normal incidence on room temperature targets /44/. The measurements have been performed by weight gain at ion doses of $3 \cdot 10^{19}$ to $5 \cdot 10^{20} cm^2$. In order to correct for the sputtered material this was also measured by collection on Al probes around the targets, which were subsequently analysed by Rutherford backscattering. As some hydrogen atoms are directly backscattered, trapping is always smaller than one. Backscattering yields determined from these measurements are in good agreement with calculated values (see Fig.3). The scatter in the points below 1.5 keV is attributed to an oxide layer on the surface which may have a different thickness for different targets used. As H does not diffuse well in the oxide trapping is decreased if a major part of the incident gas ions is stopped in the oxide layer.

If the _solubility_ of the incident gas ions in the solids is _low_, as for He bombardment of most metals and for hydrogen bombardment of several metals as for example stainless steel or molybdenum, the ions come to rest at interstitial positions in the solid and they will diffuse until they become trapped relatively stable at damage sites, predominantly vacancies /47/.For different doses this is shown in Fig.7 for ^{3}He implanted into polycristalline Nb /50/. The profiles have been obtained using the ^{3}He$(d,p)^4$He nuclear reaction. However, also several other techniques are possible for such measurements /51 to54/. At the lowest dose the implantation profile follows within the statistics of the measurements mostly the calculated distributions. At higher doses a saturation concentration of∼50 atomic % is reached after which the distribution tends to a rectangular profile within the range of the incident ions.

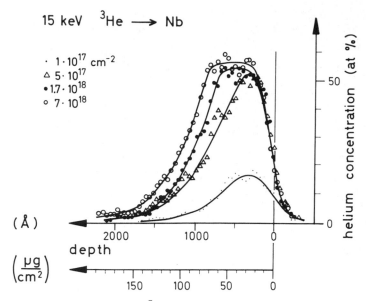

FIG.7. *Implantation profile of ^{3}He in niobium for different bombardment doses (ref.50,56).*

As the range of the ions increases with the incident energy, the total amount of trapped gas also increases with energy /55/. This is shown in Fig.8 again for helium bombardment of Nb. The maximum concentration reached within the range of the ions is independent on the implantation energy, but decreases with the target temperature. The gas emitted after saturation is   presumably released with an energy corresponding to the temperature of the first wall.

ION AND NEUTRAL SPUTTERING

Physical sputtering is the removal of surface atoms from a solid via a collision cascade initiated in the near surface region by incident energetic particles as ions, neutrons or electrons. Thus sputtering can be regarded

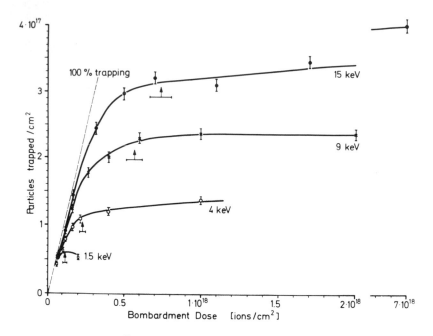

FIG.8. Trapping of ^{3}He in niobium at room temperature as a function of the bombarding dose at different energies (ref.55).

as radiation damage in the surface. During the spread of the cascade the surface stays cold contrary to surface erosion by evaporation.

The sputtering yields (atoms per incident particles) are proportional to the energy deposited by the incident particles in the surface layer in nuclear motion and inversely proportional to the surface binding energy /57/. These yields have been investigated since more than 100 years /57 to 63/, however, nearly no data are available for the parameters of interest in fusion research /3,10,13,14/. This has basically two reasons: The yields for low energy light ion bombardments are low and intense ion beams are difficult to obtain. In Fig.9 and in Fig.10 the sputtering yields of most inte rest in fusion research are summarized. They have been

54

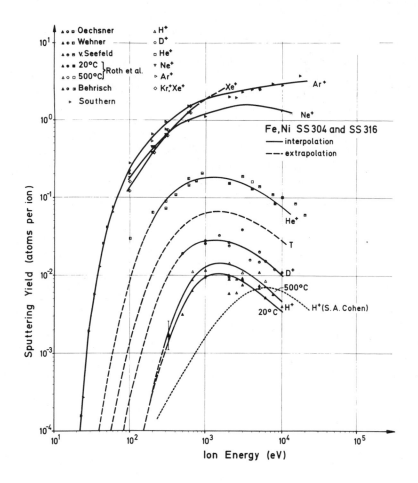

FIG.9. *Sputtering yields for normal incidence of different ions on Fe,*
Ni, SS 304 and SS 316 as measured by different authors (ref.63 to 67,
70). The values used in the plasma simulation of ref.1 (S.A.Cohen)
are also introduced.

measured for normal incidence of different ions on stainless steel,

nickel or iron as well Mo(molten)/63 to 70/. They have been obtained

partly by weight loss partly by measuring the decrease of a thin film

by Rutherford backscattering and partly by activation analysis.

55

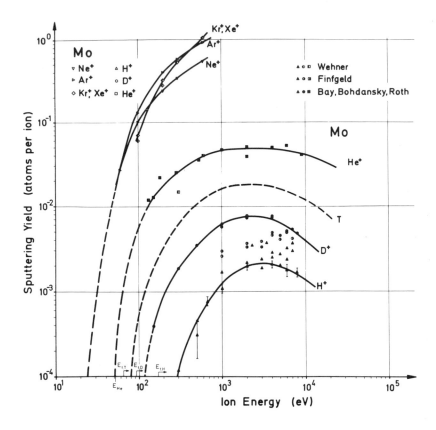

FIG.10. *Sputtering yields for normal incidence of different ion on molybdenum (molten material) as measured by different authors. /66,68,69/. The solid lines are fitted to the experimental points, while the dashed lines are extrapolations.*

The solid lines are a fit to the experimental values and the dashed lines and extrapolation. Below a threshold energy, E_t of 10 to 100 eV no sputtering occurs /71/. Yields then increase with energy to a maximum which occurs for hydrogen at an energy around 1 keV. The decrease at a further increase of bombarding energy is due to the decrease of the deposited energy in the surface region. The difference between $20^{\circ}C$ and

500°C SS is due to different surface compositions at the different temperatures /67/. The dependence of the sputtering yield on the mass of the incident ion is much larger at low energies than at high energies. The expected threshold for D bombardment is 1/2 and for ^{4}He bombardment 1/4 of the threshold energy for proton bombardment. Thus at energies where the sputtering yield for protons is zero the heavier ions as D,T and He still show considerable sputtering.

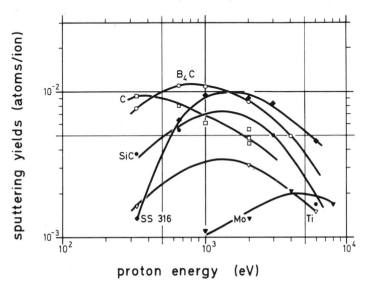

FIG.11. Sputtering yields of different low Z materials (ref.64).

Fig.11 shows sputtering yields of low Z materials which have been discussed as first wall materials in the last two years /64/. For these materials the maximum in the sputtering is considerably shifted to lower energies compared to higher Z materials.

The dependence of the sputtering yields on the angle of incidence has hardly been investigated. The few measurements performed have confirmed the theoretically expected increase of the yields with $(\cos\alpha)^{-f}$ where $1 < f < 2$ and α is the angle of incidence with respect to the normal /3/.

57

The atoms removed from the surface are predominantly neutral. Their energy and angular distribution has not yet been measured for the parameters of interest. It is expected that the mean energy will be low (1 eV to ~ 10 eV) and that the atoms will be emitted in a nearly cosine distribution /57-62/.

Results on neutron sputtering yields have achieved great attention due to extremely high yields of 0.3 atoms/neutron found in some experiments /72/. However, the latest more careful investigation at several places have let to the conclusion that neutron sputtering yields are below~10^{-4} atoms/neutron in agreement with theoretical prediction /73/.

EROSION BY CHEMICAL SPUTTERING AND BLISTERING

Chemical Sputtering

If the bombarding ions can form a volatile compound with the target material as for hydrogen bombardment of carbon, much larger yields than expected from the collisional theory may occur. At room temperature bombardment of carbon with hydrogen only small yields and no CH_4 formation could be measured. At increased target temperature however, CH_4 is found and simultansously a large increase in sputtering. This is shown in Fig.12 for different bombarding energies at normal incidence /74/. The steep increase in the yields at 400 to $500^{\circ}C$ has been explained by increased mobility of H on the surface, so that CH_4 can be formed. The decrease of the yields at temperatures above $700^{\circ}C$ can be understood by increased out diffusion of hydrogen so that the probability for the formation of CH_4 decreases.

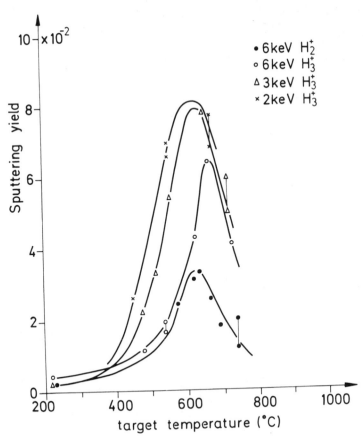

FIG.12. *Temperature dependence of the sputtering yield for pyrolytic graphite for hydrogen bombardment at different energies (from ref.74).*

Similar effects have been found for silicon carbide and boron carbide. However, in these cases the surfaces seem to become depleted of carbon after some dose, and chemical sputtering disappears /15,69/.

Finally, in all compounds and alloys a composition change due to preferential sputtering by ion bombardment is expected and has partly been observed. However, systematic investigations for the parameters of interest in respect to CTR have not yet been performed.

Blistering

If the incident gas ions are not soluble in the material, transmission electron microscope observations have shown that at bombarding doses of $\sim 10^{16}/cm^2$ corresponding to several atomic percent gas injected into the solid, the gas starts to coalesce to small bubbles of 10 to 30 Å diameter inside the implanted layer /56,76/.

If the ion bombardment is increased further, blisters i.e. bending up of a surface layer can be observed /42,56,77-82/. The critical doses for the appearance of blisters lie between 10^{17} to 10^{18} ions/cm^2. They depend on the bombarding energy, the current density and angular and energy spread of the bombardment, as well as the materials and its temperature /79 to 82/.

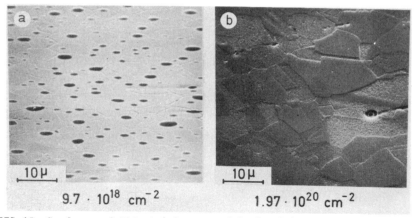

FIG.13. Surfaces of 304 stainless steel bombarded with different doses of 7.5 keV hydrogen ions at room temperature and normal incidence. Blistering is observed only at low bombarding doses (a). At higher doses the surface becomes rough and the differently oriented grains are eroded at different rates (ref.13 and 70).

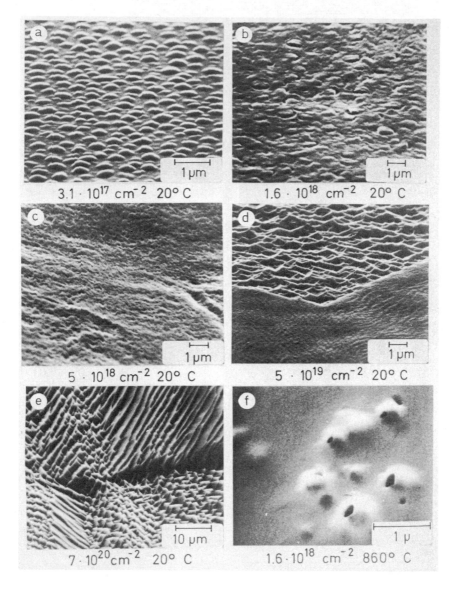

FIG.14. *Surface of polycrystalline Nb bombarded with increasing doses of 9 keV He-ions at normal incidence. It can be seen that blistering is a transient phenomenon disappearing at high doses (ref.84). Fig. 14 f (ref.85) shows a blister kind surface structure at 860° bombardment.*

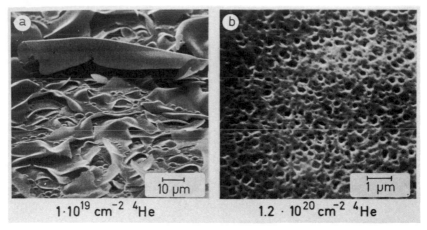

FIG.15. *Surfaces of polycrystalline Nb at room temperature bombarded with different doses of 100 keV He at normal incidence. Blisters disappear after about one deckeldicke has been removed by sputtering (ref.84).*

Blistering and the blistering mechanism has been widely investigated within the last years, because it had been regarded as the most dangerous source for first wall erosion in fusion reactors /56/. It was found that the bending-up of surface layers is basically caused by the release of lateral stress produced in the implanted surface layer /50/. The pressure in the gas bubbles formed by the injected gas in the solid also contributes,but it is not the major driving mechanism.Thus, blisters break up at the depth corresponding to the total range of the injected gas in the solid and not at the mean projected range where the maximum gas concentration occurs /50/.

This blistering mechanism has lead to the conclusion that blistering may be a transient phenomenon which may disappear if the surface adapts to the ion bombardment by forming a structure in which stress can more build

up /13,83,84/. In fact, such a new surface structure develops if more than
one deckeldicke (= thickness of the covers of the blisters) is sputtered
away. Further, blistering is considerably reduced if the bombardment is
performed at a broad energy distribution and/or a broad angular distribution of the incident ions /83,84/. The figures 13,14 and 15 show some
surface structures of stainless steel and niobium with blisters and after
disappearance of the blisters as observed in the scanning electron
microscope. Blisters disappear at sufficient high doses even at energies
as high as 100 keV leaving a sponge like structure. A picture for high
temperature blistering is also included /56,70,79-82/. The temperature
can largely modify the blister appearance. At medium temperatures (500 to
$800^{\circ}C$) blistering generally is increased and large exfoliation of the
blister covers is observed. At a further increase in temperature blistering
is decreased again.

CONSEQUENCES FOR PLASMA EXPERIMENTS   AND A LATER FUSION REACTOR

The few atomic data known about the interaction of energetic ions and
neutrals with a solid surface have now to be incorporated into plasma
simulation codes in order to find which data are the most critical ones
for plasma wall interaction and where more information is needed. However
the following general consequences can already be drawn.

Part of the ions leaving the plasma are directly backscattered from the
first wall into the plasma mostly as neutrals with energies up to close
the incident energy. They may penetrate deep into the plasma and create
energetic charge exchange particles leaving the plasma toward the first
wall. As backscattering increases with decreasing energy, low energy
neutrals (10 to 100 eV) will predominantly contribute to recycling.

63

As backscattering yields and energy distributions change with the first
wall material, this will have a noticable effect on the recycling by charge
exchange neutrals.

Due to the trapping of the implanted ions high concentrations of gas can
be built up in the surface layers of the first wall. The total amount
trapped gas in the first wall may be comparable to the amount of gas
in the plasma

Reemission from the first wall will change with bombardment dose and
target temperature. However, in any steady state operation of a fusion
reactor for reemission together with reflection a factor one may finally
be assumed (except at the holes in the CTR vessel).

The impurity introduction in todays experiments occurs predominatly by
ion, neutral, photon and electron desorption /31/ as the first walls are
generally not atomically clean and the temperature of the plasma near
the wall is too small to cause large sputtering.

At higher plasma temperatures and clean first wall surfaces sputtering by
fast charge exchange neutrals will dominate. The wall bombardment and
surface erosion will presumably not decrease substantially by introducing
a divertor.

The impurity introduction and erosion can only be kept low, if the energy
of the bombarding ions and neutrals can be kept below the threshold for
sputtering of 1 to 10 eV as it may be achieved probably by a cold gas
blanket .

Among the low Z-materials pure carbon may have to be excluded because its high chemical erosion yields.

The other low Z materials show exceptionally high yields at low energies which may compensate the advantage of a higher concentration of low Z materials in the plasma.
The yields for light ions are generally below 3×10^{-2} atoms/ion, while they are of the order of one for heavy ions. Thus heavy ions once introduced and heated in the plasma, which will come back to the first wall with energies above 100 eV, will cause considerably sputtering. Even if the impurity concentration is low, this sputtering may be higher than sputtering by hydrogen atoms and will cause a further fast increase of the impurities in the plasma.

Neutron sputtering yields are low. As neutron fluxes to the first wall will be much lower than the ion- and neutral fluxes the contribution of neutron sputtering to first wall erosion can be neglected /10,91/.

Blistering may possibly contribute to wall erosion in a fusion reactor only at the beginning of the operation, as it is a transient phenomenon which disappers at sufficiently high doses.

The angular and energy distribution of the ions bombarding the first wall may prevent blistering from the beginning. Due to the increased wall temperature an equilibrium structure at the surface may form already after the first few discharges.

Finally, the sputtering yields measured by weight loss generally have been obtained with doses well above the dose for blisters and thus include blistering.

The spongelike structure found after high dose ion bombardment will have different mechanical, thermal and electrical properties compared to the unbombarded material, which has to be taken into account in designing a fusion reactor.

ACKNOWLEDGEMENTS

It is a great pleasure to thank my colleagues especially H.Bay, J.Bohdansky, D.F.Düchs, W.Eckstein, J.Roth, B.M.U.Scherzer, H.Verbeek and H.Vernickel for several discussions about the problems on plasma wall interaction and the relevant atomic data for understanding these questions.

LITERATURE

/1/   D.M.Meade, H.P.Furth, P.H.Rutherford, F.G.P.Seidl, D.F.Düchs,

        Plasma Physics and Contr.Nucl.Fusion Research Tokyo, Proc.5th

        Int.Conf., Tokyo, 1974, IAEA Vienna 621 (1975).

/2/   E.S.Hotston, G.M.McCracken, CLM-P 455, Culham 1976 to be publ.

        in J.Nucl.Mat.

/3/   R.Behrisch, Nucl.Fusion 12 (1972) 691

/4/   B.Budger et al.UWMAK, Wisconsin, Tokamak Reactor Design, UWFDM-68,

        Nov.1973.

/5/   R.G.Mills, F.H.Tenney et al.: Princeton Reference Design Reactor,

        Princeton 1974.

/6/   Fusion Reactor Design Problems, Proc.of an IAEA Workshop,

        Culham, 1974, Nucl.Fusion Special Supplement 1974.

/7/   E.Hinnov, J.Nuclear Mat. 53, (1974), 16.

/8/   M.Kaminsky, IEEE Trans.Nucl.Sci. NS 18, 208 (1971).

/9/   H.Vernickel, Nucl.Fusion, 12, (1972) 386.

/10/   H.Vernickel, Course on the Stationary and Quasistationary

        Toroidal Reactors, Erice 1972, EUR 4999 e, p 303 (1973).

/11/   G.M.McCracken, Fusion Reactor Design Problems, Nucl.Fusion

        Special Suppl.p.471, (1974).

/12/   H.Vernickel, Proc.1st Top.Meeting on the Techn.of Contr. Nucl.

        Fusion, San Diego, Conf.740402-P2, Vol.II, 347 (1974).

/13/   R.Behrisch and B.B.Kadomtsev, Plasma Physics and Contr. Nucl.

        Fusion Research, Proc.5th Int.Conf.Tokyo 1974, IAEA, Vienna II,

        229, (1975).

/14/   B.M.U.Scherzer, J.Vac.Sci.Techn.13, 420 (1976).

/15/ R.Behrisch, "Materials Limiting Problems in Energy Production (ed.Ch.Stein), chapt.4, Fusion First-Wall Problems, Academic Press, 1976.

/16/ D.F.Düchs, H.P.Furth, P.H.Rutherford, PPL, TM-265, Princeton 1973.

/17/ D.F.Düchs, H.P.Furth, P.H.Rutherford, Proc.Europ.Conf.on Contr. Fusion and Plasma Phys.Moscow (1973) p 29 and p 326.

/18/ S.Yano, JAERI-mono. 4243 (1970).

/19/ S.Rehker, H.Wobig, Plasma physics 15 (1973) 1083

/20/ J.Hackmann, Y.C.Kim, H.Reutus, J.Uhlenbusch, LuPPG, Physical Inst.III, Universität Düsseldorf 1974.

/21/ G.Haas, Proc.Europ.Conf. on Contr.Fusion and Plasma Phys., Moscow (1973) p.

/22/ J.Hackmann, J.Uhlenbusch to be publ.J.Nucl.Mat.(1976).

/23/ R.Behrisch, W.Heiland, Proc.6th Symp.Fusion Techn.Aachen, 1970 (461).

/24/ D.M.Meade, Joint European- US Workshop on large Tokamak designs, Culham (1974).

/25/ D.Düchs, G.Haas, D.Pfirsch, H.Vernickel, J.Nucl.Mat. 53, (1974) 102

/26/ Dr.Steiner, Nucl.Appl.Techn.9 (1970), 83

/27/ R.Wienecke, Z.Naturforschung 18a (1963) 1151.

/28/ B.Lehnert, Nucl.Instr.and Meth. 129 (1975) p.31, Nucl.Fusion 13 (1973) p.781 and p.958, and references in this articles.

/29/ J.Klüber et al.Nucl.Fusion 15 (1975) 1194

/30/ D.F.Düchs, private communication

/31/ S.A.Cohen, J.Vac.Sci.Technol. 13 (1976), 449.

/32/ D.Lichtmann, J.Nucl.Mat. 53 (1974) 285.

/33/ N.Bohr, Mat.Fys.Medd. 18, No 8 (1948).

/34/ R.Weißmann and P.Sigmund, Rad.Eff. 20, 65, (1973)

/35/ J.Bøttiger and K.B.Winterbon, Rad.Eff.20, 65, (1973).

/36/ M.T.Robinson, Proc.3rd Nat.Conf.Atomic Coll.with Solids, Kiew 1954.

/37/ O.S.Oen, M.T.Robinson, Nucl.Instr. and Meth.132.(1976) 647

/38/ J.E.Robinson, Rad.Effects, 23, 29, 1974, 677 (1976).

/39/ T.Ishitani, R.Shimizu, K.Murata, Jap.J.Appl.Phys.11, 125 (1972)

/40/ C.F.Barnett and J.A.Ray, Nucl.Fusion 12, 65 (1972).

/41/ F.E.P.Matschke, W.Eckstein, H.Verbeek, Verh.DPG IV, 10, 47
(1975)and to be publ.

/42/ W.Eckstein, F.E.P.Matschke, H.Verbeek, to be publ.in J.Nucl.Mat.
(1976).

/43/ G.Sidenius, Phys.Lett.49A, 409 (1974), and Nucl.Instr. Meth., 132,
673 (1976).

/44/ J.Bohdansky, J.Roth, M.K.Sinha, W.Ottenberger, Proc.9th Symp.
Fus.Techn., Garmisch, June (1976).

/45/ J.Lindhard, M.Scharff, H.Schiøtt, Mat.Fys.Medd.33, No 14, (1963).

/46/ H.Verbeek, J.Appl.Phys,46, 2981 (1975).

/47/ E.Kornelsen, Rad.Eff.13, 227 (1972).

/48/ G.M.McCracken, Rep.Progr.Phys.38, 241 (1975).

/49/ O.C.Yonts, R.A.Strehlow, J.Appl.Phys.33,, 2902 (1962).

/50/ R.Behrisch, J.Bøttiger, W.Eckstein, U.Littmark, J.Roth and
B.M.U.Scherzer, Appl.Phys.Lett.27, 199 (1975).

/51/   R.S.Blewer, Appl.Phys.Lett.53, 593, 1973.

/52/   D.A.Leich, T.A.Tombrello, Nucl.Instr.Meth.108, 67 (1973).

/53/   J.Bøttiger, S.T.Picraux, N.Rud.Ion Beam Surface Layer Analysis,
       p.811   (1975).

/54/   J.Roth; R.Behrisch, B.M.U.Scherzer, Appl.Phys.Lett.25, 643, (1974).

/55/   R.Behrisch, J.Bøttiger, W.Eckstein, J.Roth and B.M.U.Scherzer,
       J.Nucl.Mat.56, 365 (1975).

/56/   J.Roth, Application of Ion Beams to Materials, Inst. of Physics,
       Conf.Series Number 28, p.280 (1975).

/57/   P.Sigmund, Phys.Rev. 184, 383 (1969).

/58/   G.K.Wehner, Ad.Electr.Ectro.Phys.7, 239 (1955).

/59/   R.Behrisch, Erg.exakt Nature 35, 295, 1964.

/60/   M.Kaminsky, Atomic and Ionic Impact Phenomena on Metal Surfaces,
       Academic Press, Springer 1965.

/61/   G.Carter, J.S.Colligon, Ion Bombardment of Solids, Elsvere Publ.
       Comp.1968.

/62/   N.W.Pleshivtsev, Cathode Sputtering, Atomisdat., Moskau, 1968.

/63/   H.v.Seefeld, H.Schmidl, R.Behrisch, B.M.U.Scherzer, to be publ.
       J.Nucl.Mat. (1976).

/64/   J.Bohdansky, J.Roth, M.K.Sinha, Proc.9th Symp.Fusion Techn.,
       Garmisch, (1976).

/65/   H.Oechsner, Thesis, University of Würzburg 1963.

/66/   J.Rosenberg u.G.K.Wehner, J.Appl.Phys.35, (1964) 1842.

/67/ J.Roth, J.Bohdansky, W.O.Hofer, J.Kirschner to be published.

/68/ C.R.Finfgeld, Final report No OKD-3557-15, Contract NO AT-(40-1)
3557, AEC Wash and Ruane Coll.Salem (1975).

/69/ H.Bay, J.Bohdansky, J.Roth, priv.Comm.

/70/ R.Behrisch, J.Bohdansky, G.H.Oetjen, J.Roth, G.Schilling,
H.Verbeek, J.Nucl.Mat. 60, 321 (1976).

/71/ E.Hotston, Nucl.Fusion 15, 544 (1975).

/72/ M.S.Kaminsky, J.Peavey, S.Das, Phys.Ref.Lett.32, 599 (1974).

/73/ R.Behrisch, Nucl.Instr.Meth.132, 293, 1976.

/74/ J.Roth, J.Bohdansky, W.Poschenrieder, M.K.Sinha, to be publ.
in J.Nucl.Mat.(1976).

/75/ J.Roth, J.Bohdansky, M.K.Sinha, private comm.

/76/ G.M.McCracken, see ref./6/ and private comm.

/77/ W.Primack, J.Appl.Phys.34, 3630 1963.

/78/ M.S.Kaminsky, Adv.Mass.Spectrom.3, 69 (1964).

/79/ S.K.Erents, G.M.McCracken, Rad.Eff.18, 191 (1973).

/80/ M.S.Kaminsky, S.K.Das, Rad.Eff.18, 245, 1973.

/81/ W.Bauer, G.J.Thomas, Proc.Int.Conf. on Defects and Defect Clusters
in bcc Metals and their Alloys, p 255 (1973).

/82/ J.Roth, R.Behrisch, B.M.U.Scherzer, J.Nucl.Mat.53, 147 (1974).

/83/ J.Roth, R.Behrisch, B.M.U.Scherzer, J.Nucl.Mat.57, 365 (1975).

/84/ R.Behrisch, B.M.U.Scherzer, M.Risch,J.Roth, Proc.9th Symp.Fusion
Techn., Garmisch, June (1976).

/85/ J.Roth, T.Picraux, W.Eckstein, J.Bøttiger, R.Behrisch,
J.Nucl.Mat.to be publ.(1976).

71

COMPUTER PREDICTIONS FOR FUTURE TOKAMAKS

D.F. Duechs

Max-Planck-Institut für Plasmaphysik,
8046 Garching bei München, Germany

ABSTRACT

Proceeding from a reasonable agreement with existing experimental
results, this lecture presents radial particle and energy transport
computations which extrapolate to large (up to reactor dimensions) future
Tokamaks. Special consideration is given to the behaviour of alpha-
particles, the influence of high-z impurities, and the thermal stability
of the plasma.

"The paper was not available at the time of publishing the book".

NUMERICAL STUDIES OF IMPURITY PHENOMENA

D.F. Duechs and R. Behrisch

Max-Planck-Institut für Plasmaphysik,
8046 Garching bei München, Germany

ABSTRACT

The various effects of impurities on hydrogenic plasmas are reviewed. The problems involved in impurity diffusion are explained, and the simplifications necessary for including impurity diffusion in computer studies are discussed, several models are presented as examples. The different contributions to the impurity particle fluxes are compared. Finally, the concept of a "Cold plasma blanket" for impurity flux control is investigated.

"The paper was not available at the time of publishing the book".

COMPUTER STUDIES

D.F. Duechs

Max-Planck-Institut für Plasmaphysik,
8046 Garching bei München, Germany

ABSTRACT

Plasma theory (i.e. formulae) is available only for isolated single phenomena, e.g. particle diffusion, bremsstrahlung, ionization, instabilities etc. It will be shown to what extent, at present, such theories can be combined to a more consistent description of Tokamak plasma behaviour by means of computer studies. Such codes should serve as scaling devices. As an example, the codes on radial transport of particles and energy are discussed in some detail. The relative importance of the effects included is illustrated (for present-day Tokamak parameters).

"The paper was not available at the time of publishing the book".

77

Review of Experimental Results I

H. P. Furth

Plasma Physics Laboratory, Princeton University,
Princeton, New Jersey 08540, USA

ABSTRACT

An illustrative survey of operating tokamak
devices is given. The ohmic heating of tokamak
plasmas is reviewed, and the limitations imposed
by MHD instability phenomena are described. A
modification of the MHD theory, including finite-
resistivity effects, appears to give a reasonable
fit of experimental results. The combined requirements of ohmic heating, MHD stability, and
radiation cooling define the experimental parameters of present-day tokamak devices.

1. Tokamak Devices

The basic elements of the tokamak confinement scheme [1,2]
are illustrated in Fig. 1. Metal discharge chambers are used —
typically incorporating sections of bellows or ceramic voltage
breaks. A toroidal magnetic field B_t is generated by a set of
external coils. An air-core (or iron core) transformer winding
is used to induce the plasma current I , which generates a
poloidal field component B_p and thus imparts a weak helicity
to the magnetic field lines. Force balance along the major
radius R is provided by addition of a basically vertical
magnetic field B_v , which can be feedback-controlled or pre-
programmed to locate the plasma at a desired point in R . For
moderately convex curvatures of B_v , the equilibrium position
is stable; in addition, a massive copper shell can be used to
guarantee positional stability. The plasma edge is defined by
a metal limiter, typically a diaphragm or a pair of horizontal
rails made of tungsten.

Some representative tokamak parameters are given in Table
I; the major and minor radii, the maximum toroidal field, the

79

[14] S. von Goeler, in Controlled Fusion and Plasma Physics
(Proc. 7th Europ. Conf. Lausanne, 1975) II, EURATOM-CEA
(1975) 71.
[15] S. von Goeler, W. Stodiek, N. Sauthoff, Phys. Rev. Lett.
33 (1974) 1201.
[16] I. N. Golovin, et al., in British Nuclear Energy Society
Nuclar Fusion Reactor Conference (Proc., Culham, 1969) 194.

Table I.

Parameters of Illustrative Tokamak Devices

	R (cm)	a_{max} (cm)	B_{tmax} (kG)	I_{max} (kA)	I_{typ} (kA)
PLT	130	45	50	1500	600
T-10	150	37	50	1000	400
TFR	98	20	60	400	300
ORMAK	80	23	26	230	200
Alcator	54	10	100	300	150
Pulsator	70	12	30	150	80
TM-3	40	8	40	70	50

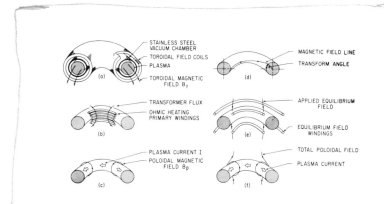

Fig. 1. Outline of the tokamak confinement
scheme. (PPPL 753361)

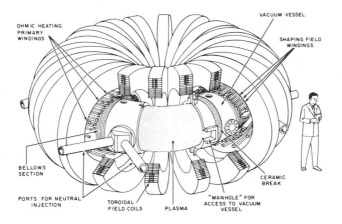

Fig. 2A.   Schematic of the PLT device at PPPL.
(PPPL 723111)

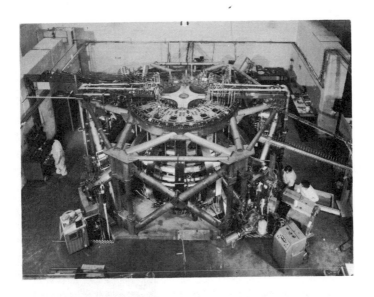

Fig. 2B.   The PLT device on completion of fabrication.   (PPPL 764321)

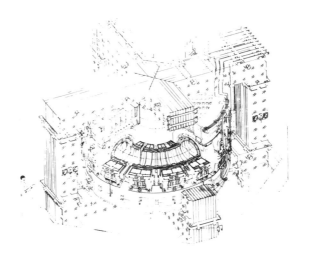

Fig. 3A. Schematic of the T-10 device at the
I. V. Kurchatov Institute. (PPPL 753274)

Fig. 3B. The T-10 device on completion of fabrication. (PPPL 753865)

maximum current capability and the typical operating current are quoted. Different types of detailed tokamak design are illustrated in Figs. 2-6. The PLT and T-10 are the largest operating tokamaks in the world. The recently completed PLT device [3] (Fig. 2) uses an air-core transformer and an electronically controlled vertical-field, and is outfitted for neutral-beam injection (cf. Lecture II). The T-10 (Fig. 3), which has been in operation somewhat longer [4], exemplifies traditional tokamak design features: an iron transformer core and a copper shell for plasma positioning. The Adiabatic Toroidal Compressor or ATC (Fig. 4) uses electronic position control to compress the discharge in major radius [5]. The Alcator (Fig. 5) operates with very high magnetic field strengths and small major radius, thus maximizing ohmic heating [6]. The Doublet II A (Fig. 6) studies tokamak discharges with noncirular minor cross sections [7]; for this purpose, a more elaborate coil system is needed to shape the externally applied portion of the poloidal field.

The world tokamak population has risen from ~10 in 1971 towards ~10^2 in 1976, counting minor facilities. Interestingly there appears to be a very strong family resemblance among tokamak discharges, so that the smallest (a ~ 3 cm) and weakest (B ~ 1 kG) exhibit essentially the same characteristics as the largest (a ~ 45 cm) and strongest (B ~ 80 kG). This continuity of phenomena is encouraging: if tokamak devices of reactor parameters were to preserve the same stability patterns and scaling laws that characterize present-day experiments, the basic feasibility of the tokamak reactor approach would be assured.

83

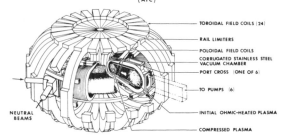

Fig. 4. The Adiabatic Toroidal Compressor (ATC) at PPPL. (PPPL 733328)

Fig. 5. The Alcator at MIT. (PPPL 763719)

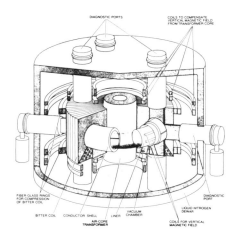

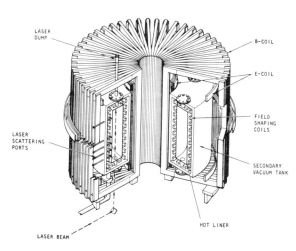

Fig. 6. The Doublet II A at General Atomic Corporation. (PPPL 753275)

2. The Tokamak Discharge

The toroidal discharge must be initiated at moderately high ohmic heating power (i.e., high voltage — typically ~100 V for R ~ 100 cm) in order to ionize the initial gas filling and force the electron temperature T_e through the 20-50 eV "radiation barrier," where light impurities such as oxygen have maximum radiation power. Thereafter, the electron temperature rises easily to the keV level, while the discharge voltage drops to a few volts. At the same time, a characteristic "black center" appears (Fig. 7A) when the minor plasma cross section is viewed photographically: the light impurities continue to radiate strongly only from the surface of the plasma toroid, where they can remain partly stripped.

If the ohmic heating power input is high compared with the power that can be carried from the plasma edge by atomic processes such as radiation and charge exchange, the discharge will expand until it touches the limiter. In this type of standard tokamak operation, more than half the input power flows to the limiter by plasma transport. Alternatively, if the input power is weak or there is a high level of hydrogen or impurity influx, the discharge channel shrinks from limiter contact, as illustrated in Fig. 7A and B.

When seeking to achieve high plasma density in a tokamak, it is advisable to minimize impurity influx and to build up the hydrogen gas feed gradually, in such a manner as to reduce the voltage requirement for initial T_e-breakthrough and to avoid shrinking of the discharge channel thereafter. The consequence of excessive initial voltage is the formation of electron runaway populations in the MeV range; the consequence of discharge

85

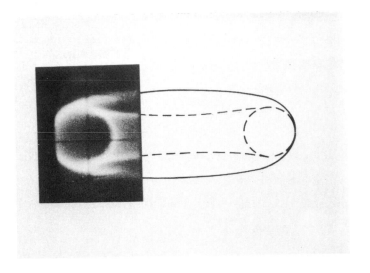

Fig. 7A. Photograph in visible light of a tokamak plasma in ATC; the camera is aimed tangentially, in the horizontal midplane. (PPPL 733510)

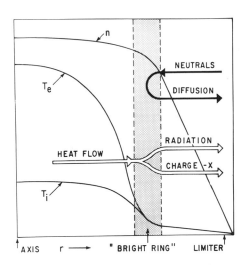

Fig. 7B. Schematic of the particle and energy flow in a tokamak profile such as that of Fig. 7A. (PPPL 743446)

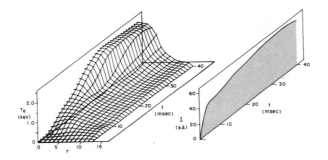

$(n_{max} = 2 \cdot 10^{13} \; cm^{-3} \; at \; 40 \; msec)$

Fig. 8A. T_e-profiles measured by Thomson scattering in the ST tokamak. (PPPL 733051)

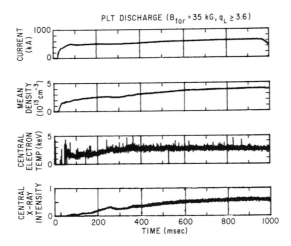

Fig. 8B. A representative discharge on PLT. The T_e-trace is an approximate measurement, based on x-ray foil absorption. The late broading of the x-ray intensity trace is due to the $m = 1$ "sawtooth." (PPPL 763287)

shrinking, as we shall see, is gross destabilization of the plasma. Elaborate techniques have been developed for "cleaning" the vacuum wall and for programming the gas feed, but up to the present time this field of endeavor has remained more of an art than a science.

The evolution of a representative small-tokamak discharge at the 60-kA level is shown in Fig. 8A. The electron temperature profiles were obtained by Thomson scattering on the ST device [8]. Discharges at 600 kA in PLT (Fig. 7B) show a remarkably similar pattern, including the flattening of the central portion of the T_e-profile, and the absence of any marked outward peaking of T_e during the current rise, which would have characterized a prolonged skin-current phase. Extensive T_e-profile measurements carried out on PLT by x-ray and Thomson scattering techniques will be reported in Ref. [3].

3.   Gross Stability

In "normal operation" the tokamak discharge is grossly stable. Only when the insufficiency of ohmic heating power causes the discharge channel to shrink (Fig. 7) or, alternatively, when the plasma current is made excessively large, does one find evidence of the sort of spectacular malfunctions that are identified as instabilities in pinch experiments. Grossly unstable tokamak behavior, which has been given the name "disruptive instability," is illustrated in Fig. 9. Typically one sees a slowly growing helical magnetic perturbation, leading into a sudden expansion of the discharge channel (as shown in Fig. 9, these expansions seem to come in two distinct sizes: small and large). The exterior regions of the poloidal

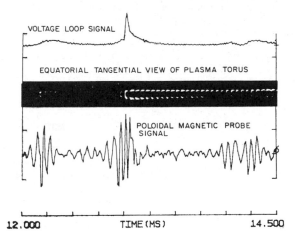

VOLTAGE LOOP SIGNAL

EQUATORIAL TANGENTIAL VIEW OF PLASMA TORUS

POLOIDAL MAGNETIC PROBE SIGNAL

12.000   TIME (MS)   14.500

Fig. 9. Minor and major disruptions in the ATC. A negative voltage signal is shown correlated with the framing camera record and a magnetic pick-up loop trace. (PPPL 733355)

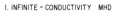

TOKAMAK KINK MODES

I. INFINITE – CONDUCTIVITY MHD

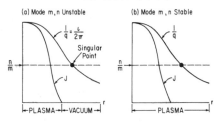

(a) Mode m,n Unstable   (b) Mode m,n Stable

2. FINITE — RESISTIVITY MHD

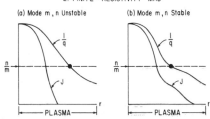

(a) Mode m,n Unstable   (b) Mode m,n Stable

Fig. 10. Review of the tokamak kink instability theory. (PPPL 763741)

flux are pushed outward during the expansion, giving a characteristic negative voltage spike (shown inverted in Fig. 9).

The onset of this disturbance is readily understandable from the theoretical point of view. Defining $\iota(r)$ as the transform angle (cf. Fig. 1) during passage once around the torus, and $q(r) = 2\pi/\iota = B_t r/B_p R$ as the "MHD safety factor," one can compute a clear-cut set of rules for the permissible profile of ι (or q). The basic principles are contained in ideal MHD stability theory [9], where the plasma is assumed to have perfect conductivity. For magnetically-driven helical perturbations (kink modes) characterized by the form $\exp i(m\theta - n\phi)$, where θ and ϕ are the angles the short and long way around the torus, one finds that a plasma region wherein $q(r) < 1$ is obeyed will be unstable against the fundamental mode $m = 1$, $n = 1$. As illustrated in Fig. 10, Case 1, higher-harmonic kink modes ($m \geq 2$) can be unstable only if the singular point $m/n = q(r_s)$ falls outside the plasma. In actual tokamak experiments, such higher-harmonic modes are generally observed even when r_s falls well within the plasma; hence it is necessary to use the finite-resistivity theory [10] to describe these instabilities. Infinite-conductivity and finite-resistivity modes have exactly the same stability conditions for similar configurations (Fig. 10, Cases 1A and 2A); the presence at r_s of resistive plasma, rather than vacuum, merely decreases the growth rate of the instability. A feature of the experiments that can be described only by finite-resistivity theory is that in general the current density $J(r)$ is nonzero at r_s, and thereby exerts a significant stabilizing influence [11].

The observed destabilization of tokamak discharges by excessive
gas and impurity influx, or by an excessively high total current
that forces the singular point $q(r_s) = 2$ (or 3, 4,....) to approach the limiter, is apparently due to the suppression of $J(r)$
near r_s by local cooling of the plasma. (Another possible destabilization would be due to the suppression of the local plasma
pressure gradient [12], which provides a high-temperature correction to the simple finite-resistivity theory.) When the stability condition is violated, one can observe the appropriate mode
conveniently as a magnetic oscillation [13], since the perturbation tends to be swept along in the θ-direction by the electron diamagnetic drift. Low-level perturbations ($\lesssim 1\%$ in B_p)
are commonly seen to persist in quasi-steady state without
causing disruption; their structure has been mapped in some
detail by means of pinhole x-ray detectors [14] (Fig. 11). At
this point, one can consider the understanding and avoidance
of the onset of kink-type perturbations to be fairly well in
hand. Limiter q-values of $\gtrsim 3$, combined with well-scrubbed
vacuum walls and moderate rates of density build up will insure
a grossly stable discharge. When kink modes are allowed to
grow, however, the exact mechanism of the resultant disruptive
instability still remains veiled in the mysteries of nonlinear
MHD theory.

The fundamental $m = 1$, $n = 1$ kink mode of the tokamak, was
not discovered until sometime after the others, since it appears
only in the region $q(r) < 1$, so that its manifestations are
confined to the central part of the discharge. The first clue
to its existence was the observed flattening of T_e-profiles, as
in Fig. 8A. Examination of the central plasma region with

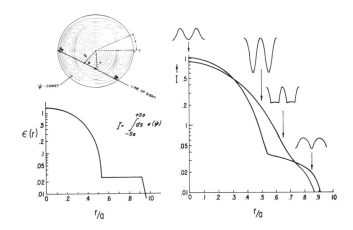

Fig. 11. Distortion of the tokamak flux
surfaces by a resistive m = 2, n = 1 mode, and
expected modulation of a pinhole x-ray signal;
such signals were observed in ST.[14] (PPPL
743889)

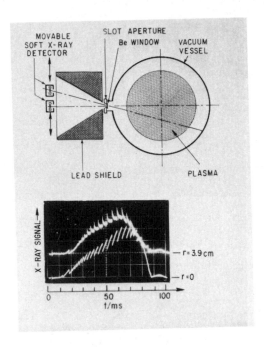

Fig. 12A. Use of a faster sweep and
high-frequency filter reveals that the
"sawtooth" pattern has m=1, n=1 fluctuations superimposed. (PPPL 743481)

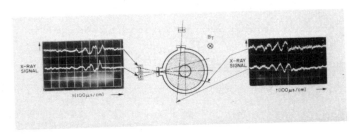

Fig. 12B. Pinhole x-ray observations of the
"sawtooth" effect in ST[15]; the fluctuations are
due mainly to T_e. (PPPL 743480)

pinhole x-ray detectors [15] revealed that the flattening phenomenon was being produced by a sawtooth-like fluctuation (Fig.
12A), a kind of minidisruption associated with the growth and
nonlinear relaxation of the $m = 1$, $n = 1$ mode (Fig. 12B).

An important by-product of the discovery of the $m = 1$
regime on the ST tokamak was the observation that discharges
operating in this manner had the highest plasma energy content
and the longest energy confinement times. When the $m = 1$
mode appeared, the $m = 2$ mode would fade away; as indicated
in Fig. 13, this observation is consistent with MHD stability
theory. A central discharge channel with $J(r)$ constant and
$q(r) \equiv 1$ is expected to be marginally stable against modes in
the range $1 \lesssim m/n \lesssim 2$ and positively stable against others.
If $q(0) < 1$, then $m = 1$ is unstable, but all the higher harmonics can become positively stable; if $q(0) > 1$, the higher
harmonics tend to be destabilized. In the latter case, the
$m = 2$, $n = 1$ mode is found to dominate — apparently with a more
damaging effect on overall energy confinement than is produced
by the centrally localized $m = 1$, $n = 1$ mode.

4.   Limitations on Ohmic Heating

In the simplest model of a tokamak reactor plasma [16]
the safety factor $q(a)$ at the limiter is taken to characterize
$q(r)$ throughout the plasma. The ohmic heating power density is
then given by $P_{OH} = J^2 \eta$, where $J = B_t / 2\pi Rq$, and $\eta \propto T_e^{-3/2}$.
Cooling by hydrogenic Bremsstrahlung alone would impose the
upper limit

$$nT_e < 2.4 \cdot 10^{12} \, J \, keV/A \, cm \quad . \tag{1}$$

For magnetic field strengths B_t in the range of practical interest (40-70 kG), the large physical size requirement of a reactor ($R \gtrsim 500$ cm) depresses this upper limit on ohmic heating capability to $nT_e \lesssim 3\text{-}5 \cdot 10^{14} q^{-1} cm^{-3}$ keV.

If we were to assume that the usual limiter value $q(a) \sim 3$ is the appropriate q-value to use in Eq. (1), the restriction on n would be very severe. As discussed in the previous section, the choice $q \sim 1$ is more appropriate. Even so, D-T ignition temperatures can be contemplated only for rather low densities $n < 5\text{-}8 \cdot 10^{13} cm^{-3}$ — neglecting all energy losses by plasma transport processes (i.e., assuming $n\tau_E = \infty$). As we shall see in Lecture II, energy confinement against plasma transport is, however, rather critical; its optimization is expected to depend mainly on the maximization of density ($n\tau_E \propto n^2$) — as will the achievement of economically attractive fusion power density ($P_F \propto n^2$). Accordingly, there is a strong incentive to raise the plasma density of a reactor plasma to the MHD β-limit. That this cannot be done in an ohmic-heated reactor plasma can be seen by rewriting Eq. (1) in the form

$$\beta_{pe} < \frac{800 \text{ kA}}{I} \tag{2}$$

where $\beta_{pe} = 8\pi nT_e/B_p^2$. Tokamak reactor currents must necessarily be in the range >3 MA (just to confine the D-T alpha particles satisfactorily), whereas the MHD-limit on $\beta_p \sim 2\beta_{pe}$ is well above unity (cf. Lecture II).

As has been noted earlier, the operation of present-day tokamak experiments already depends on a balance between ohmic heating and radiation cooling, but the ingredients are somewhat

95

different. The main objective is to make the ohmic-heating input power at least several times greater than the line radiation from impurity ions, so as to avoid the shrinking and MHD-instability of the discharge channel. An empirical formula that gives a rough idea of the appropriate operating density is

$$n < 2 \cdot 10^{11} \; B_t/R \; \text{gauss}^{-1} \text{cm}^{-2} \tag{3}$$

If we identify $q \sim 1$, $J \sim B_t/2\pi R$, this becomes $n < 10^{11} \; J \; \text{cm}^{-1} \text{A}^{-1}$ — a limit that is about 20 times more severe than Eq. (1) for tokamak plasmas of average $T_e \sim 1$ keV.

A lower bound on the experimental operating density is set simply by the ability of the plasma to carry the discharge current. Defining the drift velocity $v_{d\parallel} = cJ/en$, it follows identically that

$$\frac{v_{d\parallel}}{v_{th,e}} = \left(\frac{2}{n\pi r^2 \beta_{pe} r_{e\;class}} \right)^{1/2} \tag{4}$$

where $v_{th,e}$ is the electron thermal velocity and $r_{e\;class}$ is the classical electron radius $e^2/m_e c^2 = 3 \cdot 10^{-13}$ cm. In order to avoid the excitation of current-driven microturbulence — typically accompanied by electron runaway phenomena — one desires $v_{d\parallel}/v_{th,e}$ to be small [e.g., of order $(m_e/m_i)^{1/2}$ or less]. For devices with $a \sim 10$ cm, $\beta_{pe} \sim 1$, this implies $n \gtrsim 10^{13} \text{cm}^{-3}$.

A very rough schematic idea of present plasma-heating trends in tokamaks is conveyed by Fig. 14. Since the original "tokamak breakthrough" by the T-3 experiment in late 1960's, ohmic heating has been used to raise T_e and T_i somewhat;

96

the principal parameter extension has been the order-of-magnitude increase in density achieved on the Alcator experiment — by virtue of its outstanding value of B_t/R (cf. Table I). The prospects of raising the plasma temperature further, towards levels of practical reactor interest, appear to rest on the successful development of non-ohmic heating methods — in particular, neutral-beam injection, which will be discussed in Lecture II.

ACKNOWLEDGEMENT

This work supported by U.S.E.R.D.A. Contract E(11-1)-3073.

REFERENCES

[1] L. A. Artsimovich, Nucl. Fus. 12 (1972) 215.
[2] H. P. Furth, Nucl. Fus. 15 (1975) 487.
[3] W. Stodiek, et al., in Plasma Physics and Controlled Nuclear Fusion Research (Proc. 6th Int. Conf., Berchtesgaden, W. Germany, 1976) paper A2, to be published.
[4] A. B. Berlizov, et al., to be published in JEPT Letters; A. B. Berlizov, et al., in Plasma Physics and Controlled Nuclear Fusion Research (Proc. 6th Int. Conf. Berchtesgaden, W. Germany, 1976) paper A1, to be published.
[5] K. Bol, et. al., Phys. Rev. Lett. 29 (1972) 495; E. Mazzucato, et al., in Plasma Physics and Controlled Nuclear Fusion Research (Proc. 6th Int. Conf. Berchtesgaden, W. Germany, 1976) paper A11, to be published.
[6] R. R. Parker, et al., in Plasma Physics and Controlled Nuclear Fusion Research (Proc. 6th Int. Conf. Berchtesgaden, W. Germany, 1976) paper A5, to be published.
[7] R. L. Freeman, T. Tamano, et al., in Plasma Physics and Controlled Nuclear Fusion Research (Proc. 6th Int. Conf. Berchtesgaden, W. Germany, 1976) paper A10-3, to be published.
[8] D. Dimock, et al., Nucl. Fus. 13 (1973) 271.
[9] See Lecture by D. Pfirsch.
[10] H. P. Furth, P. H. Rutherford, H. Selberg, Phys. Fluids 16 (1973) 1054.
[11] A. H. Glasser, H. P. Furth, P. H. Rutherford, submitted to Phys. Rev. Letts.
[12] A. H. Glasser, J. M. Greene, J. L. Johnson, Phys. Fluids 19 (1976) 567.
[13] S. V. Mirnov, I. B. Semenov, JETP 33 (1971) 1134.

97

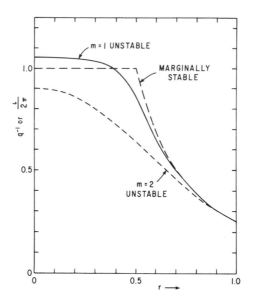

Fig. 13. Ideal marginally stable
tokamak profile, and two "realistic"
neighboring profiles, unstable against
the m = 1 and the m = 2 mode, respectively. (PPPL 763572)

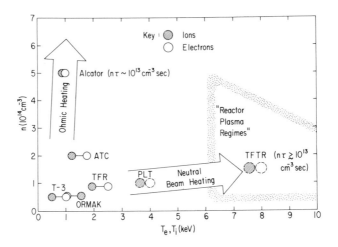

Fig. 14. Historical trends in tokamak plasma
heating. Parameters refer to the central plasma
region. (PPPL 763370)

Review of Experimental Results II

H. P. Furth

Plasma Physics Laboratory, Princeton University,
Princeton, New Jersey 08540, USA

ABSTRACT

The experimental results on energy and particle confinement in tokamaks are reviewed. MHD-
instability effects on confinement can be distinguished when they are strong, but information on
microscopic loss mechanisms is still rudimentary.
Under the constraint of dominant ohmic heating,
the true physical scaling laws of the tokamak
transport coefficients cannot be determined. Non-
ohmic heating methods, notably neutral-beam injection, are now proving successful and should improve
our knowledge of tokamak confinement physics. Non-
ohmic heating combined with refinements in tokamak
geometry offers the prospect of attaining practical
tokamak reactor parameters.

1. Plasma Energy Transport Processes

The basic advantage of a toroidal reactor is that the fuel

ions can collide many times without being lost; this feature

permits one to envisage tokamak reactor operation at the 10 keV

level, even though the ratio of Coulomb scattering to fusion

cross section exceeds 10^5. In tokamaks the energy transport

due to ion scattering is described by the so-called neoclassical

theory, which has been discussed in an earlier Lecture [1]. The

confinement predicted by this theory for D-T plasmas is better

than actually required in typical reactor designs, by an order

of magnitude or more.

As illustrated in Fig. 1, the experimental results for T_i

conform reasonably well with the neoclassical prediction, which

is based on ion heating by collisions with the electrons and on

cooling by ion thermal conduction in the plateau regime. At

the high-temperature end of the scale, and especially for

99

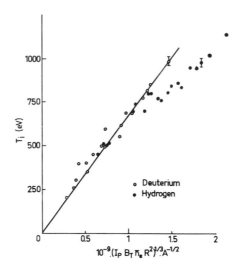

Fig. 1. Comparison of TFR experimental points with neoclassical (plateau) theory [2]. (The quantity A is 1 for hydrogen and 2 for deuterium; in absolute terms, T_i is higher for hydrogen.) (PPPL 763767)

hydrogen, one would have expected an upturn of T_i, due to entry into the collisionless regime; the opposite is, however, observed. A far greater anomaly characterizes the electron transport: whereas energy confinement is supposed to be much better for the electrons than for the ions [by $\sim (m_i/m_e)^{1/2}$], it is generally found to be somewhat worse, and its scaling has no obvious relation to classical prediction. The total plasma energy content of the tokamak, as characterized by $\beta_p = 8\pi \langle n(T_e + T_i)\rangle /B_p^2(a)$, (cf. Fig. 2) is actually not far from neoclassical expectation, i.e., typically $\beta_p = O(1)$; but the experimentally observed predominance of the electrons — rather than the ions — in the thermal transport process is fundamentally at variance with classical ideas.

The direct physical evidence for anomalous electron transport mechanisms in tokamaks has remained remarkably sketchy until recently. The reason is evidently the difficulty of diagnosing dense keV-range plasmas by noncontact methods. The development of pinhole x-ray diagnostics (cf. Lecture I) has now clarified the MHD-related transport processes inside tokamaks. In grossly stable tokamaks, it is clear that the "sawtooth" oscillations of the $m = 1$, $n = 1$ mode can produce very high transport rates near the plasma center, and that the magnetic islands of the $m = 2$, $n = 1$ mode can do the same in the vicinity of the point $q(r) = 2$. Furthermore, it seems likely that MHD activity is involved in the anomalous suppression of the skin effect in large tokamaks (cf. Fig. 3). One would expect that the interior magnetic surfaces of tokamaks become fairly ergodic in the presence of substantial MHD activity; this is confirmed by the rapid escape of runaway electrons that is observed under

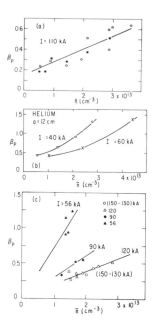

Fig. 2. Typical variations of β_p in (a) T-4 [3], (b) ST [4] and (c) ORMAK [5]. (PPPL 753311)

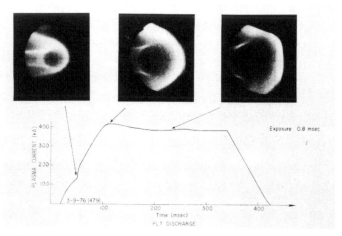

Fig. 3. Evolution of the plasma minor cross section, photographed in visible light on PLT. During current rise, the plasma edge region remains cold, thus minimizing skin effect. (Figure provided by K. Young.) (PPPL 763302)

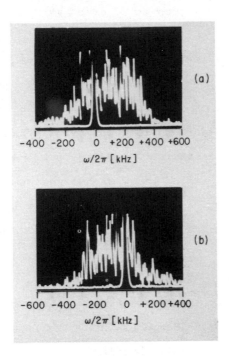

Fig. 4. Density fluctuation spectra
in ATC, observed by scattering of a 70-GHz
microwave beam [7]. The toroidal field
has opposite directions in the two cases
(a) and (b); the spectrum is shifted in
the direction of electrom diamagnetic
drift. (PPPL 763582)

such conditions. On the other hand, the fairly good (though always imperfect) confinement of runaways in MHD-quiescent cases implies the existence at least of substantial bands of high-integrity flux surfaces.

Since stellarators, multipoles, etc., are well known to exhibit large anomalous transport rates in the absence of MHD instability, it would not be astonishing if the anomalous transport of tokamaks were found to arise mainly from microinstabilities [6]. A direct examination of the small-scale density fluctuation pattern inside tokamaks has recently been carried out on ATC [7]. Similar results are reported in Ref. [8]. The diagnostic method consists in scattering a microwave or infrared beam from the plasma, and unfolding the spectrum of density fluctuations from the modulated beam spectrum (Fig. 4). The observed fluctuations are consistent with theory in that they exhibit the appropriate range of wavelengths ($\lesssim 1$ cm) and phase velocities (on the order of the electron diamagnetic drift, and with the appropriate direction). The magnitude ($\Delta n/n \sim 1\%$) is also about right to explain the observed transport rates. As we know so well from stellarator and multipole experiments, however, there are instances of anomalous transport in the presence of a theoretically appropriate fluctuation spectrum, where the transport is found to persist unabated when the fluctuations have been suppressed. In these cases, it often turns out that quasi-static convective cells [9] connected with device irregularities are mainly responsible for the losses. It seems possible that magnetic defects produced by weak MHD instabilities in the tokamak may also be able to generate such convective patterns, but in this case direct experimental evidence will not readily become available.

Plasma energy can be transported either by heat conduction or by a net outflow of hot particles. The latter mechanism tends to be less important in tokamaks, since the particle confinement time τ_p is typically several times the energy confinement time τ_E . Like the electron transport, the observed particle transport substantially exceeds the neoclassical prediction.

In evaluating τ_p , it is customary to average over the whole discharge; the actual particle replacement time, however, is a strong function of radius. In quasi-stead state, the particle outflow is balanced by the ionization of the local neutral gas $h_h(r)$. The replacement time is given for typical tokamak parameters by $\tau_p = \tau_{ionize} \sim 5.10^7 \ cm^{-3} / h_h$. The density of "warm" neutrals (~ 10 eV) at the plasma edge is typically one or two orders of magnitude greater than the density in the central plasma region, which contains only n'th-generation hot charge-exchange neutrals. Near the plasma center, $\tau_p(r)$ thus tends to be much greater than τ_E .

An interesting anomaly of particle transport has been found in experiments with high-density tokamaks, such as Alcator. In this case, the plasmas are so opaque to charge-exchange neutrals that one cannot readily account for the observed rate of central-density rise on the basis of the ionization of local neutral gas. In computer simulations of such discharges, the plasma density peaks at the edge and then diffuses inward; in the experiments, such peaking is seen only transiently: thereafter, the central density continues to rise, even though the particle transport should be outward.

2.   Empirical Scaling Laws

In the absence of direct physical evidence concerning tokamak energy transport mechanisms, a good deal of attention has focussed on empirical scaling laws for the energy confinement time τ_E. In the experiments, this quantity is usually defined as the plasma energy content in quasi-steady state, divided by the input power. When deriving empirical scaling laws, the usual practice is to interpret the experimentally measured τ_E-values in terms of plasma transport from the hot center to the cold edge: as indicated in Fig. 7 of Lecture I, radiation cooling usually constitutes an important additional energy loss channel only near the plasma edge. This interpretation of τ_E is consistent with usage in reactor studies, where τ_E refers only to plasma transport, the radiation cooling being treated explicity.

The endeavor to find an empirical τ_E-scaling law has led to a great diversity of models — ranging from the old-fashioned "pseudoclassical" [10,11] $\tau_E \propto n^{-1}T_e^{1/2}a^2$ to the contemporary "collisional" scaling [12] $\tau_E \propto na^2$ and including such formulations [13,14] as $\tau_E \propto aI$ and $\tau_E \propto nT_e^{1/2}a^2/I$. The difficulty is that essentially all tokamak data thus far have been taken under conditions where ohmic heating is the main power input, so that there has been a built-in constraint,

$$\frac{nT}{\tau_E} = \eta J^2 ,$$ (1)

linking the plasma temperature and the confining poloidal field. As a result, the true physical parameter dependences of τ_E have not been uniquely determinable from the data.

106

The basic experimental input comes in the form of the β_p-curves shown in Fig. 2. For a given device, the dimensions a and R are roughly fixed and the safety factor q is made high enough so as to avoid depressing τ_E by enhanced MHD activity (cf. Lecture I). If, for simplicity, one neglects dependences on the effective ionic charge Z_{eff} and on the relative magnitudes of T_e and T_i, there remain the three variables n, I (or B_p) and T. We can then take as our experimental input:

$$\beta_p = X(n,I) \tag{2}$$

or, using the definition of β_p,

$$T \propto \frac{I^2 X(n,I)}{n} \quad . \tag{2a}$$

When Eq. (2a) is combined with the ohmic heating constraint, Eq. (1), in the form

$$\tau_E \propto \frac{nT^{5/2}}{I^2}$$

one obtains

$$\tau_E \propto \frac{I^3 X^{5/2}(n,I)}{n^{3/2}} \quad . \tag{3}$$

Since Eq. (3) was derived for a set of ohmic-heating regimes characterized by the validity of Eq. (2a), it conceals all dependences on the parameter $I^2 X(n,I)/nT$. This ambiguity in

Eq. (3) can be taken into account by introducing an unknown functional dependence $F[I^2X(n,I)/nT]$. The general formulation that underlies all "empirical scaling laws" for ohmic-heated tokamaks is thus:

$$\tau_E = \frac{I^3 X^{5/2}(n,I)}{n^{3/2}} F[I^2X(n,I)/nT] \tag{4}$$

where F is any function.

Some examples are:

A. Pseudoclassical

$$X \propto 1 \quad,$$

$$T \propto I^2/n \quad,$$

$$\tau_E = \frac{I^3}{n^{3/2}} F\left(\frac{I^2}{nT}\right) \quad. \tag{5}$$

B. "Collisional"

$$X \propto n/I^{1+\varepsilon} \quad,$$

$$T \propto I^{1-\varepsilon} \quad,$$

$$\tau_E = n F(I^{1-\varepsilon}/T) \quad. \tag{6}$$

In the pseudoclassical model, the conventional formulation is $F(I^2/nT) \propto (I^2/nT)^{-1/2}$, so that $\tau_E \propto I^2 T^{1/2}/n$. In the "collisional" model, the convention is $F = $ constant; The quantity ε is typically said to lie in the range $-0.2 < \varepsilon < 0.2$.

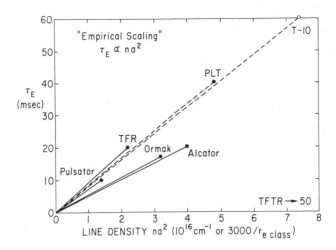

Fig. 5. Empirical scaling of confinement in ohmic-heated tokamaks, as a function of mean density and size. Linear scaling up to the maximum points shown has been documented in detail for ORMAK and Alcator. (PPPL 763290)

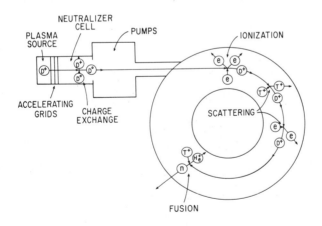

Fig. 6. Basic concepts of neutral-beam heating. (PPPL 753441)

Since tokamak experiments have by now accumulated a good deal of data where I and n are varied independently (as in Fig. 2), the pseudoclassical model of τ_E can be fairly well ruled out. The "collisional" model fits the trend of all the data at least crudely, and fits the data of some tokamak devices very well — at least up to the vicinity of their high-density operating limits. The individual variations may be connected at least partly with the Z_{eff} -dependences of ohmic heating and of τ_E , which have been neglected in the preceding analysis.

When comparing data from tokamak devices of substantially different dimensions, it is possible to infer the a-dependence of τ_E . As indicated in Fig. 5, the "collisional" empirical model fits reasonably well with the scaling

$$\tau_E \propto na^2 . \tag{7}$$

Having arrived at Eq. (7), one may ask whether it offers any new insight into the physical transport mechanisms of tokamaks. In this connection, one is forced to note, first of all, that Eq. (7) is unlikely to be valid as a true physical scaling law, since it makes no reference at all to the strength of the confining magnetic field. One would expect there to be at least a linear increase of τ_E with B (meaning either B_t or B_p), as in the case of Bohm diffusion — or even a quadratic increase with B, as in the case of classical diffusion or trapped-particle-mode transport. The structure of the unknown function F in Eq. (6) tells us that, in these two cases we would have either

110

$$\tau_E \propto na^2 \frac{B}{T^{1/1-\varepsilon}} \tag{8a}$$

or

$$\tau_E \propto na^2 \frac{B^2}{T^{2/1-\varepsilon}} \ . \tag{8b}$$

The existence of such adverse temperature dependences is plausible from a general theoretical point of view. Happily these dependences are not nearly as strong as the theoretically predicted T-dependence for trapped-particle-mode transport ($\tau_E \propto na^4 B^2 T^{-7/2}$).

The latter remark brings us to the real physical dilemma of the so-called "collisional" empirical scaling: If the observed confinement had anything to do with "collisional" and "collisionless" transport by plasma-gradient-driven drift-waves and trapped-particle modes, one would expect to see the scaling $\tau_E \propto 1/n$, not $\tau_E \propto n$, in the present range of experimental parameters — and especially in the regimes discussed in Ref. [12]. Conversely, if the relevant physical scaling parameter is not collisionality in the conventional sense, then the observed n-dependence must be related to some physical parameter other than collisionality, in which case there is no basis for the belief that Eq. (7) will continue to hold in a set of regimes that is characterized only by the same degree of collisionality.

An example of a non-collisional interpretation of Eq. (7) would be the excitation of some sort of current-driven drift wave; in that case the key parameter is $v_{d\parallel}/v_{th,e}$ (cf. Lecture I), which we have seen to diminish with rising na^2. This

interpretation tends to produce very weak τ_E-dependences on
B and T , and strongly favorable dependences on a . Still
another possibility would be that $\tau_E \propto n$ relates to the classical electron heat conductivity <u>along</u> magnetic field, with the
destruction of magnetic surfaces by low-level MHD modes providing the requisite degree of connectedness between the plasma interior and edge regions. In that case the scaling would be
$\tau_E \propto a^2 n T^{-5/2}$, with strong additional dependences on geometric
factors relating to MHD stability (cf. Lecture I).

Profile effects are clearly important, not only for MHD-
related transport, but also for microscopic modes and atomic
effects. The state of the art in tokamak data-fitting has advanced far beyond the business of attempting to fit simple
algebraic formulas to τ_E-data. One would expect, of course, that
even a one-dimensional computer code emphazing an elaborate set
of transport and atomic mechanisms would not give reasonable
predictions in the absence of the right "vital ingredient." On
the other hand, it is possible that there <u>is</u> no single vital ingredient — only a set of complex transport mechanisms operating
simultaneously in different regions of the discharge profile,
which add up to give ostensibly simple scaling laws for τ_E .
If so, the quest for physical understanding of tokamak transport
will be challenging — and the potential rewards, in terms of
the optimization of confinement, will be great.

3. Non-Ohmic Heating

Present uncertainties concerning the empirical scaling of τ_E
will, hopefully, be narrowed a good deal by the application of
intense non-ohmic heating. A number of rf heating methods are
giving promising results, especially ion and electron cyclotron

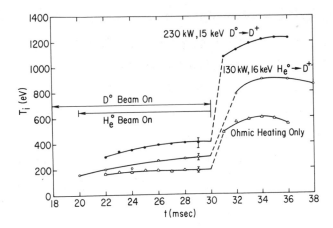

Fig. 7. Ion heating by neutral-beam injection and compression in ATC [22]. (The peak final density is 10^{14}cm^{-3} for ohmic heating only and $2 \cdot 10^{14}$cm^{-3} for the $D^O \rightarrow D^+$ case.) (PPPL 753841)

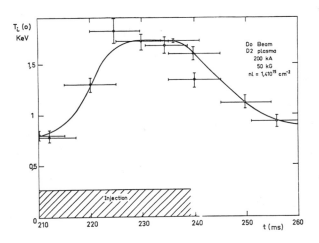

Fig. 8. Ion heating by injection of a 400-kW, 35-keV D_O beam into TFR [23]. (PPPL 763766)

heating [15,16]. Experiments of the latter type in TM-3 have
already yielded information on the local electron thermal conductivity at the plasma center, with the remarkable result that
confinement improves with rising T_e [16]. Neutral-beam injection [17] has been used to raise T_i in a number of tokamaks
(ATC, ORMAK, TFR, etc.), with the result that the total energy
confinement is found to be insensitive to T_i, while the ion
energy confinement follows the roughly classical lines of ohmic-
heated regimes.

Near-term experimental prospects of exceeding the ohmic-
heating power by non-ohmic means appear to rest mainly on neutral-
beam injection. The basic concept is illustrated in Fig. 6. An
intense ion beam is accelerated by multi-aperture grids, and then
neutralized by charge-exchange in a gas cell. In its neutral
state, the beam passes into the tokamak plasma and is captured
there by ionization and charge exchange, in the form of a high-
energy ion component. The injected ions slow down by collisional
processes [18], in excellent agreement with classical theory,
dividing their energy between the bulk plasma electrons and ions.
For $T_e \sim 5$ keV, the slowing-down time τ_s is characterized by
$n\tau_s \sim 10^{13} cm^{-3} sec$. A potentially useful incidental feature of
the slowing-down process, shown in Fig. 6, is that 100-200-keV
range deuterons injected into a tritium plasma of $\gtrsim 5$ keV can reproduce their injection power in the form of direct D-T fusion
power, not even counting thermal reactions in the bulk plasma [19]
Some results of present-day neutral-beam heating experiments are
shown in Figs. 7 and 8. The outstanding feature of present-day
experiments is "Eubank's Law": the ion temperature increase is

114

roughly linear with injected power, and amounts to about an eV per kW.

In applying neutral-beam injection to tokamak reactor experiments [20], one can utilize the reactions between the injected ions and the bulk plasma ions to achieve reactor-like conditions of fusion power density and energy multiplication, while requiring only $n\tau_E \gtrsim 10^{13} cm^{-3}$ for the tokamak plasma. The ultimate goal of tokamak reactor experiments, however, is to attain ignition; this will require $n\tau_E$-values above $3 \cdot 10^{14} cm^{-3}$sec, and will impose severe additional conditions on the neutral-beam injection system. In particular, if one assumes a scaling law of the type $n\tau_E \propto n^2 a^2$, it follows that adequate neutral-beam penetration will depend on raising the injection energy roughly according to $W_0 \propto (n\tau_E)^{1/2}$. To meet this requirement by a straightforward extrapolation of the scheme of Fig. 6 would be impractical, since the beam neutralization efficiency drops precipitously for deuteron energies beyond 100 keV. A promising approach to the solution of the penetration problem is the so-called "ripple-assisted penetration [21]," where the injection is vertical, and the injected particles continue to move toward the plasma center by ∇B-drift if they have been ionized after partial penetration. This type of orbit can be realized by means of a "ripple" in the toroidal field that is sufficiently strong near the injection point to mirror-trap the energetic ions, and diminishes toward the plasma center so as to untrap them. (The apparent violation of Liouville's theorem by this scheme is resolved by noting that most of the injected ions become thermalized before they can find the escape orbits that necessarily accompany the introduction of the ripple-injection orbit.)

115

4. Tokamak Beta Limits

As we have seen in Lecture I, ohmic heating will not raise problems of MHD β-limitation in large plasmas, since radiation cooling sets a more severe upper limit. Even in small plasmas, the MHD β-limitation cannot be reached in quasi-steady state, because of cooling by plasma transport, which generally gives β_p-values of order unity or less (cf. Fig. 2). One of the most interesting aspects of powerful non-ohmic heating will be the possibility of exploring the real β-limits of tokamaks.

First of all, there is a need to provide MHD-equilibrium. Returning to the simple picture of Lecture I, Fig. 1, it is not surprising to find that the B_v-field must be raised for higher β_p-values, to provide major-radius equilibrium. Eventually, B_v becomes comparable to B_p; this happens for $\beta_p \sim R/a$. At this point the tokamak minor cross section necessarily becomes non-circular; a separatrix makes its appearance and sets an outer bound to the plasma confinement region. The conventional wisdom is, therefore, that $\beta \equiv \langle n(T_e + T_i) \rangle / B_t^2$ is limited by equilibrium constraints to

$$\beta = \beta_p a^2 / q^2 R^2 < a/Rq^2 \tag{9}$$

which is commonly estimated as less than 5%.

One obvious way to escape from this constraint on β is to elongate the tokamak minor cross section vertically, since β_p is then made larger for given q, so that β approaches closer to β_p. A potential difficulty with this approach is that the MHD-stability threshold of the safety factor q may also rise, so that the net benefit for β may be modest. The

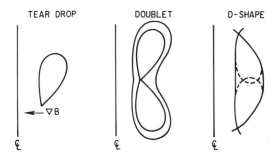

Fig. 9. Noncircular tokamak minor cross sections. The basic tear-drop shape has modestly advantageous MHD-properties, which are retained in two composites: the Doublet and D-shape. (PPPL 743870)

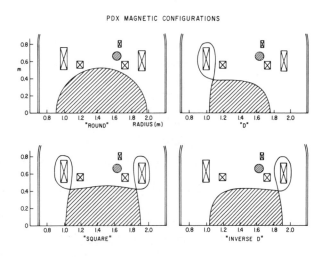

Fig. 10. Minor cross sections of various tokamak configurations designed for the Poloidal Divertor Experiment (PDX) at PPPL. (PPPL 753910)

117

best theoretical configurations for achieving stable elongation are illustrated in Fig. 9. Experimentally, some encouraging results have been obtained for noncircular configurations
on the Doublet II A experiment (cf. Lecture I, Fig. 6), but
these relate to the advantages of noncircularity in respect to
ohmic heating power — not to the MHD limitations on β. The
actual β-values obtained thus far in noncircular configurations
are about the same as those in standard tokamaks (cf., a central
β-value of about 2% in ATC).

A second possibility for raising β has recently been
stressed in Ref. [24]: the very appearance of an outer separatrix at high β_p-values can be used to advantage, since it tends
to force the plasma-edge q-value toward infinity. This means
that a very high plasma current density can flow near the plasma
edge, without producing dangerously low q-values. As a result,
very high local values of vertical field can be made to appear
where they are most needed (i.e. on the large-R side of the
plasma), even without resorting to substantial plasma elongation.
In this way, β_p can exceed its conventional limit, and average β-values of order 20% or more can be produced. In the experiments, such configurations could arise in a natural way if
a well-conducting plasma were heated non-ohmically to high β:
the high-conductivity constraint would then tend to preserve
the original q-value on each flux surface, and so would generate
high-β equilibrium configurations of the desired type [24].

The main source of concern about schemes involving high
β_p is the prospect of instability against localized flute modes
(ballooning modes). This consideration imposes a conventional
limit on β_p that is essentially the same as the equilibrium

limit (i.e., $\beta_p < R/a$), and therefore leads to the same constraint on β, i.e., Eq. (9). It is difficult to see how ballooning mode instabilities could be circumvented at high β_p.

Finally, a hopeful note arises from the MHD kink-mode considerations discussed in Lecture I. The most relevant q-value to be used in Eq. (9) is clearly not a limiter q-value of $\gtrsim 3$, since one finds $q(r) \sim 1$ over much of the profile, even in the case of present-day experimental tokamaks. In large, hot tokamaks of the future, it may be possible to provide kink-stable configurations with $q(r) \sim 1$ over most of the plasma region [25]; in that case, substantial β-values would result even for moderate β_p. The outlook for the attainability of average tokamak β's of order 10% by one of the methods described above — or, more likely, by an optimal combination — seems favorable. An experimental demonstration can be expected to emerge only in the course of the next several years.

5. Impurity and Divertors

In ohmic-heated experiments, the presence of impurity ions has at least one redeeming feature: the plasma resistivity increases as a function of z_{eff} (roughly linearly). The increase in radiated power with z_{eff} is even more rapid, but more importantly, the effect on plasma transport does not appear to be unfavorable. As a result, the highest T_e-values of tokamaks (~3 keV in ST and TFR) have been achieved in fairly impure discharges, and T_e has been observed to diminish when impurities are removed, e.g., by gettering in ATC [26]. The main advantage of high plasma purity for ohmic-heated tokamaks consists in raising the upper limit on the plasma density, which is imposed by radiation cooling (cf. Lecture I).

119

In neutral-beam heated tokamaks, on the other hand, the effect of impurities on the input power will be unfavorable: the most sensitive problem in this area is the depth of neutral-beam penetration into large, dense plasmas; for 100-keV range beams, this penetration will be reduced roughly in inverse proportion to Z_{eff}, due to impact ionization of the injected neutrals.

In present-day tokamaks the impurity influx consists mainly of lightly bound atoms, which are easily sputtered from the vacuum chamber wall, and redeposited there after each discharge. By means of assiduous discharge cleaning this migrant population of impurity atoms can be largely removed from the system. Another fortunate circumstance is that this population consists mainly of light atoms, which radiate strongly only from the plasma edge (cf. Lecture I, Fig. 7), where radiation cooling can actually play a helpful role in sparing the limiter. In future tokamak experiments, with — hopefully — much higher plasma ion temperatures, the mechanical sputtering of metallic impurities by charge-exchange neutrals should become the most serious problem. In this case, scrubbing of the vacuum chamber will not suffice to prevent impurity influx. With neutral-beam injection, there is the particular danger of an unstable feedback process, where reduced beam penetration heats the plasma edge preferentially, thus increasing the sputtering and further reducing the beam penetration [27]. In assessing the severity of this problem, a key experimental question is, whether impurities injected into the plasma edge will tend to be drawn towards the plasma interior by collisions with the hydrogen ion outflow (the classical prediction) and thus accumulate preferentially, or whether they will be lost from the plasma edge by anomalous transport processes along with the hydrogen. At present, there is evidence for the existence of both phenomena, but the factors governing their relative strength have not been established.

The existence of fairly stringent impurity limits for the attainability of ignition, especially in the 10-keV range, is well recognized. In order to maintain quasi-steady state reactor burns, it seems likely that special techniques must be found for blocking impurity influx; the most promising of these is the use of a magnetic divertor [28] (Fig. 10). The object here is to drain the tokamak edge plasma along the separatrix into a separate pumping chamber, so that the main tokamak plasma will not be subjected to a direct influx of neutral hydrogen and impurity atoms from plasma recombination on a mechanical limiter. As illustrated in Fig. 10, axisymmetric divertor geometries require noncircular minor cross sections. The separate considerations of high-β MHD-stability and plasma impurity control may well lead to a combined optimum design — for example a plasma D-shape with divertor loops at the top and bottom tips.

The past decade of tokamak experimental research has concentrated on exploiting the natural survival mechanisms of the tokamak discharge (ohmic self-heating, spontaneous feedback-stabilization of the $m = 1$ kink mode, discharge cleaning, etc.). The next decade will be characterized by various efforts to improve on nature (non-ohmic heating, control of radial profiles and cross sections, divertor action, etc.). There are many possibilities for improving an already favorable performance, and

121

the prospects of arriving at a practical end product seem good. The success of this second-stage experimental effort will depend a great deal, however, on the quality of our understanding of basic tokamak physics; success along this line is yet to be won.

ACKNOWLEDGEMENT

This work supported by U.S.E.R.D.A Contract E(11-1)-3073.

REFERENCES

[1] See Lecture by D. Pfirsh.
[2] TFR Group, to be published in Nuclear Fusion.
[3] L. A. Artsimovich, et al., in Plasma Physics and Controlled Nuclear Fusion Research (Proc. 4th Int. Conf. Madison, Wisconsin, USA, 1971) 1, IAEA (1971) 433.
[4] D. Dimock, et al., in Plasma Physics and Controlled Nuclear Fusion Research (Proc. 4th Int. Conf. Madison, Wisconsin, USA, 1971) 1, IAEA (1971) 451.
[5] J. F. Clarke, Oak Ridge National Laboratory Report ORNL-TM-4585 (1974).
[6] See Lecture by B. Coppi.
[7] E. Mazzucato, et al., in Plasma Physics and Controlled Nuclear Fusion Research (Proc. 6th Int. Conf. Berchtesgaden, W. Germany, 1976) paper All, to be published.
[8] A. Samain, F. Koechlin, in Plasma Physics and Controlled Nuclear Fusion Research (Proc. 6th Int. Conf. Berchtesgaden, W. Germany, 1976) paper A8, to be published.
[9] S. Yoshikawa, et al., in Plasma Physics and Controlled Nuclear Fusion Research (Proc. 3rd. Int. Conf. Novosibirsk, USSR, 1968) 1, IAEA (1969) 403.
[10] L. A. Artsimovich, JETP Lett. 13 (1971) 70.
[11] S. Yoshikawa, Phys. Rev. Lett. 25 (1970) 353.
[12] D. R. Cohn, D. L. Jassby, R. R. Parker, Nucl. Fus. 16 (1976) 31.
[13] E. P. Gorbunov, S. V. Mirnov, V. S. Strelkov, Nucl. Fus. 10 (1970) 43.
[14] C. Daughney, Nucl. Fus. 15 (1975) 967.
[15] Symposium on Plasma Heating in Toroidal Devices (Proc. 3rd Symp., Varenna, 1976) to be published.
[16] V. V. Alikaev, et al., Moscow Institute of Atomic Energy Report IAE-2564 (1975); Oak Ridge National Laboratory Report ORNL-TR-4080 (1976).
[17] D. R. Sweetman, Nucl. Fus. 13 (1973) 157.
[18] L. A. Berry, et al., in Plasma Physics and Controlled Nuclear Fusion Research (Proc. 5th Int. Conf. Tokyo, Japan, 1974) 1, IAEA (1975) 113.

[19] J. M. Dawson, H. P. Furth, F. H. Tenney, Phys. Rev. Lett. 26 (1971) 1156.
[20] A. H. Spano, Nucl. Fus. 15 (1975) 909.
[21] D. L. Jassby, R. J. Goldston, Princeton Plasma Physics Laboratory Report MATT-1206 (1976), to be published in Nuclear Fusion.
[22] R. A. Ellis, et al., Princeton Plasma Physics Laboratory Report MATT-1202 (1976), to be published in Nuclear Fusion.
[23] J. P. Girard, in Plasma Physics and Controlled Nuclear Fusion Research (Proc. 6th Int. Conf. Berchtesgaden, W. Germany, 1976) paper A4-2, to be published.
[24] J. F. Clarke, Oak Ridge National Laboratory Report ORNL-TM-5429 (1976).
[25] A. H. Glasser, H. P. Furth, P. H. Rutherford, submitted to Physical Review Letters.
[26] P. E. Stott, C. C. Daughney, R. A. Ellis, Jr., Nucl. Fus. 15 (1975) 431.
[27] D. E. Post, et al., Princeton Plasma Physics Laboratory Report MATT-1262 (1976), submitted to Nuclear Fusion.
[28] L. Spitzer, Phys. Fluids 1 (1958) 253.

MHD Equilibrium and Stability

D. Pfirsch

Max-Planck-Institut für Plasmaphysik, 8046 Garching bei München, Federal Republic of Germany

ABSTRACT

The first part of this lecture discusses the influence of current profiles and noncircular cross-sections on the maximum ß obtainable in a Tokamak from the MHD equilibrium point of view. The second part treats limitations on such MHD equilibria resulting from various MHD instabilities like external and internal kinks, localized and nonlocalized modes- and axisymmetric instabilities.

1. Equilibrium

In a Tokamak there is an axisymmetric toroidal field B_t generated by external coils. In this field one has the plasma, which carries a toroidal current. This current leads to a poloidal field B_p around the minor circumference of the plasma torus. Both fields together form a screw-like field. In addition, one needs weaker vertical fields, generated by toroidal coils.

There is, in principle, no axisymmetric equilibrium without having a toroidal current and the corresponding poloidal field. In this case the plasma would, so to speak, slip through the toroidal field. On the other hand, there exist equilibria with no toroidal field present. Such configurations would, however, be strongly unstable. To avoid at least the most dangerous instability, one has to apply a toroidal field of such strength that the screw-like field winds at most once around the minor circumference when going once around the major circumference. With

$$\frac{1}{q} = \frac{\text{number of turns the short way}}{\text{number of turns the long way}}$$

it is necessary that $\frac{1}{q} < 1$ or $q > 1$

for this instability not to occur. This is the famous Kruskal-Shafranov condition, which will be derived below by some rough arguments.

The larger q the more likely it is that one can avoid instability. q is therefore called the safety factor. Using this quantity q, we can write

$$\frac{B_t(a)}{B_p(a)} = \frac{R_o q}{a} = Aq \ , \ A = \frac{R_o}{a}$$

with R_o = major plasma radius, a = minor plasma radius and A = aspect ratio. From this relation we see that, unfortunately, much higher external fields are necessary than those generated by the plasma current.

The quantity

$$\beta = \frac{<p>}{B^2/8\pi},$$ which for economic reasons should be of the order of 10% in a reactor,

can then be decomposed into

$$\beta = \beta_p \frac{a^2}{R_o^2 q^2} = \beta_p \frac{1}{A^2 q^2} \text{ with } \beta_p = \frac{<p>}{B_p^2(a)/8\pi} \ .$$

β_p is called the poloidal β.

For $\beta_p \leq 1$ the equilibrium is caused by the poloidal field and the toroidal current as in a z-pinch. This is the regime in which most present-day Tokamaks are operated.

A reactor in this regime would have, for example, A = 3 and q = 2,5, which for β_p = 1 would result in

$\beta = 1,8\%.$

This would be much too small for a reactor to be economic.

In Alcator (MIT) and Pulsator (Garching), for example, one has $\beta_p > 1$. In this case it is mainly the toroidal field which confines the plasma. β_p values of about A/2 represent, however, the upper limit beyond which only strongly desturbed equilibria are possible, if at all.

The physics in this regime is as follows: Since B_{tor} is approximately proportional to 1/R, where R is the distance from the axis of symmetry, the poloidal current density j_p necessary for the equilibrium according to the equation

$$\frac{1}{c} \underline{j}_p \times \underline{B}_t = \nabla p$$

must be larger for large R than for small R. This results in a net current in the vertical direction, and this current has to flow back in a force-free way, i.e. along the lines of force. The net current density in the vertical direction is of the order $-\frac{r}{R} j_p$; the current density along the lines of force having a component in the vertical direction which is the negative of r/R j_p must be $\frac{B_t}{B_p}$ times as large as this quantity and therefore $j_{\parallel} \approx q \, j_p$.
A more exact expression is

$$j_{\parallel} = 2q \, j_p \sin \Theta,$$

where Θ is the angle with the vertical direction in a meridional plane.

The β_p limit mentioned above is essentially given by

$$j_{\parallel} \approx j_t \approx \frac{c}{4\pi} \frac{B_p}{r} :$$

127

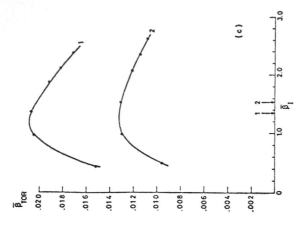

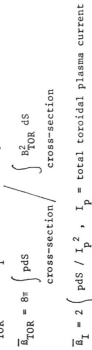

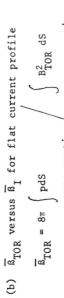

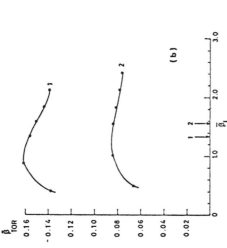

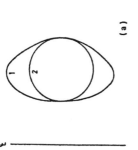

Fig 1: (a) 1: D-shaped; 2: circular cross-section; aspect ratio $\approx 2,4$

(b) $\bar{\beta}_{TOR}$ versus $\bar{\beta}_I$ for flat current profile

$$\bar{\beta}_{TOR} = 8\pi \underset{\text{cross-section}}{\int} pdS \bigg/ \underset{\text{cross-section}}{\int} B^2_{TOR}\, dS$$

$$\bar{\beta}_I = 2 \underset{\text{cross-section}}{\int} pdS \bigg/ I^2_p, \quad I_p = \text{total toroidal plasma current}$$

(c) $\bar{\beta}_{TOR}$ versus $\bar{\beta}_I$ for peaked current profile

In (b) and (c) arrows indicate the $\bar{\beta}_I$ values beyond which there is current reversal.

128

with $\nabla p \approx p/r$ we obtain $j_p \approx c \, p/B_t$ and therefore from $j_\parallel \approx j_t$

$$2q \; cp/B_t r \approx \frac{c}{4\pi} B_p/r$$

or $\beta_p = p/B_p^2/8\pi \approx \frac{1}{q} \frac{B_t}{B_p} = \frac{R}{r} = A.$

Using $\beta_p = \frac{1}{2} A$ as upper limit, we obtain for the maximum value of β

$$\beta_{max} = \frac{1}{2} \frac{1}{Aq^2} \; .$$

If we insert again $A = 3$ and $q = 2,5$, we get

$$\beta_{max} = 2,7 \; ,$$

which is still too small for an economical reactor.

In general, one has

$$q = \oint \frac{B_t}{B_p} \frac{dl}{2\pi R}$$

minor circumference on magnetic surface

and therefore roughly

$$q = \frac{B_t}{B_p} \cdot \frac{\text{minor circumference}}{\text{major circumference}} \; .$$

For a rectangular cross-section with half-axis a_z in the vertical and a_R in the horizontal direction, one has for $a_R < a_z$ roughly

$$\nabla p \approx p/a_R \; , \; j_t \approx \frac{c}{4\pi} \, B_p/a_R \; , \; j_\parallel \approx j_p \, a_R/R \cdot B_t/B_p, \; j_p \approx \frac{cp/a_R}{B_t} \; ,$$

$$\oint dl = 2(a_R + a_z)$$

and therefore

$$\beta_p \lesssim \frac{1}{2} R/a_R$$

and $\beta_{max} = \beta_p \dfrac{B_t^2}{B_p^2} \approx \dfrac{1}{2q^2 R/a_R} \dfrac{1}{4} \left(1 + \dfrac{a_z}{a_R} \right)^2 .$

Thus β increases by a factor of $2,25$ for $a_z/a_R = 2$ or by a factor of 4 for $a_z/a_R = 3$.

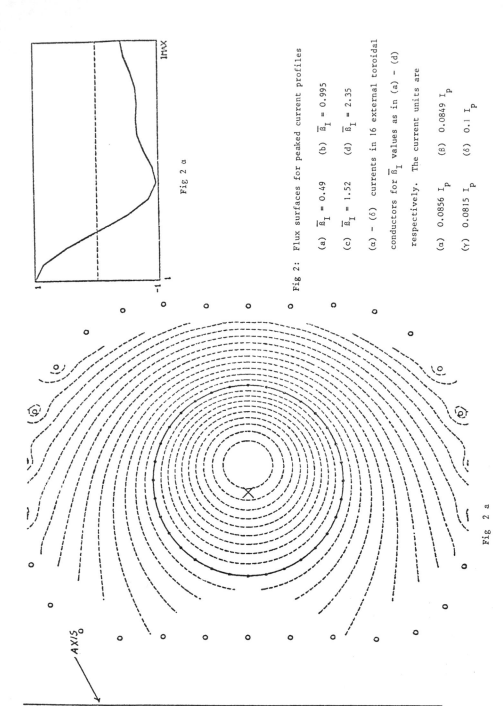

Fig 2α

Fig 2: Flux surfaces for peaked current profiles

(a) $\bar{\beta}_I = 0.49$ (b) $\bar{\beta}_I = 0.995$

(c) $\bar{\beta}_I = 1.52$ (d) $\bar{\beta}_I = 2.35$

(α) – (δ) currents in 16 external toroidal conductors for $\bar{\beta}_I$ values as in (a) – (d) respectively. The current units are

(α) 0.0856 I_p (β) 0.0849 I_p

(γ) 0.0815 I_p (δ) 0.1 I_p

Fig 2 a

In reality, things are, of course, somewhat more complicated.
Figs. 1 and 2 give characteristics of toroidal equilibria with
flat and peaked current profiles, where peaked means
$j_{tor} \sim \psi \sim (1 - \frac{r^2}{a^2})$. [1] With the D-shaped cross-section the ratio
of the half-axes is about 1.7. There are two remarkable results:

1. There is no further increase in β_{tor} for $\beta_{pol} > 1$, which is
   due to the definition of ß as

   $$\beta = \frac{\int p \; dS}{\int B_t^2/8\pi \; dS}$$

   together with the distortion of the magnetic surfaces. This
   implies that only part of the cross-section is really filled
   up with plasma, and that correspondingly the total plasma
   current decreases, keeping q on axis equal to 1.

2. There is a very strong influence of the current distribution
   on ß. In a reactor it is more likely to have a peaked current
   profile because of α-particle heating, and therefore ß values
   beyond 3% will hardly be obtained, even with moderately noncircular cross-sections.

K. Lackner [2] has calculated the series of equilibria shown in
Figs. 2 a-d and 3 a-d. In these equilibria a roughly circular
cross-section of the plasma boundary is prescribed. It can be seen that
the deformation of the magnetic surfaces increases with increasing β_{pol}.
Figs. 2 α-δ and 3 α-δ show the currents in 16 external toroidal
conductors, which are necessary to obtain the almost circular
cross-sections. The sum of the absolute values of these currents
reaches the order of the toroidal plasma current for $\beta_p > 1$,
which might cause some technical problems in a reactor.

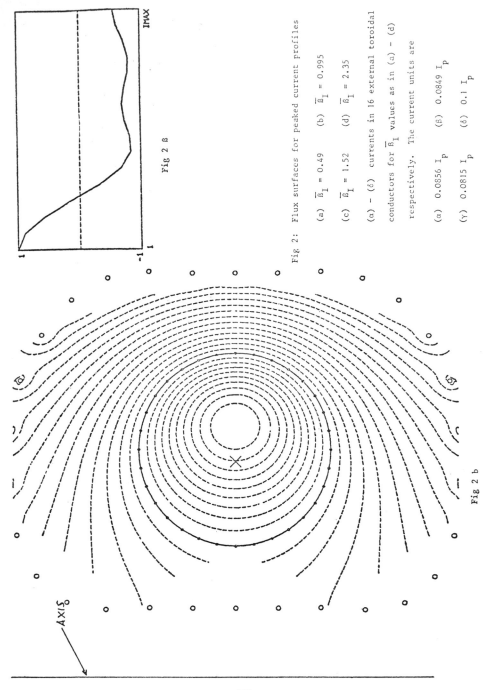

Fig 2 β

Fig 2 b

Fig 2: Flux surfaces for peaked current profiles

(a) $\overline{\beta}_I = 0.49$ (b) $\overline{\beta}_I = 0.995$

(c) $\overline{\beta}_I = 1.52$ (d) $\overline{\beta}_I = 2.35$

(α) - (δ) currents in 16 external toroidal

conductors for $\overline{\beta}_I$ values as in (a) - (d)

respectively. The current units are

(α) $0.0856 \, I_p$ (β) $0.0849 \, I_p$

(γ) $0.0815 \, I_p$ (δ) $0.1 \, I_p$

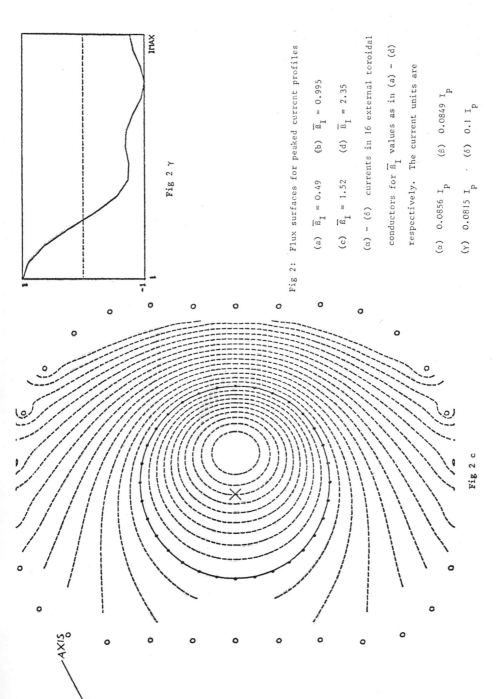

Fig 2 γ

Fig 2: Flux surfaces for peaked current profiles

(a) $\bar{\beta}_I = 0.49$ (b) $\bar{\beta}_I = 0.995$

(c) $\bar{\beta}_I = 1.52$ (d) $\bar{\beta}_I = 2.35$

(α) - (δ) currents in 16 external toroidal
conductors for $\bar{\beta}_I$ values as in (a) - (d)
respectively. The current units are

(α) 0.0856 I_p (β) 0.0849 I_p

(γ) 0.0815 I_p · (δ) 0.1 I_p

Fig 2 c

AXIS

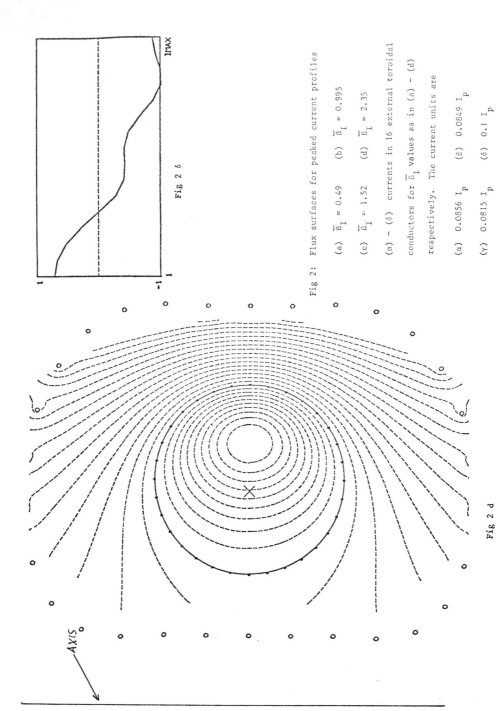

Fig 2 δ

Fig 2: Flux surfaces for peaked current profiles

(a) $\overline{\beta}_I = 0.49$ (b) $\overline{\beta}_I = 0.995$

(c) $\overline{\beta}_I = 1.52$ (d) $\overline{\beta}_I = 2.35$

(α) – (δ) currents in 16 external toroidal conductors for $\overline{\beta}_I$ values as in (a) – (d) respectively. The current units are

(α) 0.0856 I_p (β) 0.0849 I_p

(γ) 0.0815 I_p (δ) 0.1 I_p

Fig 2 d

AXIS

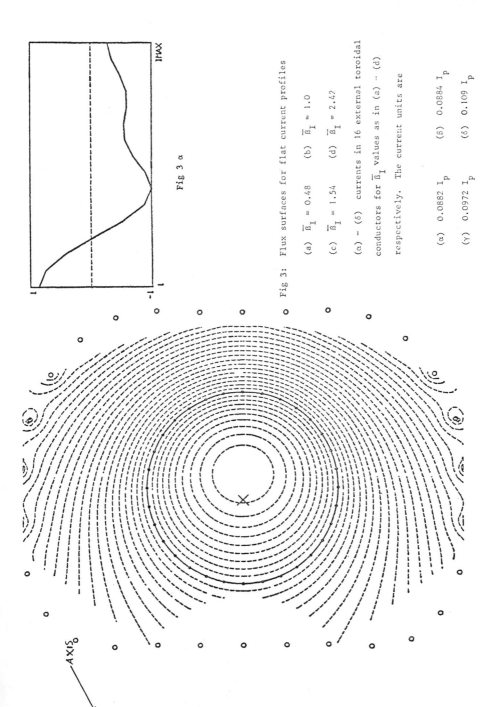

Fig 3 α

Fig 3 a

Fig 3: Flux surfaces for flat current profiles

(a) $\bar{\beta}_I = 0.48$ (b) $\bar{\beta}_I = 1.0$

(c) $\bar{\beta}_I = 1.54$ (d) $\bar{\beta}_I = 2.42$

(α) – (δ) currents in 16 external toroidal conductors for $\bar{\beta}_I$ values as in (a) – (d) respectively. The current units are

(α) 0.0882 I_p (β) 0.0884 I_p

(γ) 0.0972 I_p (δ) 0.109 I_p

AXIS

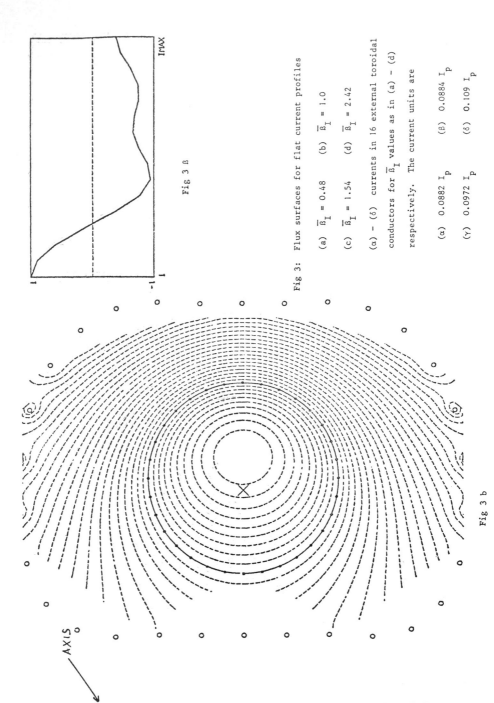

Fig 3 β

Fig 3 b

Fig 3: Flux surfaces for flat current profiles

(a) $\overline{\beta}_I = 0.48$ (b) $\overline{\beta}_I = 1.0$

(c) $\overline{\beta}_I = 1.54$ (d) $\overline{\beta}_I = 2.42$

(α) – (δ) currents in 16 external toroidal conductors for $\overline{\beta}_I$ values as in (a) – (d) respectively. The current units are

(α) 0.0882 I_p (β) 0.0884 I_p

(γ) 0.0972 I_p (δ) 0.109 I_p

AXIS

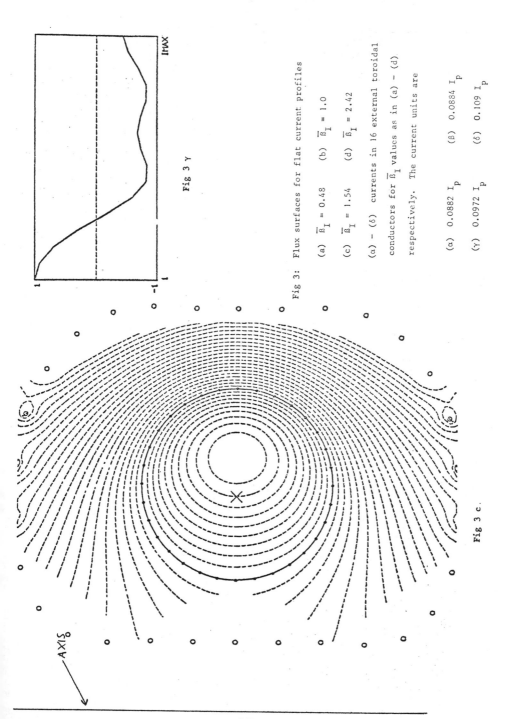

Fig 3 γ

Fig 3 c.

Fig 3: Flux surfaces for flat current profiles

(a) $\bar{\beta}_I = 0.48$ (b) $\bar{\beta}_I = 1.0$

(c) $\bar{\beta}_I = 1.54$ (d) $\bar{\beta}_I = 2.42$

(α) – (δ) currents in 16 external toroidal conductors for $\bar{\beta}_I$ values as in (a) – (d) respectively. The current units are

(α)   0.0882 I_p (β)   0.0884 I_p

(γ)   0.0972 I_p (δ)   0.109 I_p

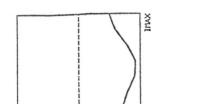

Fig 3 δ

Fig 3: Flux surfaces for flat current profiles

(a) $\overline{\beta}_I = 0.48$ (b) $\overline{\beta}_I = 1.0$

(c) $\overline{\beta}_I = 1.54$ (d) $\overline{\beta}_I = 2.42$

(α) – (δ) currents in 16 external toroidal
conductors for $\overline{\beta}_I$ values as in (a) – (d)
respectively. The current units are

(α) $0.0882\ I_p$ (β) $0.0884\ I_p$

(γ) $0.0972\ I_p$ (δ) $0.109\ I_p$

AXIS

Fig 3 d

It is sometimes said that equilibria having current reversal cannot be kept quasi-stationary if finite conductivity is taken into account. This, in the author's opinion, is not true. Of course, the externally induced electric field alone cannot sustain such current distributions. There is, however, no difficulty with such distributions if plasma convection is allowed for. This question is discussed in more detail in the author's lecture on classical diffusion in toroidal systems.

To conclude the discussion on Tokamak equilibria, a few analytical results for flat current profiles (j_t independent of ψ) shall be given:

Flux function:

$$\psi = \frac{2\pi \, |p'|}{1+\alpha^2} \left[z^2 (R^2 - \gamma) + \frac{\alpha^2}{4} (R^2 - R_o^2)^2 \right] \, , \quad p' = \frac{dp}{d\psi} = \text{const}$$

Separatrix:

$$R_s^2 = \gamma \quad , \qquad -\frac{\alpha}{\sqrt{2}} \sqrt{R_o^2 - \gamma} < z < \frac{\alpha}{\sqrt{2}} \sqrt{R_o^2 - \gamma}$$

and $R_s^2 = 2R_o^2 - \gamma - \frac{4}{\alpha^2} z^2$

Pressure profile:

$$p = P_{max} - |p'|\psi \, , \qquad P_{max} = |p'|\psi_s = \frac{\pi}{2} |p'|^2 \frac{\alpha^2}{1+\alpha^2} (R_o^2 - \gamma)^2$$

with $\psi_s = \psi$ on separatrix, therefore

$$p = P_{max} \left[1 - \frac{4}{\alpha^2} \frac{z^2 (R^2 - \gamma)}{(R_o^2 - \gamma)^2} - \frac{(R^2 - R_o^2)^2}{(R_o^2 - \gamma)^2} \right] \, .$$

Volume average of p inside separatrix:

$$<p> = \frac{16}{35} P_{max}$$

Ratio of half-axes close to magnetic axis:

$$\left(\frac{a_z}{a_R}\right)_o^2 = \alpha^2 \frac{R_o^2}{R_o^2 - \gamma} = \alpha^2 \frac{A^2+1}{2A}$$

$$A = \text{aspect ratio} = \frac{R_{s\ max} + R_{s\ min}}{R_{s\ max} - R_{s\ min}}$$

Safety factor q_o on magnetic axis:

$$q_o^2 = \frac{4}{35} \left(1 + \left(\frac{a_z}{a_R}\right)_o^2 \frac{2A}{A^2+1}\right) \frac{1}{<\beta>} \frac{2A}{A^2+1}$$

with $\quad <\beta> = \dfrac{<p>}{B_t^2(o)/8\pi}$

Mean plasma β for $q_o = 1$:

$$<\beta> = \frac{4}{35} \left(1 + \left(\frac{a_z}{a_R}\right)_o^2 \frac{2A}{A^2+1}\right) \frac{2A}{A^2+1} \quad .$$

With $A = 3$, $\quad a_z/a_R = 2 \quad$ one obtains $<\beta> = 23\%$, or

with $A = 3$, $\quad a_z/a_R = 1 \quad$ one obtains $<\beta> = 11\%$.

These would be reasonably high values of β.

As already mentioned, however, a more realistic peaked current profile will reduce these values by a factor of roughly 10.

2. MHD Instabilities

We can distinguish 4 classes of instabilities:

a) external kinks

b) internal kinks

c) localized modes

d) slip modes and general axisymmetric modes.

External kinks are global modes influencing also the vacuum region between the plasma boundary and a conducting wall.

Internal kinks are extended over a larger fraction of the plasma region but leave the plasma surface and therefore also the vacuum region unperturbed. Localized modes are localized around certain magnetic surfaces within the plasma. Slip modes are axisymmetric modes which leave the toroidal field unperturbed.

2 a) External kinks:

These are the most severe modes. The most dangerous of these are the $m = 1$ modes, where m is the mode number around the minor circumference defined in the following way:
In a cylindrical coordinate system a, θ, z perturbations of a circular cylinder with periodicity length $2\pi R$ along the z direction are proportional to exp $(im\theta - in \frac{z}{R})$ or to cos $(m\theta - n \frac{z}{R} + \phi(r))$.

Thus $m = 1$ means only a shift of the plasma without deformation, whereas $m \geq 2$ means a deformation without a shift. If the geometrical structure of an $m = 1$ perturbed plasma coincides with the geometrical structure of the magnetic field outside

141

the plasma but inside a conducting wall, then the plasma can slip through the magnetic field as if there were no "toroidal" field in the z direction, and this is an unstable configuration.

Going once around the major circumference is equivalent to increasing z by $2\pi R$. Constant perturbation therefore means a change of θ by $2\pi \frac{n}{m}$, or the number of turns the short way per turns the long way is just $\theta/2\pi = n/m$. The geometrical structure of the magnetic field therefore coincides with the geometrical structure of a possible perturbation if $q = m/n$ somewhere between the plasma boundary and the wall. We therefore easily obtain instability boundaries: If a is the plasma radius and b the wall radius, then we find from $q(r) = \frac{rB_z}{RB_\theta} \sim r^2$, because $B_z \approx$ const and $B_\theta \sim 1/r$, in the vacuum region

for $\qquad q(r=a) = \frac{m}{n} \qquad$ that $\qquad q(a) = \frac{m}{n}$

and for $q(r=b) = \frac{m}{n} \qquad$ that $\qquad q(a) = \frac{m}{n}\frac{a^2}{b^2}$.

We would therefore have instability for

$$m\frac{a^2}{b^2} < nq(a) < m .$$

But this is true only of perturbations without deformation of the cross-section, i.e. for $m = 1$. Thus we have instability of $m = 1$ modes if it holds that

$$\frac{a^2}{b^2} < nq(a) < 1 .$$

Most critical is obviously the case $n = 1$, which leads to

$$\frac{a^2}{b^2} < q(a) < 1 \quad \text{for instability.}$$

This is the famous Kruskal-Shafranov condition.

The relation for q(a) giving instability which is also valid
for $m \neq 1$ is, for $a/R = \varepsilon \ll 1$, $B_\theta/B_z = O(\varepsilon)$, $\beta_{pol} = O(1)$,
$q = O(1)$ and hence for $\beta = O(\varepsilon^2)$, $p = O(\varepsilon^2)$, which is called
the usual Tokamak scaling, and for j_z = const, as follows:

$$ m - 1 + \left(\frac{a}{b}\right)^{2m} < nq < m . $$

This is more restricted than the former condition

$$ m\left(\frac{a}{b}\right)^2 < nq < m . $$

It coincides with the exact condition only for $m = 1$ (or $a = b$)
since only in the latter case does no deformation work have to
be done. The dependence $(a/b)^{2m}$ instead of $(a/b)^2$ results from
the short range of higher m-fields, for which wall stabilization
is therefore weak. Taking all m-modes we have the following
instability pattern:

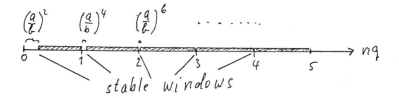

Of course, j_z = const and cylindrical approximation is not
realistic. Toroidal effects, for instance, lead to broadening
of the unstable regions. [7] More realistic results concerning
current profile effects are obtained mainly from numerical
calculations. Such results as found by Wesson [3] are shown
in Fig. 4. These results exhibit a stabilizing effect of current
peaking. Unfortunately, this means, as we have seen before,

143

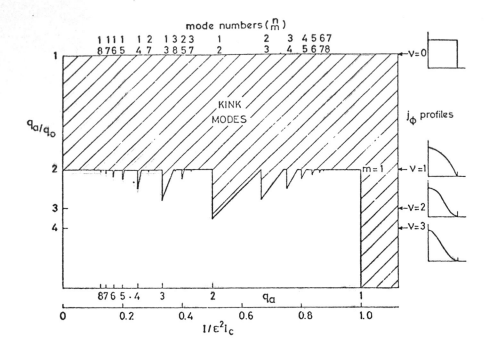

Fig 4: Stability diagram for kink modes. The shaded region is unstable.

$\varepsilon = a/R$, I_c = toroidal field current

ν = exponent in formula for the toroidal current profile:

$$j_t = j_{to} \, (1 - r^2/a^2)^{\nu}$$

144

lowering of ß if $q_o > 1$ is required for reasons of internal instability.

The growth rates of such instabilities can be estimated in the following way: Since a long Θ-pinch is marginally stable, the instability of slender tori is mainly due to the longitudinal current j_ϕ and the corresponding poloidal field B_p. Thus the growth rates are of the order

$$\gamma \approx v_{Ap}/a \quad , \qquad v_{Ap} = \sqrt{\frac{B_p^2}{4\pi\rho}} \quad ,$$

or, written with the full Alfvén speed $\quad v_A = \sqrt{\frac{B^2}{4\pi\rho}} \quad ,$

$$\gamma^2 \approx \frac{v_A^2}{a^2} \cdot \epsilon^2 \quad , \qquad \epsilon = \frac{a}{R} \quad .$$

We speak in this case of current driven instabilities.

There are a few papers discussing toroidal effects and the influence of noncircular cross-sections on the stability behaviour of such plasmas. Calculations are usually done with flat current profiles. They show no major changes compared to the results for linear configurations with circular cross-sections. Plasmas with noncircular cross-sections appear to be rather more unstable (Kerner [4], Laval, Pellat, Soulé [5]). There is therefore obviously no hope to find stable configurations with flat current profiles, making it more difficult to find configurations with reasonable ß values.

There is an indication that a pressureless plasma between the plasma and the wall can increase ß by, say, a factor of two [6]. But, owing to the finite conductivity of the wall and plasma, this would still imply weaker instabilities, which would have to be feedback stabilized.

To conclude this review of external modes, a few words shall be said on the so-called peeling mode, an external mode localized close to the boundary. It was first discussed by Frieman et al. [7] A necessary condition for stability for general equilibria was derived by Lortz [8]. For the cylindrical case it requires decreasing rotational transform for stability and a pressure gradient which must not be too large:

$$- \frac{4\pi}{B_t^2} \frac{rdp}{dr} < \begin{cases} \frac{1}{8} \kappa^2 \, , & -4 < \kappa < 0 \\ -\kappa - 2 \, , & R < -4 \end{cases} \quad , \quad \kappa = \frac{d\ln\iota}{d\ln r} < 0$$

There is another necessary condition valid in cylindrical geometry found by Wesson [3] which is stronger than Lortz's condition and states that $j_\phi \neq 0$. Thus flat profiles would certainly be unstable.

2 b) Internal modes:

There is a similarity to external kinks in that a q-value of $\frac{m}{n}$, now inside the plasma, again allows unstable perturbations of the same geometrical structure as the magnetic field, i.e. the perturbations are nearly constant along the field. If a mode is localized around a q = m/n surface, Mercier's criterion for Tokamaks with circular cross-sections and large aspect ratios

146

yields stability for

$$\frac{8\pi}{B_t^2} p'(1-q^2) + \frac{r}{4}\left(\frac{q'}{q}\right)^2 > 0 .$$

Since p' < 0, only q > 1 guarantees stability; otherwise, at
least for r → 0, the condition is violated. The factor $1 - q^2$
is a toroidal effect resulting from the toroidal magnetic well.
It is missing in the corresponding Suydam criterion for linear
configurations. Applying Mercier's criterion to configurations
with noncircular cross-sections having certain combined elliptical
and triangular deformations yields critical q-values below 1.
However, Mercier's criterion is only necessary for stability.
Calculations by Laval [9] using a step profile for the current
density and Strauss [10] using a peaked current profile show that
there are strongly unstable nonlocalized internal modes as soon
as q on axis falls below 1. Nonlinear computer calculations
by Strauss with $a_z/a_R = 2$ and $q_o = 0,82$ show that small magnetic
perturbations are accompanied by substantial plasma convection.

Using the sufficient criterion of Lortz [11] for internal modes,
Lortz and Nührenberg [12] found stable configurations up to
ß-values of 1%; but such configurations, which include current
peaking, elliptical and triangular deformations, would have an
aspect ratio of 14.6 . For an aspect ratio of 5 they could only
find ß's of 0.2%. Whereas Laval and Strauss have assumed infinite
aspect ratio, there is no assumption concerning A in the treatment
of Lortz and Nührenberg.

147

For circular cross-sections the internal instabilities are
pressure driven. This is obvious for the localized modes
from Mercier's criterion. The growth rates are therefore
described by the sound velocity instead of the Alfvén
velocity, i.e.

$$\gamma^2 \approx \frac{v_s^2}{a^2} \cdot \varepsilon^2 = \beta \frac{v_A^2}{a^2} \cdot \varepsilon^2 \sim \varepsilon\, 4 \quad \text{since} \quad \beta \sim \varepsilon^2 \quad .$$

This means that, in general, toroidal effects influence such
instabilities. With noncircular cross-sections the growth
rates are similar to those for external modes. Toroidal
effects should therefore be of minor importance in these cases.

2 c) Axisymmetric modes:

An unfavourable effect of noncircular cross-sections is the
existence of unstable slip motions, discussed mainly by Laval,
Pellat, Soulé, Lackner, McMahon, Rebhan, Salat and Rosen [13].
Slip motions are defined as leaving the toroidal field unaltered.
Thus, increasing this field has no stabilizing effect. Special
cases are rigid vertical displacements. The most general slip
perturbation was found by Rebhan [14]. The results show that
only very moderate deformations are allowed, say a ratio of the
half-axes of about 1.4 for A = 3, while larger A reduces this
value. Wall stabilization can reduce the growth rates, but
still makes feedback stabilization necessary.

All this shows that it is not possible at present to say that
noncircular cross-sections would yield an interesting improvement
with regard to β over circular cross-section Tokamaks. It appears

therefore highly desirable to do experimental configuration studies with medium-sized Tokamaks.

On the whole it can be stated that the equilibrium and stability situation is satisfactory as far as physics experiments are concerned. For reactors, however, the situation is rather critical.

References

[1]   B.J. Green, J. Jacquinot, K. Lackner and A. Gibson
Nuclear Fusion 16, 521 (1976)

[2]   K. Lackner, private communication

[3]   J.A. Wesson, Proc. VIIth European Conference on Controlled
Fusion and Plasma Physics, Lausanne 1975, Vol. II p. 102

[4]   W. Kerner, Nuclear Fusion 16, 643 (1976)

[5]   G. Laval, R. Pellat, J.L. Soulé, Phys. Fluids 17, 835 (1974)

[6]   G. Freidberg in D.J. Sigmar, MIT Report PRR 7526 p. 7 (1975)

[7]   E. Frieman, J. Greene, J. Johnson, K. Weimar
Phys. Fluids 16, 1108 (1973)

[8]   D. Lortz, Nuclear Fusion 15, 49 (1975)

[9]   G. Laval, Phys. Rev. Letters 34, 1316 (1975)

[10]   H.R. Strauss, Phys. Fluids 19, 134 (1976)

[11]   D. Lortz, Nuclear Fusion 13, 817 (1973)

[12]   D. Lortz, J. Nührenberg, Proc. of 5th IAEA Conference on
Plasma Physics and Controlled Nuclear Fusion Research,
Tokyo (1974), Vol. I p. 439

[13]   For references see e.g. E. Rebhan, A. Salat
Max-Planck-Institut für Plasmaphysik, Report IPP 6/148, June 1976

[14]   E. Rebhan, Nucl. Fusion 15, 277 (1975)

Collisional Transport

D. Pfirsch

Max-Planck-Institut für Plasmaphysik, 8046 Garching bei München,
Federal Republic of Germany

ABSTRACT

Collisional particle and heat transport is treated in plane
and toroidal geometry. In particular, temperature gradient effects
on impurity diffusion - so-called temperature screening - are
considered for the different collisional regimes. The existence of
quasi-stationary self-consistent Tokamak equilibria with finite
resistivity and a possible limitation of the maximum ß caused by
particle diffusion is discussed.

A) Plane Geometry

The case of plane geometry serves, on the one hand, as a

reference case and, on the other, it can be of interest for Tokamaks

with highly elongated cross-sections.

1) Ambipolarity

In this section the motion of single particles in a magnetic

field $\underline{B}$ which is constant to lowest order in the gyro-radius is

considered, and the influence of collisions on this motion is

investigated. Let $\underline{B}$ have only a z component in a rectangular

coordinate system x,y,z, $B = (0,0,B)$, then the unperturbed motion

of a particle is described by

$$m\ddot{x} = + \frac{e}{c} \dot{y} B, \qquad m\ddot{y} = - \frac{e}{c} \dot{x} B \quad .$$

For $B \approx$ const to lowest order in the gyro-radius we obtain

with $w = x + iy$

$$\ddot{w} = - i \omega_g \dot{w} , \quad \dot{w} = \dot{w}_o e^{-i\omega_g t} , \quad w = w_o - \frac{1}{i\omega_g} \dot{w}_o e^{-i\omega_g t} ,$$

where w_o is the complex gyro-centre position.

If the particle undergoes a collision with a sudden change in w but with w unchanged, one obtains a change in w_0 given by

$$\Delta w_0 = \frac{1}{i\omega_g}\, \Delta \dot{w}_0\, e^{-i\omega_g t} = \frac{1}{i}\, \frac{\Delta \dot{m w} \cdot c}{e B} \ .$$

Using vector symbols $\underline{r}_0 = (x_0, y_0)$, $\underline{v} = (\dot{x}, \dot{y})$, we obtain

$$\Delta \underline{r}_0 = c\, \frac{\Delta m \underline{v} \ x \ \underline{B}}{e B^2} \ .$$

If we define the mean centre of charge of the colliding particles by $\ \Sigma e \underline{r}_0 / \Sigma |e| \ ,$

where the summation goes over the two colliding particles, then we get for its change during collision

$$\Sigma e \Delta \underline{r}_0 / \Sigma |e| = c\, \frac{\Sigma \Delta m \underline{v} \ x \ \underline{B}}{B^2\, \Sigma |e|} = 0 \quad \text{since } \Sigma \Delta m \underline{v} = 0 \ .$$

Thus, to lowest order in the gyro-radius there is ambipolar diffusion due to momentum conservation.

This holds very generally for all situations in which the change of the mean position of a particle during collision in the direction of the pressure gradient is proportional to the gyro-radius.

2) Classical Diffusion

From $\nabla p = \frac{1}{c}\, \underline{j} \ x \ \underline{B}$, $\eta \cdot \underline{j} = \underline{E} + \frac{1}{c}\, \underline{v} \ x \ \underline{B}$, and $\underline{E} = 0$ one finds

$$\nabla p = \frac{1}{\eta c^2}\, (\underline{v} \ x \ \underline{B}) \ x \ \underline{B} = \frac{1}{\eta c^2}\, (\underline{B}(\underline{v} \cdot \underline{B}) - \underline{v}\, B^2) \ ,$$

from which

$$\underline{v} = -\frac{\eta c^2}{B^2}\, \nabla p \equiv \underline{v}_{cl}$$

results. This is so-called classical diffusion. For $T = \text{const}$, which, in reality, is assumed here from the beginning, this can be written as

$$\underline{v} = - \frac{nc^2}{B^2} T \nabla n \equiv - \frac{D}{n} \nabla n$$

with the diffusion coefficient

$$D = nT\eta c^2/B^2.$$

The resistivity η can be expressed by the electron collision frequency

$$\eta = \nu_{ei} \frac{m}{e^2 n} \quad .$$

With this expression we obtain

$$D = \nu_{ei} \frac{mTc^2}{e^2 B^2} = \nu_{ei} \, r_{ge}^2 \quad .$$

Thus, the diffusion can be described as a stochastic process with the gyro-radius as the step size of a single process.

The magnetic diffusion coefficient describing the skin effect is

$$D_m = \frac{\eta c^2}{4\pi} \quad ,$$

which leads to the relation

$$D = \frac{4\pi nT}{B^2} D_m = \frac{1}{2} \, \beta \, D_m \quad , \qquad \beta = \frac{nT}{B^2/8\pi} \quad .$$

This means that particle diffusion is small compared with magnetic diffusion as long as the local β is smaller than 1. Using the expression for the Spitzer resistivity

$$\eta = 1,15 \times 10^{-14} \frac{Z \ln \Lambda}{T_{eV}^{3/2}}$$

and taking $Z = 1$, $T_{eV} = 10^4$ ($= 10$ keV) ,

we obtain

$$D_m = 0,82 \ \ln \Lambda \ \ cm^2/sec$$

$$15 \ cm^2/sec \quad .$$

To get a confinement time of 1 sec for $\beta = 1$, it would be enough to have a half-thickness of the plasma of the order of 5 cm. A half-thickness of 5m would lead to particle confinement times of 10^4 sec.

For comparison, Bohm diffusion $D_B = \frac{cT}{16eB}$ for $T = 10$ keV and $B = 40$ kG would give $D_B = 1.56 \; 10^6$ cm^2/sec.

3) Classical Heat Conduction Perpendicular to $\underline{B}$

The transport of heat perpendicular to $\underline{B}$ in the direction of ∇T can be described by a stochastic process similar to diffusion. Thus

$$\underline{q} = - \kappa \nabla T = - D_{particle} \; n \nabla T \; ,$$

where $D_{particle}$ now also contains contributions from equal particle collisions. Since $\nu_{ii} = \frac{m_i}{m_e} \nu_{ie}$ and electrons and ions give the same contribution to D as far as collisions between these two species are concerned, only ion-ion collisions are of interest. Thus, the time-scale for heat conduction is shorter than for diffusion by a factor of m_e/m_i , which is 1/61 in the case of deuterium and 1/74 for tritium.

4) Impurity Diffusion

Qualitatively, we have the following situation: Since the impurities interact much more strongly with the hydrogenic ions than with the electrons, the mean relative motion between impurities and ions is small compared with that between electrons and the heavy particles. We can therefore go into a system of reference in which all the heavy particles are at rest simultaneously. These particles are then not confined by a magnetic field but by an electrostatic space charge potential U(x) with, say, U(0) = 0, where x = 0 means the centre of the plasma. Only the electrons are confined magnetically. Then, since we have almost thermal equilibrium,

$$n_i(x) = n_i(0) \; e^{- \frac{e_i U(x)}{T_i}}$$

154

or
$$\frac{n_i(x)}{n_i(0)} = \left(\frac{n_k(x)}{n_k(0)}\right)^{\frac{e_i}{T_i}\frac{T_k}{e_k}} .$$

For $e_k = e$ (hydrogenic ions) and $e_i = Ze$, $T_i = T_k = T$ we have

$$\frac{n_Z(x)}{n_Z(0)} = \left(\frac{n_H(x)}{n_H(0)}\right)^Z ,$$

a result first obtained by J.B. Taylor [1]. Since $Z \geq 2$, all kinds of impurities are more concentrated in the centre of the plasma than the hydrogenic ions as far as this formula is applicable. A remarkable change of this result occurs if one takes into account a temperature gradient. One can do a very simple perturbation theory with respect to a temperature gradient in a kinetic equation of the Bathnager-Krook type, leading to a perturbation of a Maxwellian distribution f_M caused by gradients and electric fields:

$$\delta f^{(Z)} = \frac{-\nu_Z v_x + \omega_Z v_y}{\nu_Z^2 + \omega_Z^2}\left[\frac{\partial}{\partial x} - \frac{ZeE_x}{T}\right] f_M^{(Z)} , \quad \omega_Z = \frac{ZeB}{m_Z c} .$$

ν_Z is the collision frequency of impurities of charge Z with hydrogenic ions. Doing the corresponding calculation for the hydrogenic ions and imposing the condition of momentum conservation during collisions in the y direction, one can eliminate the electric field and obtains ambipolar diffusion in the x direction. The flux of the impurities calculated from $\delta f^{(Z)}$ is then

$$\Gamma_Z = \langle n v_x \rangle_Z = -\frac{\nu_Z n_Z}{m_Z \omega_Z^2}\frac{3}{2\pi} TZ\left[\frac{1}{Z}\frac{1}{n_Z}\frac{\partial n_Z}{\partial x} - \frac{1}{n_H}\frac{\partial n_H}{\partial x} + \frac{1}{2}\left(1 - \frac{1}{Z}\right)\frac{\partial T}{\partial x}\right].$$

If there were no temperature gradient, the stationary state would require that

$$\frac{1}{Z}\frac{1}{n_Z}\frac{\partial n_Z}{\partial x} = \frac{1}{n_H}\frac{\partial n_H}{\partial x}$$

or $\qquad n_Z \sim n_H^Z$

as we had found before. To find out what the influence of the temperature gradient is, it is convenient to discuss temperature profiles related to the hydrogenic density profile by

$$T(x) \sim n_H^\alpha \quad .$$

For n_Z we write

$$n_Z \sim n_H^\beta \quad .$$

The stationary state is then obtained for

$$\frac{\beta}{Z} - 1 + \frac{\alpha}{2}\left(1 - \frac{1}{Z}\right) = 0$$

or

$$\beta = Z - \frac{\alpha}{2}(Z-1) \quad .$$

For $\alpha < 2$ we have $\beta > 1$, assuming $Z > 2$, and for $\alpha > 2$ we have $\beta < 1$. Thus, up to $\alpha = 2$ there is impurity influx. For $\alpha > 2$ there would be impurity outflux, an effect which is called temperature screening. The physics of this effect in plane geometry is, as can easily be seen from the calculations, the v-dependence of the collision frequency ν_Z. If ν_Z were independent of v, impurities could never be thrown out by temperature gradients.

B) Toroidal Geometry

1) Pfirsch-Schlüter Regime

In toroidal geometry we have to distinguish whether collisions are frequent or rare. The macroscopic theory applies if the mean free path of the particles is shorter than the connection length qR after which a field line has appreciably changed its meridional position on a magnetic

156

surface. In this case, the only problem in establishing an equilibrium
is that, unlike in the plane geometry, currents parallel to $\underline{B}$ also have
to be driven against the plasma resistance. This cannot be done
directly by diffusion through a $\frac{1}{c} \underline{v} \times \underline{B}$ term in Ohm's law since such a
term acts only perpendicularly to $\underline{B}$. We therefore need, in addition,
an electric field to drive the parallel currents. Such fields arise
automatically by charge separation because the diamagnetic current is
not divergence-free. But such space charge fields have to be pure
poloidal fields because of axial symmetry. The projection of these
fields on the field lines must drive $j_{\parallel}$ and is therefore given by
$n_{\parallel} \, j_{\parallel} \approx n_{\parallel} \, q \, j_{diam}$. The poloidal field therefore has the strength
$n_{\parallel} \, q \, j_{diam} \cdot \frac{qR}{r}$. The component of this field perpendicular to $\underline{B}$
has to be transformed away by a $\frac{1}{c} \underline{v} \times \underline{B}$ term otherwise it would drive
a current $\frac{n_{\parallel}}{n_{\perp}} q \, \frac{qR}{r} \, j_{diam} \gg j_{diam}$.

The velocity, being of the $\underline{E} \times \underline{B}$ type, has to be in the poloidal and
mainly outward direction (Fig. 1) and is obviously of the order

$$v_p \approx \frac{n_{\parallel}}{n_{\perp}} \, 2 \, \frac{q^2 R}{r} \, v_{c1} \quad .$$

To ensure an almost divergence-free motion of the plasma with a
stationary density profile, the plasma has to flow back along the lines
of force, the necessary velocity being

$$v_{\parallel} \approx \frac{qR}{r} \, v_p \quad .$$

But a fraction r/R of the poloidal particle flux leaks out, giving rise
to a diffusion velocity

$$v_{PS} \approx \frac{r}{R} \, v_p \approx \frac{n_{\parallel}}{n_{\perp}} \, 2q^2 \, v_{c1} = q^2 v_{c1} \quad .$$

This velocity means that the expansive energy corresponding to v_{pS} not
only goes into Joule dissipation of the diamagnetic current but also

157

into Joule dissipation of the secondary current, which is a factor $\frac{n_{\parallel}}{n_{\perp}} 2q^2$ larger than the first one.

This so-called Pfirsch-Schlüter factor q^2 is also present in heat conduction. The physics there, however, is very different from that for diffusion. It has to do with a heat flow

$$\underline{q}_{k\perp} = \frac{5}{2} \frac{n_k T_k}{e_n B^2} (\underline{B} \times \nabla T_k)$$

which is perpendicular to ∇T. Since it is not divergence-free, it leads to a heat flow parallel to $\underline{B}$ and, consequently, to a poloidal temperature gradient. With this gradient the above formula yields a radial heat flow which is then of the order of q^2 times the plane value.

The above discussion of particle diffusion and plasma convection does not lead to a condition that current reversal in the toroidal direction should not be allowed. Strictly speaking, however, the theory applies only to low-ß stellarators. There might arise severe changes if one treats Tokamak equilibria in a self-consistent way. The latter shall be done now.

In a cylindrical coordinate system R, ϕ, Z one can describe an axisymmetric field configuration by

$$\underline{B}_p = \nabla \psi \times \nabla \phi \quad , \quad \underline{B}_t = F(\psi) \nabla \phi \ .$$

The pressure p is a function of the flux function ψ alone, and because of the large heat conduction of the electrons parallel to $\underline{B}$ it is reasonable to assume that the temperature is also a function of ψ alone. For simplicity, we neglect anisotropy effects in the resistivity η . It thus holds that

158

$$p = p(\psi), \quad \eta = \eta(\psi), \quad n = n(\psi),$$

where n is the particle density.

The question is now what functions $p(\psi)$ and $F(\psi)$ are possible for self-consistent equilibria.

The answer follows from the question how the currents necessary for the equilibria can be driven. The current densities are

$$\frac{4\pi}{c} j_t = Rp' + \frac{1}{R}\frac{1}{2}(F^2)', \quad \frac{4\pi}{c} j_p = F'\nabla\psi \times \nabla\phi = F'\underline{B}_p.$$

The toroidal current can be driven partly by a loop voltage $2\pi U$ or the corresponding electric field U/R, partly by plasma convection. It thus follows that

$$\eta j_t = U/R + \frac{1}{c}(\underline{v} \times \underline{B})_\phi.$$

With

$$(\underline{v} \times \underline{B})_\phi = (\underline{v}\times(\nabla\psi \times \nabla\phi))_\phi = \left[-\nabla\phi(\underline{v}\cdot\nabla\psi) + \nabla\psi(\underline{v}\cdot\nabla\phi)\right]_\phi = -\frac{1}{R}\underline{v}\cdot\nabla\psi,$$

we find

$$\frac{1}{c}\frac{1}{R}\underline{v}\cdot\nabla\psi = U/R - \eta j_t = U/R - \eta(Rp' + \frac{1}{R}\frac{1}{2}(F^2)')\frac{c}{4\pi}.$$

If $-\dot{N}(\psi)$ is the total number of particles diffusing through a magnetic surface ψ = const per unit time, then

$$n(\psi)\int_{\psi=\text{const}} \underline{v}\cdot d\underline{S} = -\dot{N}.$$

It is

$$\int_{\psi=\text{const}} \underline{v}\cdot d\underline{S} = \oint_\psi 2\pi R \frac{\underline{v}\cdot\nabla\psi}{|\nabla\psi|} dl = 2\pi \oint_\psi \frac{\underline{v}\cdot\nabla\psi}{R} R \frac{dl}{B_p}.$$

The volume enclosed by a flux surface is

$$V(\psi) = \int\int 2\pi Rdl\left|\frac{d\psi}{\nabla\psi}\right| = \int^\psi d\psi \oint_\psi \frac{2\pi dl}{B_p}.$$

159

and therefore

$$V'(\psi) = 2\pi \oint_\psi \frac{dl}{B_p} \quad .$$

Defining averages by

$$< G > \equiv \frac{2\pi \oint_\psi G \frac{dl}{B_p}}{2\pi \oint_\psi \frac{dl}{B_p}} = \frac{1}{V'(\psi)} \; 2\pi \oint_\psi G \frac{dl}{B_p} \quad ,$$

we obtain

$$- \frac{\dot{N}}{nV'} = cU - \frac{nc^2}{4\pi} \left[p' <R^2> + \frac{1}{2} (F^2)' \right]$$

or

$$p' = \frac{1}{\eta <R^2>} \left[U + \frac{\dot{N}}{nV'} - \frac{\eta}{2} (F^2)' \right] \quad .$$

The poloidal current can partly be driven by a poloidal electric potential field $\underline{E} = - \nabla\Phi$ and partly again by convection. It thus follows that

$$\eta j_p = - \nabla\Phi + \frac{1}{c} (\underline{v} \times \underline{B})_p \quad .$$

For the last term here we find

$$(\underline{v} \times \underline{B})_p = (\underline{v} \times (\nabla\psi \times \nabla\phi))_p + (\underline{v} \times F\nabla\phi)_p = \nabla\psi \underline{v} \cdot \nabla\phi + F(\underline{v} \times \nabla\phi) \quad .$$

Therefore, it holds that

$$\frac{c\eta}{4\pi} F' \underline{B}_p = - \nabla\Phi + \nabla\psi \frac{1}{c} \underline{v} \cdot \nabla\phi + F \frac{1}{c} (\underline{v} \times \nabla\phi) \quad .$$

Uniquenes of the potential requires that

$$\oint_\psi \frac{B_p \cdot \nabla\Phi}{B_p} \, dl = 0 \quad .$$

With

$$\underline{B}_p \cdot (\underline{v} \times \nabla\phi) = (\underline{v} \times \nabla\phi) \cdot (\nabla\psi \times \nabla\phi) = \frac{1}{R^2} \underline{v} \cdot \nabla\psi$$

this condition yields

$$\oint dl \left[\frac{c\eta}{4\pi} F' B_p - \frac{1}{R^2} \frac{1}{B_p} \frac{1}{c} \underline{v} \cdot \nabla\psi \right] = \oint \frac{dl}{B_p} \left[\frac{c\eta}{4\pi} F' B_p{}^2 - \frac{1}{R^2} \frac{1}{c} \underline{v} \cdot \nabla\psi \right] = 0 \quad .$$

160

Inserting the expression for $\underline{v}\cdot\nabla\psi$ and using the definition of averages, we get

$$\frac{cn}{4\pi} F' <B_p^2> - <\frac{F}{R}\left[U/R - \frac{cn}{4\pi}(Rp' + \frac{1}{R}\frac{1}{2}(F^2)')\right] = 0$$

or

$$\frac{cn}{4\pi} F' <B_p^2> - FU<\frac{1}{R^2}> + \frac{cn}{4\pi} p'F + \frac{cn}{4\pi}\frac{1}{2}(F^2)'F<\frac{1}{R^2}> = 0$$

Multiplication with F and insertion of p' yields

$$\frac{cn}{4\pi}\frac{1}{2}(F^2)'<B_p^2> - F^2<\frac{1}{R^2}>\left[U - \frac{cn}{4\pi 2}(F^2)'\right] + F^2\frac{1}{<R^2>}\left[U + \frac{\dot{N}}{nV'}, \frac{cn}{4\pi 2}(F^2)'\right] = 0$$

or

$$\frac{cn}{4\pi 2}(F^2)'<B_p^2> = \frac{cn}{4\pi} p'F^2\left[<R^2> <\frac{1}{R^2}> - 1\right] + \frac{\dot{N}}{cnV'}, F^2<\frac{1}{R^2}>.$$

From this we obtain with

$$F^2<\frac{1}{R^2}> = <(\frac{F}{R})^2> = <B_t^2>$$

$$\dot{N} = \frac{nV'}{<B_t^2>}\frac{c^2n}{4\pi}\left[\frac{1}{2}(F^2)'<B_p^2> - p'<B_t^2>\left[<R^2> - <\frac{1}{R^2}>^{-1}\right]\right].$$

Multiplication with $\frac{\nabla\psi\cdot\nabla\psi}{|\nabla\psi|^2}$ yields with $|\nabla\psi|^2 = R^2 B_p^2$,

$$nV'\nabla\psi = nV , \quad \dot{N} = - n\underline{v}_D\cdot\nabla V , \quad p'\nabla\psi = \nabla p \text{ etc.}$$

$$\underline{v}_D = - \frac{n}{<B_t^2>}\left[\frac{1}{2}\frac{\nabla R^2 B_t^2}{R^2} - \nabla p\frac{B_t^2}{B_p^2}\frac{1}{R^2}\left[<R^2> - <\frac{1}{R^2}>^{-1}\right]\right].$$

For $\frac{1}{2}\frac{\nabla F^2}{R^2} \approx - \nabla p$ one has just the Pfirsch-Schlüter formula and it obviously does not matter how large $p/B_t^2/2$ is. Thus there is no condition that would not allow current reversal. There was only one relation between F and p which can be solved for p':

$$p' = - \frac{1}{2}(F^2)'<\frac{1}{R^2}>\left[1 + \frac{<B_p^2>}{<B_t^2>}\right] + \frac{4\pi cU}{c^2 n}<\frac{1}{R^2}>.$$

Using this, the toroidal current density becomes

$$j_t = R\left(\frac{1}{2}(F^2)'\left(\frac{1}{R^2} - \langle\frac{1}{R^2}\rangle\right) - \frac{1}{2}(F^2)'\langle\frac{1}{R^2}\rangle\frac{\langle B_p^2\rangle}{\langle B_t^2\rangle} + \frac{4\pi cU}{c^2\eta}\langle\frac{1}{R^2}\rangle\right).$$

U has to be chosen so as to guarantee q_o on axis greater 1. If the magnetic surfaces close to the magnetic axis have circular cross-sections, then

$$\frac{4\pi}{c}j_{to} = \frac{2B_t^o}{R_o q_o} \quad\text{and therefore}\quad \left[\frac{4\pi cU}{c^2\eta}\langle\frac{1}{R^2}\rangle\right]_o = \frac{4\pi j_t^o}{cR_o} = \frac{2B_t^o}{R_o^2 q_o}.$$

Current reversal can occur if $\frac{4\pi cU}{c^2\eta}\langle\frac{1}{R^2}\rangle$ becomes of the order of $\frac{1}{2}(F^2)'(\frac{1}{R^2} - \langle\frac{1}{R^2}\rangle)$ somewhere.

Since $\frac{1}{R^2} - \langle\frac{1}{R^2}\rangle \sim \frac{2r}{R}\frac{1}{R^2}$, $\frac{1}{2}(F^2)' \approx R^2 p'$,

we get $\frac{1}{2}(F^2)'(\frac{1}{R^2} - \langle\frac{1}{R^2}\rangle) \approx p'\frac{2r}{R} = \frac{|\nabla p|\frac{2r}{R}}{RB_p} \approx \frac{\frac{2r}{a^2}p_o\frac{2r}{R}}{RB_p}$.

The ratio of this over the above quantity at r is therefore

$$\frac{\frac{2r}{a^2}p_o\frac{2r}{R}}{RB_p} \cdot \frac{R^2 q_o\eta}{2B_t\eta_o} = \frac{2r^2}{a^2}\frac{q_o p_o\eta}{B_p B_t\eta_o} = \frac{r^2}{a^2}\beta_p q_o\frac{B_p}{B_t}\frac{\eta}{\eta_o} .$$

This is larger than 1 for $\beta_p > \frac{q}{q_o}\frac{R}{r}\frac{a^2}{r^2}\frac{\eta_o}{\eta}$.

Assuming $q/q_o = 2,5$, $T/T_o = \frac{1}{4}$, $r = a$, it then holds that $\beta_p > 0.31\frac{R}{a}$, and there is no problem with the existence of such solutions.

To conclude this discussion, the relation for p is written by using vector quantities. We obtain after multiplying by $\nabla\psi$

$$\nabla p = -\langle\frac{1}{R^2}\rangle\frac{1}{2}\nabla(R^2 B_t^2)\left[1 + \frac{\langle B_p^2\rangle}{\langle B_t^2\rangle}\right]R\langle\frac{1}{R^2}\rangle\frac{4\pi cU}{c^2\eta}(R\nabla\phi \times \underline{B}_p) .$$

162

2) The Banana Regime

Even if there were no collisions at all, not all the particles in a Tokamak plasma could move freely around the torus along the field lines. Since the magnetic field strength varies along a field line over a length after which a field line has appreciably changed its meridional position on a magnetic surface, i.e. after a length of the order qR, a particle sees magnetic mirrors at a distance qR. The strength of the mirrors $\Delta B/B$ is given by the inverse aspect ratio

$$\Delta B/B \approx a/R.$$

There are therefore particles trapped between such mirrors according to the law of energy conservation

$$\mu B + \frac{1}{2} mv_\|^2 = \text{const} \quad \text{or} \quad \mu \Delta B + \Delta \frac{1}{2} mv_\|^2 = 0 \; ,$$

for which it holds that

$$\left| \Delta \frac{1}{2} mv_\|^2 \right| = \left(\frac{1}{2} mv_\|^2 \right)_{\text{max}} \; .$$

Since $\mu = \frac{1}{2} mv_\perp^2/B$, this means that

$$\frac{v_\|^2}{v_\perp^2} = \frac{\Delta B}{B} = \frac{a}{R} \ll 1 \; .$$

Such particles drift essentially in the vertical direction with a velocity

$$v_{\text{Drift}} = c \, \frac{mv_\perp^2/R}{eB} \; .$$

During the time the particle flies from one to the other mirror, that is the time $qR/v_\|$, the particle moves a distance

$$\delta = v_{\text{Drift}} \, qR/v_\| = \frac{mv_\perp c}{eB} \, q \, \frac{v_\perp}{v_\|} = r_g \, q \, \frac{R}{a}$$

out of a magnetic surface in the vertical direction. This is the thickness of the banana-like orbits (Fig. 2).

163

If there are few collisions, the first to occur is a reversal of $v_\parallel$, since $v_\parallel$ is much smaller than $v_\perp$. This means that an inner part of a banana orbit becomes an outer one or vice versa, i.e. the particle does steps of the order of the banana thickness δ . The banana thickness therefore replaces the gyroradius in plane geometry. The time to reverse $v_\parallel$ is, however, not the usual mean free time but is shorter than this by the factor $\frac{v_\parallel^2}{v^2}$. Thus we have to use a trapped particle collision frequency

$$\nu_t = \frac{v^2}{v_\parallel^2} \nu \approx \frac{R}{a} \nu \quad .$$

The number of such trapped particles is proportional to the $v_\parallel$ interval given by the trapping condition, i.e.

$$n_t = n \frac{v_\parallel}{v} = n \sqrt{\frac{a}{R}} \quad .$$

A stochastic process with δ as step size then yields the diffusion coefficient

$$D_B = \delta^2 \nu_t \frac{n_t}{n} = r_g^2 q^2 \frac{v_\perp^2}{v_q^2} \frac{v^2}{v_\parallel^2} \nu \frac{v_\parallel}{v} = r_g^2 \nu q^2 (\frac{R}{a})^{3/2} \quad .$$

$$= D_{PS} (R/a)^{3/2} .$$

This derivation is valid as long as particle trapping is not inhibited by collisions, i.e. for $\nu_t q R/v_\parallel < 1$

or $\nu \frac{v^2}{v_\parallel^3} qR = \frac{qR}{\lambda} \left(\frac{v}{v_\parallel}\right)^3 = \frac{qR}{\lambda} A^{3/2} < 1$ or $\lambda > A^{3/2} qR,$

where λ is the mean free path.

Thus there is a regime left:

$$qR < \lambda < A^{3/2} qR \quad .$$

Since D_B, $D_{PS} \sim \frac{1}{\lambda}$,

one has

$$D_B(\lambda = A^{3/2} qR) = D_{PS}(\lambda = qR).$$

One therefore expects the following behaviour of D shown in **Fig. 3**. The inner part forms a plateau and is therefore called the plateau regime. In reality, there is a smooth transition from the banana to the Pfirsch-Schlüter regime.

Since the stochastic process including banana orbits is only modified by geometrical factors compared to that with gyro-orbits, ambipolar diffusion again holds. There are two accompaning effects which are of importance

1) the so-called bootstrap current,

2) the Ware effect.

There is, so to speak, an induction effect of the high diffusion velocity leading to a current density in the toroidal direction

$$j_B = B \frac{v_B}{p\,c} = - \frac{c}{B_p} \frac{dp}{dr} \left(\frac{r}{R}\right)^{1/2} .$$

The corresponding current in the poloidal direction does not occur because of a special kind of viscosity. An important consequence of this current is as follows: Insertion of this current into Maxwell's equation yields

$$\frac{1}{r} \frac{d}{dr} r\, B_p = \frac{4\pi}{c} \frac{c}{B_p} \frac{dp}{dr} \left(\frac{r}{R}\right)^{1/2}$$

or

$$\beta_p = \frac{<p>}{B_p^2/8\pi} \approx A^{1/2} .$$

165

I want to mention here some preliminary considerations of Borrass and the author. In these considerations it is stated that the diffusion velocity should never exceed the magnetic field diffusion velocity in a plasma with finite resistivity. There are a number of arguments for this. Assuming this to be true, one easily obtains a relation for an upper limit of ß in the following way: We had found above the relation valid in plane geometry

$$v_{cl} = \frac{1}{2} \beta \, v_{magn} \quad .$$

The magnetic diffusion described by v_{magn} does not depend on the geometry. Thus, from the condition

$$v_{magn} \overset{>}{} v_D$$

one obtains

$$\beta < \frac{2 v_{cl}}{v_D} \quad .$$

For the banana regime this yields $\beta < \frac{1}{A^{3/2} q^2}$ or, with $\beta = \frac{1}{q^2 A^2} \beta_{pol}$, one gets $\beta_{pol} < A^{1/2}$, in agreement with the result just derived. Pseudo-classical diffusion is given by $v_D \sim q^2 A^2 v_{cl}$, and therefore $\beta_{pol} < 1$, as is already well known. Pfirsch-Schlüter diffusion is expressed by $v_D \approx q^2 v_{cl}$, and therefore $\beta_{pol} < A^2$, which would be much larger than often assumed, and which is probably not in contradiction to an exact theory as outlined in section 1a. Bohm diffusion at 10 keV and 40 kG would give $\beta < 10^{-5}$. If this ß relation is correct, it would lead to extremely severe restrictions as to the permissible diffusion rates if these are not caused by a corresponding anomalous resistivity.

The Ware effect is the pendant to the bootstrap current in the sense of Onsager's relation and states that the usual E/B drift is replaced by

$$v_E = c \frac{E}{B_{pol}} \quad .$$

3) Impurity Transport

For impurities in the Pfirsch-Schlüter regime the impurity transport is similar to that in plane geometry but is modified by a Pfirsch-Schlüter factor [6] [7] [8], i.e. there is impurity influx as long as the temperature profile is flatter than given by $T \sim n^2$. If $n_z Z^2/n_i Z_i^2 > 1$, the ion flux is enhanced by a factor of $\sqrt{m_i/m_e}$ over the pure plasma case and the ion heat conduction becomes about $A^{3/2}$ times the ion heat conduction in the pure case [7], but the electron flux remains unchanged.

Calculations in which the impurities are also assumed to be in the banana or plateau regime were made by Hinton and Moore [9].

If the impurities are in the banana regime, there is temperature screening as in the Pfirsch-Schlüter regime. If, however, the impurities are in the plateau regime, no temperature screening occurs. This might be a special problem for the α-particles. But for low Z and low mass impurities theories usually do not apply, and so one has to consider the situation as being open for such species.

References

[1] J.B. Taylor   Phys. Fluids $\underline{4}$, 1142 (1961)

[2] D. Pfirsch and A. Schlüter   Max-Planck-Institut für Physik und Astrophysik, München, MPI/PA/7/62 (1962)

[3] M.D. Kruskal and R.M. Kulsrud   Phys. Fluids $\underline{1}$, 265 (1958)

[4] A.A. Galeev and R.Z. Sagdeev   Zh. êksp. teor. Fiz. $\underline{53}$, 348 (1967)

[5] K. Borrass and D. Pfirsch   to be published

[6] P.H. Rutherford   Phys. Fluids $\underline{17}$, 1782 (1974)

[7] S.P. Hirshman   MIT Report PRR-7513 (1975)

[8] A. Samain and F. Werkoff   to be published in Nucl. Fusion

[9] F.L. Hinton and T.B. Moore   Nucl. Fusion $\underline{14}$, 639 (1974)

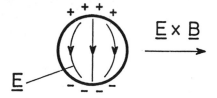

Fig 1: $\underline{E} \times \underline{B}$ drift caused by charge separation

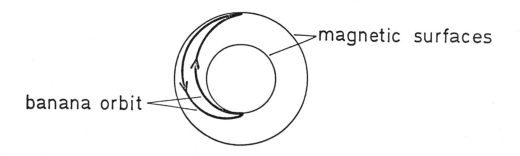

Fig 2: Banana orbits

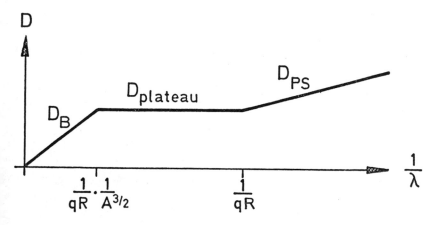

Fig 3: Diffusion coefficient as function of the inverse mean
free path (idealized).

Magnetic Divertors

M. Keilhacker

Max-Planck-Institut für Plasmaphysik, Garching, Germany

ABSTRACT

The different needs for divertors in large magnetic confinement
experiments and prospective fusion reactors are summarized,
special emphasis being placed on the problem of impurities.
After alternative concepts for reducing the impurity level
are touched on, the basic principle and the different types
of divertors are described. The various processes in the
scrape-off and divertor regions are discussed in greater detail. The dependence of the effectiveness of the divertor
on these processes is illustrated from the examples of an
ASDEX/PDX-size and a reactor-size tokamak. Various features
determining the design of a divertor are dealt with. Among
the physical requirements are the stability of the plasma
column and divertor throat and the problems relating to the
start-up phase. On the engineering side, there are requirements on the pumping speed and energy deposition, and for a
reactor, the need for superconducting coils, neutron shields
and remote disassembly.

I.   INTRODUCTION

It was as long as 25 years ago that Professor Lyman Spitzer
at Princeton described and proposed the concept of a magnetic divertor /1/: a coil configuration that magnetically di-
verts an outer layer of the magnetic flux confining the plasma
and conducts this flux - and with it the surface plasma - to
an external chamber. This concept was then tested experimentally
on the B-65 Stellarator in 1957 and on the Model C Stellarator
in 1963, where it proved to be very effective, reducing impurities by an order of magnitude and decreasing the recycling.

171

Despite this successful operation no further divertor experiments were proposed until a few years ago when the attempt to reach the collisionless plasma regime in high-power-level tokamaks such as T-4 and TFR failed owing to an increased evolution of impurities. In addition, in these machines high-current discharges sometimes even damaged limiters and walls of the vacuum chamber. One of the most promising solutions to these problems appears to be the magnetic divertor. In a divertor the plasma boundary is defined by a magnetic limiter (separatrix) and contact with material walls is removed to a separate chamber. The divertor reduces contamination of the confined plasma by cutting down the recycling of cold gas (unloading action) and - under certain conditions - by screening off the influx of impurities from the vacuum chamber wall (shielding action). Furthermore, the divertor is suited to solve the following additional problems related to a fusion reactor: to pump the particle through-put of 10^{22}-10^{23} particles/sec, to take up the energy flux of several 100 MW associated with this particle through-put, and to remove fuel (D, T) and ash (He).

In the following discussion of magnetic divertors emphasis is more on the physical aspects of a divertor than on its engineering problems. The paper starts by outlining the needs and possible methods of impurity control (section 2). Then the basic principle of a divertor, the stability requirements on the divertor field configuration and problems relating to the start-up of the plasma current are described (section 3). The main part of the paper (section 4) deals with the various processes in the scrape-off and divertor regions that determine the effectiveness of a divertor. Section 5 finally discusses engineering requirements of a divertor.

II.   SIGNIFICANCE OF IMPURITIES IN LARGE FUSION EXPERIMENTS

II. 1. Need for Impurity Control

Impurities have a number of deleterious effects on large magnetic confinement experiments and prospective fusion re-

actors. These effects have already been widely discussed
(e.g. /2/ - /6/) and will only be briefly summarized here:

i)   Enhanced radiation losses due to line radiation (par-
     tially stripped impurities) and recombination and brems-
     strahlung radiation (fully stripped impurities).

ii) Changes in temperature and current profiles: Edge cooling
     by low-Z impurities may cause the plasma to shrink and
     become unstable to a disruption, while high-Z impurities
     near the plasma centre may cause hollow temperature pro-
     files and unstable current.

iii) Reduction of reacting fuel ion density resulting in a
     decrease of thermonuclear reaction rates and an increase
     in ignition temperature.

iv) Reduced neutral beam penetration ($\sigma_{i,eff} \simeq z_{eff}\, \sigma_{i,i}$)
     leading to increased requirements on beam energy.

Apart from their role in present-day and next-generation tokamak experiments impurities severely affect the economy of prospective fusion reactors. Especially the presence of high-Z
impurities, because of the afore-mentioned processes i) and
iii), shifts the Lawson and ignition criteria significantly
towards higher temperatures and finally rules out ignition
altogether. A number of studies /4,5/ have indicated that the
maximum tolerable impurity concentrations, above which ignition
is no longer possible, are 5 to 10 % for low-Z (C,O), 0.5 to
1 % for medium-Z (Fe to Mo) and about 0.1 % for high-Z (W, Au,
Ta) impurities. This is illustrated in Fig.1.

II. 2. Origin of Impurities

During the steady-state phase of conventional tokamak discharges impurities are mainly produced by the following processes:

i)   Charged particle sputtering (hydrogen and impurity ions)
     of the vacuum chamber wall and limiters

ii) Neutral particle sputtering (fast charge-exchange atoms)
     of the vacuum chamber wall

173

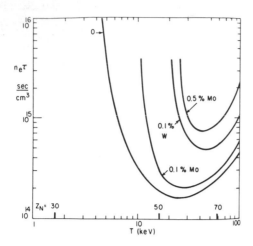

Fig.1: Ignition condition for D-T, with various percentages of molybdenum and tungsten impurity contamination /4/. The vertical bars indicate the temperatures at which ions of nuclear charge Z_N are ionized to helium-like states.

iii)   Evaporation of limiter or wall material due to over-
        heating.

iv)   Desorption from the wall surface by ions, electrons or
        photons.

Prior to the quasi-stationary phase with constant plasma
current, impurities are generated during the initial phase
of plasma breakdown and plasma current build-up. During this
period there is special danger of plasma-wall contact and of
evaporation by run-away electrons.

II. 3.   Possible Methods of Impurity Control

Various methods have been proposed to control the impurity
concentration of the plasma.These methods may be divided into three groups depending on the point at which the process
of plasma contamination is interrupted:

i)   Methods to suppress the sputtering and thus the formation
     of impurities itself

ii)   Methods to shield the hot plasma core against the influx
      of impurities

iii) Methods to retard or even reverse the accumulation of im-
      purities in the centre of the plasma (purification).

i) The most obvious method to suppress sputtering seems to be
the use of special wall material resistant to sputtering. In
addition, charge-exchange sputtering of the wall material can
be strongly reduced if the temperature of the plasma edge is
kept low, either by radiation cooling or by injecting hydrogen
gas. The first method could be used in a controlled way by
making walls and limiter out of suitable low-Z material. In
the second case - injection of hydrogen gas - a cold, dense
outer plasma region is formed that protects, at least transiently, the wall from charge-exchange-neutral bombardment while,
at the same time shielding the hot plasma core against the influx of cold neutral gas and impurities. To get this favourable
effect on the sputtering problem, it must be ensured that the
mean energy of the resultant charge-exchange neutrals stays below the sputtering threshold of the wall material. This concept of a cold-gas (cold-plasma) blanket has been proposed by
H. Alfvén and E. Smårs /7/ and is discussed, e.g. in ref./7 - 10/.
A viable solution to this problem, of course, is a magnetic divertor, reducing both the plasma ion and charge-exchange neutral
sputtering, as will be discussed later.

ii) The influx of sputtered impurity atoms can be attenuated
in the outer plasma layer by a magnetic divertor or possibly
by a cold gas blanket. In the divertor, part of the incoming
impurity atoms are ionized in the scrape-off layer and swept
into the divertor by the outflowing plasma. In the case of a
continuously refreshed cold gas blanket impurities could be
flushed away with the outstreaming gas.

iii) The most important, but least understood, problem is how
to retard or possibly reverse the inward diffusion of impurities

predicted by classical theory. Several techniques have been
proposed to give the impurities an outward momentum, e.g.
the use of plasma waves /11,12/ or local injection of hydrogen gas /13/. Also, if the impurity transport is classical,
the impurities could be forced to flow outward by reversing
the hydrogen plasma density gradient towards the plasma edge
either by neutral beam injection("profile shaping") or by means
of a cold-plasma blanket.

III.   DIVERTOR MAGNETIC FIELD CONFIGURATION

III. 1. Basic Principle and Types of Divertors

The principle of the magnetic divertor is well known and
shall only be briefly described here as a basis for the following discussion.
External currents near the plasma surface produce a separatrix that defines the size and shape of the hot plasma.
Charged particles diffusing across this separatrix can flow
along the magnetic field lines into a separate chamber where
they are neutralized at collector plates and can be pumped
off. This removal of the cold neutral reflux (the recycling)
reduces charge-exchange sputtering of the vacuum chamber
walls.
The plasma region outside the separatrix, lying on magnetic
field lines that enter the divertor, is called the scrape-off
layer. If this layer is sufficiently thick and dense, it
can shield the hot plasma against wall-originated impurity
atoms: the incoming impurity atoms become ionized and are
then swept into the divertor by the outflowing plasma.

There are various types of divertors. They may be divided into two groups depending on whether the toroidal or poloidal
field is distorted: toroidal divertors (Model C Stellarator,
1964) and poloidal divertors (FM-1, 1973; DIVA, 1974; PDX,
1978; ASDEX, 1978). The latter are also called axisymmetric

divertors since they preserve the axisymmetry of the tokamak.
A special kind of toroidal divertor is the bundle divertor
where only a small part of the toroidal flux is diverted (DITE,
1975; proposed for TEXTOR).

III. 2. Stability of Divertor Field Configurations

One of the most important criteria for choosing a particular
divertor field configuration is its MHD stability. This question has therefore been investigated intensively during the
last years at different laboratories (see, for example, ref.
/14/ and references cited therein). Here only the most pertinent results related to poloidal (axisymmetric) divertors can
be summarized in a more qualitative fashion:

i)   The divertor fields have to be localized, i.e. of
short range, in order not to affect the equilibrium, stability
and shape of the plasma. This rules out arrangements in which
the divertor field is produced by a single coil carrying current
parallel to the plasma current (Fig. 2a or, equivalently, Fig. 2b),
and calls for divertor coil triplets with zero net current
(Fig. 2c,d). More specifically, the results of hexapole and
octupole calculations presented in ref. /14/ show that the
stability against axisymmetric modes is the better the more
localized the divertor field is, thus probably ruling out a
hexapole configuration.

ii) Divertor configurations where the plasma cross-section
is "D"-shaped (Fig. 2c) are stable over a wider range of plasma parameters than those with an "inverse D" (Fig. 2d).

iii) As far as the shape of the plasma cross-section is concerned for axisymmetric modes the stability is best for a
circular cross-section, while for non-axisymmetric modes a
triangular deformation of the plasma cross-section tends to
improve the stability.

Apart from the MHD stability of the plasma column, the following points have to be considered in designing a divertor

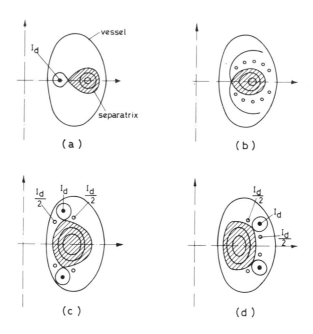

Fig. 2: Various axisymmetric divertor configurations with
parallel (•) and antiparallel (O) currents

field configuration: positional stability of the scrape-off
layer in the divertor throat, plasma breakdown near the
centre of the vacuum chamber, possibility of an expanding
magnetic limiter without programmed divertor currents (see
following paragraph).

III. 3. Expanding Magnetic Limiter for Controlling the Skin Effect

The turbulence associated with the skin effect during the
current-rise phase in large tokamaks is expected to transfer a large amount of energy to the walls and limiter, thereby boiling off impurities. To suppress this skin effect, one

would like to enforce proportionality between the plasma cross-section and the plasma current.

In a divertor tokamak this is possible by having the separatrix expand in a controlled way with the rising plasma current. In practice, however, it is not possible to enforce this perfect magnetic limiter behaviour over a large range of plasma currents by simply varying the currents in the divertor coils. An elegant way out of this difficulty, proposed by K. Lackner and discussed in refs. /15/ and /17/, is not to vary the divertor currents but rather to shift the plasma column ("plasma displacement limiter") by proper magnetic fields into the vicinity of one of the divertor stagnation points and then to let it come back to the median plane in a controlled way. This kind of expanding magnetic limiter has the further advantage that even during current build-up the plasma is connected to one of the divertors. Another concept for a plasma displacement limiter would be to strike the discharge near a mechanical limiter on the inside of the torus or at the centre of an octupole magnetic null /19/ and to let the plasma move radially outward in a controlled way as the current rises.

IV.   DIVERTOR EFFECTIVENESS

IV. 1.   Impurity Build-up with and without Divertor

The operation of a divertor system is specified mainly by the following three parameters:

C ,   the plasma capture efficiency of the divertor; a large
       C reduces the plasma ion sputtering of the limiter

R ,   the backstreaming ratio, i.e. the fraction of neutral gas
       that returns from the divertor into the vacuum chamber;
       a small R reduces the charge-exchange neutral sputtering

P ,   the shielding efficiency of the scrape-off region; a
       large P prevents wall-originated impurities from reaching
       the hot plasma core.

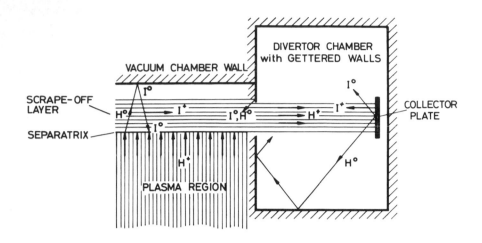

Fig. 3: Idealized representation of processes in the scrape-off and divertor regions

To see in how far a divertor is capable of reducing the level of impurities, we look at the build-up of impurities in a steady-state confined plasma due to charge-exchange and ion sputtering of the walls, limiter and collector plate as depicted in Fig. 3. In this simplified model, that neglects desorption, evaporation, blistering etc., the build-up of impurities can be described by the following relation:

$$\dot{n}_I + \frac{n_I}{\tau_I} = \frac{n}{\tau_p} \left\{ (1-C)\ (S_{HL} + \Upsilon_H\ S_{HW})\ (1-P) \right.$$

$$\left. + C \left[S_{HD}\ P_i (1-P_d) + R\,\Upsilon_{H_2}\ S_{HW}(1-P) \right] \right\} \tag{1}$$

$$+ \frac{n_I}{\tau_I} \left[(1-C) S_{IL}(1-P) + C\ S_{ID}\ P_i (1-P_d) \right]$$

Here n and n_I denote the particle densities, τ_p and τ_I the confinement times of the hydrogen and impurity ions, respectively. S_{HL}, S_{HW} and S_{HD} are the sputtering coefficients for hydrogen bombardment of the limiter, the wall and the collector plate, S_{IL} and S_{ID} the corresponding coefficients for sputtering by impurities. P_i is the ionization probability of an impurity ion in the scrape-off layer or in the plasma layer in front of the collector plate (the area density of the plasma is assumed equal in these two regions), P_d the probability that an impurity ion in the scrape-off layer is swept into the divertor before it diffuses into the confined plasma. Consequently, the product of P_i and P_d is the shielding efficiency P as defined before. The coefficients γ give the number of fast charge-exchange neutrals per neutralized plasma ion that bombard the wall.

The first line of equ. (1) describes the build-up of impurities caused by that part of the plasma that impinges on the edges of the divertor slits (which in this case are equivalent to the limiter). The second line represents the contribution by the plasma that enters the divertor and sputters the collector plate. A fraction P_i of the sputtered material becomes ionized and enters the vacuum chamber along the magnetic field lines. Of the neutralized gas a part R flows back into the vacuum chamber and leads to charge-exchange sputtering. The third line of equ. (1) finally describes the contamination due to sputtering by the impurity ions that are lost from the confined plasma.

For a qualitative discussion of divertor effectiveness one can simplify equ. (1). First, in a working divertor sputtering by impurities can be neglected. Furthermore we confine the discussion to a divertor of the unload type, i.e. with $C \simeq 1$.

This leads to the following expression for the build-up of impurities in an unload divertor (with $n_I = 0$ at $t = 0$):

$$\frac{n_I}{n} = \mathcal{D} \frac{\tau_I}{\tau_p} \left[1 - \exp\left(-t/\tau_I\right) \right]$$

(2)

with $\mathcal{D} = S_{HD}\, P_i\,(1 - P_d) + R\, \gamma_{H_2}\, S_{HW}\,(1 - P)$.

181

The build-up of impurities reaches a stationary state after a time of the order of the impurity confinement time τ_I. The absolute value of this stationary impurity concentration $c_{I,st} = n_{I,st}/n \approx \mathcal{D}\, \tau_I/\tau_p$ depends mainly on P_i and R and will typically be between 10^{-3} and 10^{-2} (for $\tau_I = \tau_p$).

For long impurity confinement times (for $\tau_I \gg \tau_p, t$) one obtains from equ. (2):

$$\frac{n_I}{n} = \mathcal{D}\, \frac{t}{\tau_p} \ ,$$

(3)

i.e. the number of impurities increases linearily with time.

Let us now, for comparison, look at the build-up of impurities in a tokamak without divertor. In this case one gets from equ. (1) (with $C = 0$, $P = 0$):

$$\frac{n_I}{n} = \frac{\mathcal{L}}{(1 - S_{IL})}\, \frac{\tau_I}{\tau_p}\, \left[1 - \exp - \left((1 - S_{IL})\ t/\tau_I \right) \right]$$

(4)

$$\text{with } \mathcal{L} = S_{HL} + \gamma_H\, S_{HW} .$$

i.e. the impurities build up exponentially without saturation if the sputtering coefficient of the impurities, S_{IL}, is larger than 1 (this may well be the case, unless the temperature of the sputtering impurities stays rather low, say below 500 eV).

For $\tau_I \gg \tau_p, t$ one gets by analogy with equ. (3)

$$\frac{n_I}{n} = \mathcal{L}\, \frac{t}{\tau_p} \ .$$

(5)

A comparison of the build-up of impurities with and without divertor shows that the impurity concentrations, $c_{I,st}$, that are reached in the stationary state (in the limiter case, of course, only for $S_{IL} < 1$) are related to each other by

$$\mathcal{R} = \frac{C_{I,\text{st Div}}}{C_{I,\text{st Lim}}} = \frac{\mathcal{D} |1 - S_{IL}|}{\mathcal{L}} . \tag{6}$$

To get a rough idea how this "figure of merit" for divertors, $\mathcal{R}$, depends on the main divertor characteristics, one may simplify the expression for $\mathcal{R}$ by neglecting S_{HL} compared to $\gamma_H \cdot S_{HW}$ (this is usually justified) and by dropping the term $S_{HD}P_i(1 - P_d)$ (which has to be checked in each case). One then gets

$$\mathcal{R} = C R (1 - P) |1 - S_{IL}| , \tag{7}$$

where C, which had been assumed to be about 1, is included again for the sake of greater generality and γ_H was assumed equal to γ_{H_2}.

Since it is not possible to design a divertor in which C, R and P are all optimized at the same time, one mainly has two modes of operation for a divertor system: an unload divertor (C $\simeq$ 1, R $\simeq$ 0, P $\simeq$ 0) or a shielding divertor (C $\simeq$ 1, 0 < R < 1, P $\simeq$ 1).

IV. 2. The Processes that Determine the Divertor Effectiveness

a) Density Profile in the Scrape-off Layer

To determine the density profile in the scrape-off region, one may consider a simple one-dimensional model /16, 18 - 20/ in which plasma diffusion across the magnetic field (diffusion coefficient $D_{\perp s}$) is balanced by plasma loss parallel to the field lines into the divertor (effective confinement time $\tau_\parallel$). Neglecting ionization in the scrape-off region is justified for a working divertor.

If x is the distance across the scrape-off layer, one has

$$\frac{d}{dx} (D_{\perp s} \frac{dn}{dx}) = \frac{n}{\tau_\parallel} . \tag{8}$$

Assuming $D_{\perp s}$ and $\tau_\parallel$ are constant across the scrape-off region (for $\tau_\parallel$ this is strictly speaking only correct for a divertor of the C-Stellarator type), the density profile is given by

$$n = n_b \exp(-x/\Delta_d), \tag{9}$$

where
$$\Delta_d = (D_{\perp s} \, \tau_{\parallel})^{1/2} \tag{10}$$

is the width of the scrape-off region.

The boundary density n_b is determined by the condition that the total plasma flow across the separatrix ($x = 0$) has to equal the total loss of confined plasma:

$$-A_p \, D_{\perp s} \, \frac{dn}{dx}\bigg|_{x=0} = \frac{\bar{n} \, V_p}{\tau_p} \, ,$$

$\bar{n}$, A_p and V_p being the average density, surface, and volume of the confined plasma, respectively. This then yields for the boundary density

$$n_b = \frac{\bar{n} \, a}{2 \, \tau_p} \cdot \frac{\Delta}{D_{\perp s}} \tag{11}$$

where a is the plasma radius.

For divertors with no magnetic mirrors, i.e. with stagnation lines on the outside of the plasma column, the plasma in the scrape-off region flows into the divertor with ion sound speed, v_s ("ion sound model"), i.e. $\tau_{\parallel}$ is simply given by $\tau_{\parallel} = L/v_s$, where L is the geometrical path length into the divertor.

The situation is different if the stagnation lines lie on the inside of the plasma core, and the plasma in the scrape-off layer therefore encounters a mirror on its way into the divertor. In this case the scrape-off region may be mostly populated by trapped particles and the plasma flows into the divertor by scattering into the mirror loss cone ("mirror confinement model"). The effective confinement time of the plasma in the scrape-off region is then approximately equal to the scattering time of trapped particles into the loss cone, τ_{sc}, since one can show that for most divertors $\tau_{sc} > L/v_s$. In the mirror model one thus has

$$\tau_{\parallel} \propto \tau_{90^\circ} \log_{10} M$$

where τ_{90° is the 90° scattering time for ions in the scrape-off region and M is the mirror ratio.

Since in the mirror model $\tau_{\parallel}$ depends on the density n, the solution of equ.(8) is slightly different from that for the ion sound model. Assuming $D_{\perp s}$ and the temperature T to be constant across the scrape-off layer the density profile in the mirror model is

$$n = n_b^* \ (1 + \frac{x}{\Delta_d^*})^{-2} \tag{12}$$

with

$$n_b^* = \frac{\bar{n} \ a}{4 \tau_p} \frac{\Delta_d^*}{D_{\perp s}} \tag{13}$$

and

$$\Delta_d^{*3} = 3.4 \times 10^8 \ \frac{A^{1/2} \ T(eV)^{3/2} \tau_p \ D_{\perp s}^2 \ \log_{10} M}{\bar{n} \ a \ \ln \Lambda} . \tag{14}$$

In calculating Δ_d^* the value given by Spitzer /21/ was used for τ_{90° and n_b^* was expressed by $\bar{n}$ by means of equ.(13).

The diffusion width of the scrape-off layer Δ_d (or Δ_d^*), as calculated above, is only meaningful as long as it is larger than the excursions Δ_b which the ions undergo on their path to the divertor. If, on the other hand, $\Delta_d < \Delta_b$ then the width of the scrape-off layer is determined by Δ_b. The order of magnitude of this so-called banana width is given by

$$\Delta_b \approx \sqrt{\frac{a}{R}} \ r_{ip},$$

where r_{ip} is the ion gyro-radius in the poloidal field.

For divertor experiments like ASDEX and PDX the banana width Δ_b is comparable to the diffusion width Δ_d (no mirrors) but smaller than Δ_d^* (with mirrors). In a divertor for a reactor the diffusion width Δ_d should prevail in any case (c.f.Table I).

185

b) Plasma Capture Efficiency of the Divertor

The plasma capture efficiency of the divertor is

$$C = 1 - \exp(-W/\Delta'),$$

where W is the width of the divertor throat and Δ' the characteristic length of the density drop in the scrape-off layer projected into the divertor throat, i.e. $\Delta' = (B_p/B_p')\Delta$, where B_p' is the poloidal field in the divertor throat and B_p is the average poloidal field. For Δ the diffusion width Δ_d or the banana width Δ_b has to be taken, whichever is larger. For $W/\Delta' = 4$, for example, a value that is easily attainable in most divertor designs, one gets $C \approx 0.98$, i.e. most of the outstreaming plasma goes into the divertor.

c) Neutral Gas Return from the Divertor

As discussed before, efficient divertor operation requires that only a small fraction of the neutral gas that is generated in the divertor by neutralizing the incoming plasma should return to the main confinement chamber. The large pumping speeds required for this purpose (e.g. 6×10^6 1/s in ASDEX /22/) can only be achieved by getter pumps. The getter pumping may be supplemented by trapping of plasma ions in the collector plates and by plasma pumping in the divertor throats (plasma pumping constitutes about 10 - 20 of the total pumping in PDX and ASDEX).

Neglecting plasma pumping, the fraction of neutral gas that returns from the divertor is given by

$$R = \frac{L(1 - f_p)}{L + S},$$

where L is the conductance of the divertor throat, f_p the trapping efficiency of the collector plates for plasma ions, and $S = f_n A$ the pumping speed (f_n is the intrinsic pumping speed, A the area of the gettering surface, respectively).

d) Shielding Efficiency of the Scrape-off Layer

The shielding process in the scrape-off layer consists of two steps: first the incoming wall-impurity atom has to be ionized and then the ion has to be swept into the divertor. In order to shield the plasma core effectively, the ionization process has to take place so far away from the separatrix that the impurity ion cannot diffuse into the confined plasma during its flight into the divertor.

Let us first consider the ionization process /2,16,18,19,20,23/. The penetration of impurity atoms into the scrape-off region is given by

$$v_{oI} \frac{dn_I}{dx} = -n_I \, n \, \langle \sigma v \rangle_i \quad , \tag{15}$$

where n_I, v_{oI} and $\langle \sigma v \rangle_i$ are the density, radial velocity and ionization rate of the incoming impurity atoms.

The solution of this equation

$$n_I(x) = n_I(\infty) \, \exp \left[- \frac{\langle \sigma v \rangle_i}{v_{oI}} \int_0^\infty n dx \right] \tag{16}$$

can be used to calculate the fraction of incoming impurities that are ionized within the scrape-off layer (up to the separatrix at $x = 0$)

$$P_i = 1 - \frac{n_I(0)}{n_I(\infty)} = 1 - \exp \left[- \frac{\langle \sigma v \rangle_i}{v_{oI}} \int_0^\infty n dx \right] . \tag{17}$$

The ionization probability, P_i, thus depends only on the area density of the scrape-off layer, $\int_0^\infty n dx$. For a given area density the ionization probability becomes larger for heavier impurities since $v_{oI} \propto (kT_I/A)^{1/2}$ and since one can assume that the impurity atoms all come off the wall with the same mean energy kT_I (c.f. the values for P_i (16) and P_i (100) in Table I).

The area density can be evaluated by using the density profiles
given by equs. (9) and (12) (which yield $n_b \Delta_d$ and $n_b^* \Delta_d^*$, respectively) or by making use of the relation

$$\int_o^\infty ndx = \frac{\bar{n}\, a}{2\, \tau_p}\, \tau_{\parallel} ,$$

(18)

where $\tau_{\parallel}$ is again the effective confinement time of the plasma in
the scrape-off region.

Relation (18) shows that in order to improve the ionization
probability and by this the screening efficiency, one has to
increase the confinement time of the plasma in the scrape-off
region $\tau_{\parallel}$. Now $\tau_{\parallel}$ can either be extended by increasing the geometrical path length into the divertor, L (torsatron, bundle
divertor) or by having a magnetic mirror in the divertor throat
(inside poloidal divertor, bundle divertor).

By increasing L, however, one also increases the confinement
time of the impurity ions in the scrape-off region, $\tau_{\parallel\, I}$, and
therefore their inward diffusion distance

$$d = (D_{\perp s, I} \cdot \tau_{\parallel\, I})^{1/2} \qquad \text{with } \tau_{\parallel\, I} = L/v_{\parallel\, I} ,$$

(19)

where $D_{\perp s, I}$ and $v_{\parallel\, I}$ are the diffusion coefficient and parallel
flow velocity of the impurity ions in the scrape-off region. In
the case that the width of the scrape-off layer, Δ, is determined
by diffusion (and therefore has the same dependence on L) one
gets

$$\frac{d}{\Delta} = \left(\frac{D_{\perp s, I} \cdot v_s}{D_{\perp s} \cdot v_{\parallel\, I}} \right)^{1/2}$$

(20)

i.e. the inward diffusion distance of the impurity atoms is
a fixed fraction of the scrape-off width, irrespective of the
geometrical path length.

Let us now consider the case in which the confinement time of the plasma in the scrape-off region is increased by a magnetic mirror in the divertor throat. In contrast to the geometrical path length, on which the confinement times of both plasma and impurity ions depend in the same way, the effect of magnetic mirrors is different for the plasma and the impurity ions. This is because the impurity ions in the scrape-off region are swept into the divertor by the plasma flow, either due to friction or due to an extended sheath potential $U_s = kT_e/e$. Since the impurities typically become highly ionized, but not heated much, during their transit time into the divertor their parallel velocity $v_{\parallel I} \simeq (Z_I kT_e/m_I)^{1/2}$ is much larger than their perpendicular velocity $v_{\perp I} = (kT_I/m_I)^{1/2}$ and they therefore pass easily through the magnetic mirror. For the protons, however, $v_{\parallel} \simeq v_{\perp}$ and they are trapped between the mirrors unless velocity-space instabilities enhance their parallel loss rate. Thus with a magnetic mirror it may be possible to increase the ionization probability for impurity atoms without impairing their parallel transport into the divertor.

e) Shielding Efficiency of Divertors for an ASDEX/PDX-Size and a Reactor-Size Tokamak

The width (Δ_b, Δ_d), plasma density (n_b) and shielding efficiency (P_i) of the scrape-off layer in the ion sound and the mirror confinement model, respectively, are illustrated for the examples of an ASDEX/PDX-size and an Experimental Power Reactor (EPR)-size tokamak in Table I. The following parameters have been used for the two reference tokamaks:

	a (cm)	R (cm)	B_t (KG)	$\bar{n}$ (cm^{-3})	τ_p (sec)
PDX/ASDEX	40	150	30	3×10^{13}	0.05
EPR	200	600	60	6×10^{13}	2

189

q(a) was assumed to be 3 and for L the relation $L \simeq q(a)\, \pi R$ was used. The calculations were carried out for a deuterium plasma $(A = 2)$. The mirror ratio M was set equal to 2. Furthermore, the diffusion coefficient of the plasma in the scrape-off layer, $D_{\perp s}$, was assumed to be 1/10 of the Bohm-diffusion coefficient.

For the electron and ion temperatures in the scrape-off region two different values, 100 eV and 1000 eV, were used. The impurities were assumed to come off the wall with a velocity corresponding to 1 eV. For the ionization rate $\langle \sigma v \rangle_i$, which is rather independent of temperature in the considered range of 10^2 to 10^3 eV, a value of $1 \times 10^{-7}\,cm^3/s$ was used for light elements (C,O,CO) and of $3 \times 10^{-7}\,cm^3/s$ for heavy elements (Nb, Mo, W) /24/.

		Δ_b	Δ_d	n_b	$\int_0^\infty ndx$	$P_i(16)$	$P_i(100)$
		(cm)	(cm)	$(10^{12}cm^{-3})$	$(10^{12}cm^{-2})$		
ION SOUND MODEL							
ASDEX PDX	100 eV	0.3	0.7	3.9	2.6	0.65	1
	1000 eV	0.9	1.2	0.7	0.8	0.27	0.91
EPR	100 eV	0.1	0.9	2.7	2.6	0.65	1
	1000 eV	0.4	1.7	0.5	0.8	0.27	0.91
MIRROR MODEL							
ASDEX PDX	100 eV	0.3	1.2	3.5	4.2	0.81	1
	1000 eV	0.9	17.6	5.1	90	1	1
EPR	100 eV	0.1	1.2	1.7	2.1	0.57	1
	1000 eV	0.4	17.6	2.5	44	1	1

Table I

190

The results shown in Table I explain themselves. Δ_d is always larger than Δ_b and was therefore used in calculating n_b and $\int n\,dx$. If the diffusion in the scrape-off layer is larger than 1/10 of Bohm-diffusion, the width of the scrape-off layer becomes larger and the density smaller.

The ionization probability in the scrape-off layer for light elements, denoted by $P_i(16)$ (for $A = 16$), is around 0.5 for the ion sound cases and between 0.6 and 1 for mirror confinement, whereas the ionization probability for heavy elements, $P_i(100)$ (for $A = 100$), is 1 or close to 1 for all cases. If the impurities come off the wall with an energy larger than the assumed 1 eV the ionization probabilities become smaller. It has further to be remembered that ionization of the impurity atoms is only the first step of the shielding process and that a considerable fraction of the impurity ions may diffuse across the separatrix before they reach the divertor throat.

f) Refuelling a Divertor Tokamak

For "stationary" operation a divertor tokamak has to be refuelled, i.e. the plasma that flows off into the divertor has to be replaced. In the case of an unload divertor ($C \simeq 1$, $R \simeq 0$, $P \simeq 0$), the effectiveness of which depends on keeping the recycling as small as possible, the gas source has to be located deep within the plasma. A shielding divertor on the other hand can either be refuelled from inside the plasma like an unload divertor ($R \simeq 0$) or from outside the plasma by operating it with surface recycling ($R \simeq 1$).

Possible refuelling methods are

- pellet injection
- cluster injection
- neutral injection
- cold gas inlet.

All these refuelling methods, which supply gas or neutral par-

ticles that have to be ionized for absorption in the plasma,
lead to additional sputtering of the vacuum chamber wall and
therefore increase the build-up of impurities. This can be taken
into account by adding to equ.(1) a term of the form

$$(1 - R) \, \frac{n}{\tau_p} \, \sum_m (\alpha_m \, Y_m) \, S_{HW} (1-P).$$

Here α_m denotes the fraction that a special method (pellet
injection, cluster injection etc.) contributes to the refuelling
and Y_m is the corresponding number of wall bombardment events
which occur by charge-exchange, incomplete absorption and so on
for every hydrogen ion, that is finally added to the plasma.

For ASDEX a combination of cluster injection, neutral injection,
and cold gas inlet is envisaged for refuelling /25/. The contribution of these three refuelling methods to the contamination of the plasma by metal impurities is discussed in /26/.

V.   ENGINEERING ASPECTS OF THE DIVERTOR

In discussing the engineering aspects of the divertor, one
might distinguish between the general requirements for a divertor design and those especially related to a fusion reactor.
Among the general requirements - connected with the considerations on divertor efficiency, plasma stability, and plasma
start-up, as discussed in the previous sections - the most
important are:

i)   A poloidal magnetic field configuration that ensures equi-
      librium and stability of the confined plasma.

ii)   A pumping system that provides the necessary pumping speed
      and is capable of taking up the particle throughput.

iii) A design of the exhaust channel and collector plates that
      is capable of taking up the energy flux associated with the
      particle throughput.

In addition, for a reactor the following are required:

iv)   Superconducting divertor coils that are shielded against the neutron flux.

v)   A design that allow remote disassembly of all components.

vi)   A possibility to recover the tritium.

As far as divertor experiments of the next generation, such as PDX and ASDEX, are concerned, the engineering requirements resulting from i) - iii) can be solved, and have already been more or less. Here we shall therefore only mention some of the consequences and possible solutions related to divertor designs for a reactor. Most of these points are discussed in more detail in some of the theoretical studies of divertors for possible fusion reactors /27 - 30/, the most recent study being a conceptual design of a divertor for a tokamak EPR jointly carried out by scientists from Princeton and the Soviet Union /31/. Figure 4 shows a cross-sectional view of this tokamak EPR divertor design.

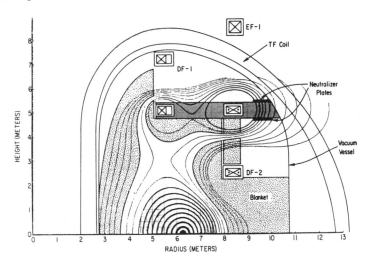

Fig.4: Cross-sectional view of the tokamak EPR divertor design of ref./31/ showing the toroidal field (TF) coil, divertor field (DF) coils and equilibrium field (EF) coil.

re ii)   Since pumping speeds in the range of $10^7 - 10^8$ liters/
sec are required, it becomes difficult to provide the necessary
pumping area. A possible solution discussed in /31/ is to pump
the escaping plasma at a higher velocity, e.g. 100 eV, instead
of room temperature.

re iii)   To take up the energy flux the scrape-off region in the
divertor has to be spread out over a sufficiently large area of
cooled neutralizer plates. The poloidal field in the divertor
chamber should therefore form long "ears" to provide extended
divertor exhaust channels. In this way, in the design of Fig.4,
the power density on the neutralizer plate could be kept down to
about 300 W/cm^2 (or even 150 W/cm^2 for a corrugated plate).

re iv)   It might be mentioned that the concept of producing an
expanding magnetic limiter by vertical displacement of the plasma
column, as discussed in section III.3., is in accord with the use
of superconducting divertor coils.

re v)   In order to render possible remote disassembly of the
inner poloidal field coils, a design might be adopted in which
these coils are in the form of 180° loops (or shorter sections,
if a modular design is aspired to) that carry positive current
half-way around the torus, crossover and return with negative
current. This design feature was first proposed for ASDEX /17/
and is also considered for a JET divertor /18/ and in the EPR
divertor design /31/.

VI.   CONCLUSIONS

A magnetic divertor appears to be capable of strongly lessening
the impurity problem and of solving some of the needs of a fusion
reactor. Even a divertor that works only in the unload mode
($C \simeq 1$, $R \leqslant 0.1$, $P \simeq 0$) should reduce the impurity level by
about a factor of 10. Beyond that a divertor should be able
to attenuate the influx of heavy metal impurities from the

wall. The shielding efficiency against light elements is more doubtful, since it depends mainly on mirror confinement of the plasma in the scrape-off region which might be impaired by loss-cone instabilities.

Alternative concepts for the control of impurities should therefore be persued in parallel.

This work was performed under the terms of agreement on association between the Max-Planck-Institut für Plasmaphysik and EURATOM.

REFERENCES

/1/   L. Spitzer, Jr., Report No. NYO-993, U.S.Atomic Energy
       Commission, Washington, D.C. (1951) and Phys.Fluids $\underline{1}$,
       253 (1958)

/2/   D. Düchs, G. Haas, D. Pfirsch, H. Vernickel, J.Nucl.
       Mat. $\underline{53}$, 102 (1974)

/3/   E. Hinnov, J.Nucl.Mat.$\underline{53}$, 9 (1974)

/4/   D.M.Meade, Nucl.Fusion $\underline{14}$, 289 (1974)

/5/   D. Eckhartt, E. Venus, JET Techn.Note $\underline{9}$ (1974)

/6/   R. Behrisch, B.B. Kadomtsev, Plasma Physics and Controlled
       Nuclear Fusion Research, Vol.II, p.229, IAEA, Vienna (1975)

/7/   H. Alfvén, E. Smårs, Nature $\underline{188}$, 801 (1960)

/8/   C.M. Braams, Phys.Rev.Lett. $\underline{17}$, 470 (1966)

/9/   B. Lehnert, Proc. 3rd Int.Symposium on Toroidal Plasma
       Confinement, Paper Cl-I, Garching (1973)

/10/   D.M. Meade, et al., Plasma Physics and Controlled Nuclear
       Fusion Research, Vol.I, p.621, IAEA, Vienna (1975)

/11/   R. Dei-Cas, A. Samain , Plasma Physics and Controlled
       Nuclear Fusion Research, Vol.I, p.563, IAEA, Vienna (1975)

/12/   T. Consoli, et al., Plasma Physics and Controlled Nuclear
       Fuison Research, Vol.I, p.571, IAEA, Vienna (1975)

/13/   T. Ohkawa, General Atomic Report GA-A13017 (1974)

/14/   G. Becker, K. Lackner, 6th International Conference on
       Plasma Physics and Controlled Nuclear Fusion Research,
       Berchtesgaden 1976, Paper B 11-3 (to be published)

/15/   K.-U.v.Hagenow, K. Lackner, Bull.Am.Phys.Soc.$\underline{19}$, 852 (1974)
       and Proc. of 7th European Conf.on Controlled Fusion and
       Plasma Physics, Lausanne 1975, Vol.I, p.19

/16/ ASDEX Proposal, Part I   (October 1973)

/17/ ASDEX Proposal, Part II (June 1974)

/18/ The JET Project, First Proposal, Report EUR-JET-R2
      (April 1974)

/19/ PDX Official Planning Document, Special Report 74/2
      (June 1974)

/20/ A.M.Stephanovsky, Report IAEA-2540, Kurchatov-Institute,
      Moskow 1975

/21/ L.Spitzer, Jr., Physics of Fully Ionized Gases, John
      Wiley & Sons, New York 1962

/22/ R. Allgeyer et al., Proc. 6th Symposium on Engineering
      Problems of Fusion Research, San Diego 1975, p. 378

/23/ A. Rogister, in Report on the Planning of TEXTOR (Novem-
      ber 1975), p.B. II/4

/24/ W. Lotz, IPP Report 1/62 (1967)

/25/ G. Haas, W. Henkes, M. Keilhacker, R. Klingelhöfer,
      A. Stäbler, Proc. 6th Symposium on Plasma Heating in
      Toroidal Devices, Varenna 1976, in press

/26/ G. Haas, M. Keilhacker, Proc. International Symposium
      on Plasma Wall Interaction, Jülich 1976, to be published

/27/ M. Yoshikawa, JAERI Report M 4494 (1971)

/28/ E.F. Johnson, Princeton Report MATT-901 (1972)

/29/ F.H. Tenney, G. Levin, Proc. 7th Symposium on Fusion
      Technology, Grenoble 1975, p. 257

/30/ R.G. Mills, International School of Fusion Reactor Tech-
      nology, Erice/Sicily 1972; Princeton Report MATT-949 (1973)

/31/ A.V. Georgievsky et al., Proc. 6th Symposium on Engineer-
      ing Problems of Fusion Research, San Diego 1975, p. 583

TOKAMAK DEVICES

THE "FINTOR 1" DESIGN

A Minimum Size Tokamak Experimental Reactor

E. Bertolini

Jet Design Group, Culham Laboratory,
Abingdon (U.K.)

INTRODUCTORY REMARKS

A co-operation in the field of Tokamak Fusion Reactor
Conceptual Design studies has been established, since early
1973, between the C.N.E.N. Laboratories at Frascati, the J.R.C.
EURATOM at Ispra and the Electrotechnical Institute of the
University of Napoli: the FINTOR (Frascati, Ispra, Napoli
Tokamak Reactor) Group has been set up.

The work carried on by the FINTOR Group has now reached
a stage where the effects of the main physics and engineering
constraints, for a minimum size Tokamak Experimental Reactor
have been clearly identified. This phase, now completed, has
allowed to produce a self-consistent design of each basic component of the reactor, FINTOR 1, [1], and to identify the more
relevant interface problems toward a further optimisation of
the reactor dimensions and characteristics to be performed in
the future (FINTOR 2).

The main parameters of FINTOR 1 are: Plasma Radius
a = 2.25 m, Major Radius R = 9 m, Toroidal Magnetic Field on
the axis B_{TO} = 3.5 T, Thermal Power P_t = 90 MW, Total Wall
Loading Q = 0.08 MW/m^2, Plasma Energy Confinemt Time τ_{Ei} = 16 s.

The idea of focussing our attention on a minimum size
reactor originated at the first Erice School on Fusion Reactors
in 1972, where the question raised by M.A. Hoffman "How small
can an experimental Tokamak reactor be ? " did not receive a
satisfactory answer.

We soon realised, however, that the question cannot be
answered in general terms; there are many "minimum size Tokamak
Reactors", according to their aims and to the plasma physics and
engineering assumptions made. Moreover, they might be very

different in design features, dimensions and thermal power output.

Therefore, in the first part of this lecture the AIMS of FINTOR 1 and the TECHNICAL AND PHYSICS ASSUMPTIONS made will be clearly stated.

In the second part the approach chosen for the determination of the FINTOR REACTOR PARAMETERS and OPERATING CONDITIONS within the framework of the assumptions made will be described. Some key results of the sensitivity analysis performed will show the effect of these assumptions on the size (in dimensions and thermal power) of the "minimum size reactor".

The main TECHNICAL FEATURES of FINTOR 1 major components (Reactor Vessel, Magnetic System, Shield and Blanket) will be presented in the third part, bearing in mind that the emphasis has been more on the individual component design and on their interface problems than on a detailed and complete self-consistent reactor design.

The lecture will end with SOME COMMENTS ON FINTOR 1 RESULTS

The detailed design of the main reactor components, the considered alternative solutions, the extensive sensitivity analysis performed on the main parameters, has allowed the fulfilment of the goals of the FINTOR 1 exercise, i.e. to identify the most relevant interface problems and to assess the relative importance of the technical assumptions made.

1.   AIMS - TECHNICAL AND PHYSICS ASSUMPTIONS

In general terms, the objective of the FINTOR 1 work has not been to produce a detailed and fully self-consistent Reactor Design, to be considered as one of the fusion power Technical Feasibility Experiments [2] to be built between the Large Tokamaks [3] under design at present (TFTR, JT-60, JET, T-20), and a Commercial Reactor, but to give consideration to the following question: "WHAT IS THE MINIMUM SIZE OF A TOKAMAK REACTOR AIMED TO FUSION POWER TECHNOLOGICAL FEASIBILITY STUDIES

AND WHAT ARE THE MAIN PROBLEMS ENCOUNTERED IN THE DESIGN, ONCE
THE ASSUMPTIONS MADE BOTH IN PHYSICS AND ENGINEERING ARE
CLEARLY DEFINED ?"

Therefore the <u>specific aims</u> of the FINTOR 1 Design have
been to assess:

A) the minimum size (in dimensions and thermal power) of such
an experimental apparatus by means of a plasma model fairly
complete, used to produce a self-consistent set of plasma
parameters.

B) the interface problems and major difficulties, in design,
construction and operation when all the main components of
a reactor are put together with their specific functions.

The key <u>technical and physics assumptions</u> made are the
following.

<u>Technical</u>:

a) Superconducting D-shaped toroidal field coils and divertor
coils made up of Nb-Ti (maximum magnetic field B_{TMAX}= 8T,
$\Delta B \leq 0.5$ T for a plasma current rise time of 10 s and
average current density in the coils $J = 10^7$ A/m^2),

b) Transformer with non-saturated iron core ($B_{TRMAX} < 2.0$ T)
and poloidal coils made up of plain copper,

c) Single null axisymmetric divertor (preferable for plasma
equilibrium).

d) Blanket sufficient to allow for a tritium breeding factor
> 1 and shield arrangement such as to limit the DPA to
< 2.0×10^{-5}/year and the nuclear energy deposition in the
superconducting coils to < 5×10^{-6} W/cm^3;   no blanket
modules assembled in the inner part of the torus,

e) Vacuum tightness moved from the first wall to the reactor
containing vessel.

<u>Physics</u>:

a) Ignition reached by ohmic and neutral injection heating
and maintained by α-particle only

b) Circular plasma cross-section

c) MHD stability margin at the plasma radius q = 2

203

d) Ratio of kinetic to poloidal magnetic pressure $\beta_p = 2$

e) Energy confinement time 10 times smaller than the neo-
    classical one, $\tau_{Ei} = 0.1\ \tau_{Ei(NC)}$

f) Plasma current rise time $\tau_{IR} = 10$ s, burning time $\tau_B = 240$ s,
    plasma current decay time $\tau_{ID} = 10$ s.

The above assumptions have been made, on purpose, somewhat
conservative on the technical side (in particular in the choice
of the superconducting material and in assuming the need of a
single null axisymmetric divertor, which both call for a relatively low toroidal magnetic field) and certainly more optimistic,
than the Tokamaks in operation at present indicate, on the
physics side (in particular in the energy confinement time
required to satisfy the plasma power balance).

2.   DETERMINATION OF THE REACTOR PARAMETERS

The FINTOR 1 parameters have been calculated by using a
rational method developed for selecting a minimum size reactor
with given key design constraints [1], [4], [5].   The method
can be readily adapted also to the case of a fusion reactor
required to produce a given (not minimum) power or wall loading.
It utilizes three models involving analyses at three different
levels of complexity:

-   Model A   Simplified plasma-engineering model for prelim-
               inary choice of parameters
-   Model B   Self-consistent analysis of quasi-steady-state
               plasma with 0-D (Zero-dimension) model
-   Model C   Self-consistent analysis of plasma with radial
               profiles (1-D model).

The utility of each of the models in providing important
input data for the more complex models should appear clear in
what follows.

Due to the present uncertainties in plasma physics,we have
chosen to compare all the results to a single reference model
of the plasma, i.e. the neoclassical model, and all calculated
results, for the required plasma confinement times to satisfy

the plasma power balance, are compared to this upper limit on the
confinement time.   However, the neoclassical scaling in no way
enters directly into the Tokamak plasma Models A and B;   that is,
any other reference confinement time, e.g. based on Bohm diffusion or trapped-ion modes or any other loss mechanism could be
used equally well to non-dimensionalize the calculated required
confinement times.

In plasma Model C, used to check the results obtained with
Model B, the transport model assumed does play a more important
role as it determines the specific form of the temperature
profiles obtained.

The three-step plasma model will be briefly described
leaving the details for the potential readers, to the references
given above.

2.1   Model A

A preliminary choice of the main reactor parameters can be
made by using three simple equations, which take into account
the plasma MHD stability condition, the blanket shield thickness requirements, as well as the toroidal coil and the transformer design constraints, i.e.:

The equation of the **plasma current**, limited by the
Kruskal-Shafranov condition:

$$(1) \quad I = \frac{2\pi}{\mu_o} \left(\frac{a}{A}\right) \times \left(\frac{B_{TO}}{q}\right)$$

where I is the plasma current, **a** the plasma radius, A the
aspect ratio of the torus (with R its major radius), B_{TO}
is the toroidal field on the minor axis, q is the MHD
safety factor and μ_o the permeability of the vacuum.

The **field-limit** equation, which takes into account
the engineering constraints imposed on the toroidal magnetic field on axis (B_{TO}) by the maximum toroidal field
(B_{TM}) permitted for the superconductor employed and/or by
mechanical stresses on the coils,

$$(2) \quad B_{TO} = B_{TM} (R - a - d_D - d_B)/R$$

205

where d_D is the additional space between the plasma and the toroidal coil on the inner side of the torus required for, e.g. a single-null divertor, and d_B is the thickness of blanket and shield in front of the coil.

The _transformer_ equation, which relates the flux swing required to induce the toroidal plasma current to the geometry and transformer constraints.

$$(3) \quad 2 \times \left[\pi \, (R-a-d_D-d_B-d_C)^2 B_{TRM} \right] = 2\pi \, R \, \ell_{PL} F_{TR} I$$

where d_C is the thickness of the toroidal coils, B_{TRM} is the transformer core maximum field, ℓ_{PL} is the plasma ring inductance per unit length and F_{TR} is a correction factor for plasma losses and for imperfect inductive coupling.

The factor of 2 on the left-hand side implies that the current in the transformer primary winding is reversed in order to give the maximum possible flux swing by varying the field from $- B_{TRM}$ to $+ B_{TRM}$.

The above three equations can be re-arranged as follows:

$$(1') \qquad a = \frac{\mu_o q}{2\pi} \frac{AI}{B_{TO}}$$

$$(2') \qquad B_{TO} = B_{TM} \left(1 - \frac{1 + \dfrac{d_B}{a} + \dfrac{d_D}{a}}{A} \right)$$

$$(3') \qquad A^3 + C_2 \, A^2 + C_1 \, A + C_o = 0$$

where C_o, C_1, C_2 are functions of $d_B, d_C, d_D, \ell_{PL}, F_{TR}, \mu_o, q, B_{TM}$ and B_{TRM}.

As a reasonable extrapolation of scientific feasibility experiments, the following input values were chosen

$B_{TM} = 8\,T, B_{TRM} = 2\,T, I \simeq 5\,MA, q \simeq 2, \ell_{PL} \simeq 10^{-7} H/m,$

$F_{TR} \simeq 3, d_B/a \simeq 0.7, d_C/a \simeq 0.5$ and $d_D/a \simeq 0.6$.

The choice of B_{TM} is imposed by present-day superconducting technology, using materials like Nb-Ti, and by the choice of keeping the stresses below about 10 kg/mm^2. $B_{TRM} \leq 2\,T$ corresponds to a non-saturated iron core. The value chosen for the plasma current, which can be considered as replacing an assumption on

206

the plasma confinement, represents a reasonable extrapolation of one of the large Tokamaks under design at present (about 3.0 MA circular cross-section), whereas the value of q is about the minimum which seems acceptable in practice. The value of d_D/a chosen is large so as to be able to accommodate a divertor with a (single) null on the inside.

For this case one obtains (Case A 1)

$$A \simeq 4.0$$
$$B_{TO} \simeq 3.5 \text{ T}$$
$$a \simeq 2.25 \text{ m}$$
$$R \simeq 9 \text{ m}$$

The implied d_B is thus about 1.6 m, which is not unreasonable for the blanket plus shield thickness (the value of d_B must be checked because it clearly does not scale with the plasma radius $\underline{a}$; i.e. the choice of $d_B/a = 0.7$ is merely a first guess and must, in general, be varied until a reasonable d_B is obtained).

2.2   Model B

Consistent global plasma and machine parameters can be determined from the following three-species 0-D model (D and T ions, electrons and alpha-particles) [4]. A self-sustained, quasi-steady state is assumed where all the plasma losses are balanced by alpha-particle heating. The space averaged ion (i) power balance per unit volume then is

$$\frac{3}{2} \frac{n_i T_i}{\tau_{Ei}} = S \Delta U_{\alpha i} + \frac{3}{2} \frac{n_i T_i}{\tau_{0(ie)}} \left(\frac{T_e}{T_i} - 1 \right) - W_{PF} \qquad (4)$$

with n denoting the density and T the temperature.

The analogous electron (e) power balance is

$$\frac{3}{2} \frac{n_e T_e}{\tau_{Ee}} = S \Delta U_{\alpha e} - \frac{3}{2} \frac{n_e T_e}{\tau_{0(ie)}} \left(\frac{T_e}{T_i} - 1 \right) - W_B - W_C \qquad (5)$$

In both equations the term on the left-hand side represents the energy loss rate due to thermal conduction to the walls (or the divertor). The characteristic times, τ_{Ei} and τ_{Ee}, are the required energy confinement times to achieve a power balance and they constitute two of the key unknowns in this model. The first terms on the right-hand side of each equation are the energy transfer rates from the alpha-particles, produced in fusion reactions, to the ions and electrons, respectively. The second terms are the collisional energy transfer rates between ions and electrons. The last term in equation (4) $W_{PF} = 3ST_i$, accounts for the disappearance of two ions in a fusion reaction and is negligible throughout. The reaction rate, appropriate for a 50:50 mixture of D and T, is given by

$$S = \frac{n_i^2}{4} < \sigma_{fus} v > \text{ with } \sigma_{fus} \text{ the cross-section for fusion}$$

processes.

The quantities $\Delta U_{\alpha i}$ and $\Delta U_{\alpha e}$ describe the energy trans-ferred from a fusion α-particle to the ion and electrons, while $\tau_{0(ie)}$ is the electron-ion energy transfer time. The terms W_B and W_C in eq. (5) refer to the bremmstrahlung and cyclotron radiation losses, respectively. In the case of cyclotron radiation partial reflection at the first wall is taken into account, including also the effect of unavoidable holes through their relative surface f_H. Line radiation losses of high-Z impurities could be taken into account by adding a term $- W_L$ to the right-hand side of eq. (5).

In order to complete the picture of the model, the follow-ing equations have been added:

$$\frac{\tau_{Ee}}{\tau_{Ei}} = 0.232 \sqrt{\frac{m_i}{m_e}} \frac{\ln \Lambda_i(n_e, T_i)}{\ln \Lambda_e(n_e, T_e)} \left(\frac{T_e}{T_i}\right)^{\frac{1}{2}} \left(\frac{n_i}{n_e}\right) \tag{6}$$

$$n_\alpha = S \tau_{N\alpha} \tag{7}$$

with $\tau_{N\alpha}$ the alpha-particle confinement time

$$n_e = n_i + 2n_\alpha + \Sigma Zn_z \tag{8}$$

208

$$P_{TOT} = n_i T_i + n_e T_e + \frac{2}{3} n_\alpha \langle U_\alpha \rangle + \Sigma \, n_z T_i \qquad (9)$$

$$\beta_p = \frac{2 \, \mu_o P_{TOT}}{B_p^2} \qquad (10)$$

$$\beta_p = \xi \, A^b , \qquad (11)$$

with $\xi = 0.5 \rightarrow 1$ and $b = \frac{1}{2}$ or 1.

$$B_p = \frac{B_{TO}}{A_q} \qquad (12)$$

for circular plasma cross section

$$I = \frac{2 \pi a B_p}{\mu_o} \qquad (13)$$

In the numerical treatment of the preceding ten equations
(4) $\div$ (13), the main input parameters are A, B_{TO}, $\underline{a}$ (e.g. ,
from Model A), T_e, q, b, ξ, $U_{\alpha o}$, $U_{\alpha m}$, n_z/n_i and Z, fusion cross-
section data and the relations for $\Delta U_{\alpha i}$, $\Delta U_{\alpha e}$, $\langle U_\alpha \rangle$ and $\tau_{N\alpha}$
from the slowing-down model for the alpha-particles.   The ten
unknowns τ_{Ei}, τ_{Ee}, n_i, n_e, n_α, T_i, P_{TOT}, β_p, B_p and I can then
be calculated.

It should be noted that this O - D Model requires no
assumptions regarding the transport mechanisms because the
calculated values of τ_{Ee} are the required ion energy confinement times necessary to achieve a steady-state power balance.
These times can then be compared to the predicted confinement
times for any Tokamak transport model employing any loss
mechanisms desired.

Although we have decided to compare them to the neoclassical confinement times, this does not imply that we
believe that neoclassical confinement will be achieved in
reactor operating conditions.   In fact, outside of the main
model consisting of equations (4) $\div$ (13), we calculate energy
confinement time reduction factors f_E as follows :

TABLE I

Plasma Model C Calculation Summary

$U_{\alpha m}$	1000 keV (regime 1)			$3/2\ T_i$ (regime 2)	
q_a	2.5		2	2.5	2
n_o/n_a	5	2	5	5	
q_o	1.02	1.02	0.82	1.02	0.82
f_{Ei}	$6.1 \cdot 10^{-2}$	$1.4 \cdot 10^{-1}$	$3.7 \cdot 10^{-2}$	$3.7 \cdot 10^{-2}$	$2.2 \cdot 10^{-2}$
$\bar{n}_e = \bar{n}_i [m^{-3}]$	$1.7 \cdot 10^{19}$	$1.7 \cdot 10^{19}$	$2.3 \cdot 10^{19}$	$1.7 \cdot 10^{19}$	$2.3 \cdot 10^{19}$
$\bar{T}_e$ [keV]	20.1	17.6	22.9	22.4	25
$T_{e,o}$ [keV]	26.9	27.9	32.4	29.9	35.3
$\bar{T}_i$ [keV]	14.6	14.5	15.8	15.2	16
$T_{i,o}$ [keV]	22.5	26.5	24.6	24.0	25.7
β_p	2.16	2.0	2.1	2.35	2.23
$\tau_{E,i}$ [s]	13	20	10	8.3	6.3
P_t [MW_t]	51	45	100	73	145

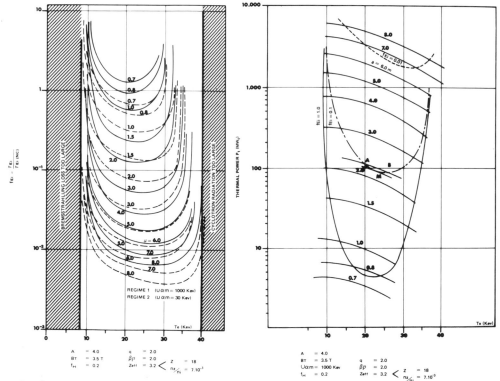

Fig.1 Required Energy Confinement due to Heat Transport losses for Different Electron Temperatures and Plasma Radii.

Fig. 2 Reactor Power and Permissible Regions for Self-Sustained Steady State Operation (Regime 1).

210

$$f_{Ei} \stackrel{\triangle}{=} \frac{\tau_{Ei}}{\tau_{Ei(NC)}} \quad ; \quad f_{Ee} \stackrel{\triangle}{=} \frac{\tau_{Ee}}{\tau_{Ee(NC)}} \qquad (14)$$

In addition to the calculation of the safety factors f_{Ei} and f_{Ee}, auxiliary equations are included in the computer code to calculate such parameters as the reactor thermal power output P_t, the neutron wall loading Q, the various radiation fluxes to the walls and so forth. For these auxiliary calculations an empirical parameter $Y \stackrel{\triangle}{=} a/a_w$ is included in the input data to account, as a first approximation, for the difference between the plasma radius $\underline{a}$ and the average minor radius a_w of the first wall. However, results from Model C which includes plasma profiles indicate that this is an acceptable approximation, except for the calculation of the thermal power.

The following reference input parameters have been chosen: $A = 4.0$, $B_{TO} = 3.5$ T, $a = 2.25$ m, $q = 2.0$, $n_z/n_i = 7 \times 10^{-3}$, $Z = 18$, as well as $\xi = 1.0$ and $b = 0.5$ (or $\xi = 2.0$ and $b = 0$) which yields a $\beta_p = 2.0$. In addition, $Y = 0.9$. Fig. 1 shows how the safety factor f_{Ei} varies with the electron temperature for Regime 1 where the alpha-particles leave the plasma at energies well above the mean ion energy ($U_{\alpha m} = 1000$ keV), and Regime 2, where the alpha-particles leave the plasma at about the mean ion energy ($U_{\alpha m} = \frac{2}{3}T_i$). One of the most interesting results is the range of temperatures over which a plasma power balance can be achieved. For the wide range of plasma radii considered below about 8 keV and above about 40 keV respectively, bremsstrahlung losses and losses due to cyclotron radiation (with $f_H = 0.2$) dominate.

Due to the limited differences between the two regimes, Regime 1, the slightly less favourable one, has been chosen for further analysis.

In Fig. 2, contours of constant f_{Ei} of 1.0, 0.1 and 0.01 are shown. This figure demonstrates that the thermal power output for the reference conditions, at $a = 2.25$ m and at

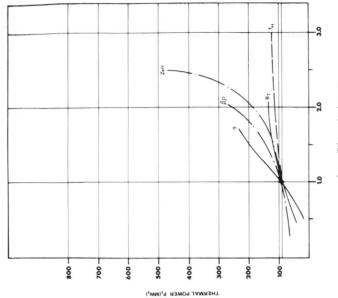

Fig. 4   Sensitivity of Minimum Reactor
Power to Main Plasma and Reactor
Parameters for $F_{Ei} = 0.1$

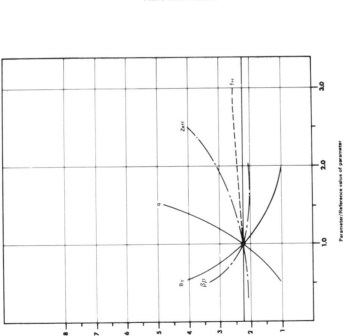

Fig. 3   Sensitivity of Plasma Radius to
Main Plasma and Reactor Parameters
for $F_{Ei} = 0.1$

T_e = 25 keV is near-minimum for about 90 MW   (Case Bl) .

For plasma transport scaling like the classical one it can be shown that the operating points to the left of the point M in Fig. 2 (such as Point A) are "thermally" unstable, while the operating points to the right (such as Point B) tend to be "thermally" stable.   Therefore, the reactor operating point will be near point B and have a temperature slightly in excess of the value of M.

A sensitivity analysis has been performed to show how some of the main input parameters might affect the dimensions and power output of the reference reactor (Figs. 3 and 4).

The variation of β_p has an effect   qualitatively similar to that of B_{TO} variations.   However, quantitatively, increases in β_p do not reduce the plasma radius very much, while P_t is increased significantly.

Comparing the sensitivity with respect to the different parameters it is seen that the minimum plasma radius depends most strongly on q and the magnetic field.   The most sensitive parameters, as regards the thermal power of a minimum size reactor, are q and β_p .   On the other hand, variations of the whole fraction f_H (that is, of the reflection properties of the wall), and, within a factor 2 of the Z_{eff} have relatively little influence.   Nevertheless at $Z_{eff} \geqslant 10$ no self-sustained steady state is possible anymore, as bremsstrahlung losses would exceed the fusion power (i.e. impurities with $Z \leq 20$ should not exceed a concentration of about 2%, and for heavy impurities, including also line radiation losses, this limit is reduced further to about 2×10^{-3} for $Z = 40$ and to 5×10^{-4} for $Z = 70$).

It is of interest to verify whether or not the plasma power balance in the reference case can be satisfied also in the presence of the trapped-ion instability .   Disregarding the influence of the density profile, this can be done with Model B comparing $\tau_{Ei(TI)}$, trapped-ion confinement with τ_{Ei}, where

213

TABLE II

Main Parameters of FINTOR 1 Reactor

Plasma radius	$a = 2.25$ m
Torus radius	$R = 9$ m
Aspect ratio	$A = 4$
Plasma Volume	$V = 900$ m^3
Toroidal magnetic field	$B_T = 3.5$ T
MHD Stability margin	$q = 2$
Poloidal β	$\beta_p = 2$
Poloidal magnetic field	$B_p = 0.44$ T
Toroidal β	$\beta = 0.03$
Plasma confinement safety factor	$f = 0.1$
Plasma current	$I = 5$ MA
Energy confinement time	$\tau_E = 16$ sec
Impurity level	$Z_{eff} = 3$
Ion temperature	$T_i = 20$ keV
Electron temperature	$T_e = 25$ keV
Ion density	$n_i = 1.7 \times 10^{19}$ m^{-3}
Electron density	$n_e = 1.9 \times 10^{19}$ m^{-3}
Wall transparency factor (cyclotron radiation)	$f_H = 0.2$
Energy transferred from α-particles	$\Delta U_\alpha = 2500$ keV
Thermal power	$P_t = 90$ MW
Total wall loading	$Q_W = 0.08$ MW/m^2
Overall radial dimension	$R_T = 18$ m
Overall height	$H_T = 27$ m

$$\tau_{Ei(TI)} = 0.228 \times 10^{-20} \frac{a^4 n_e B_{TO}^2}{T_e^{7/2}} \cdot A^{5/2} \left(1 + \frac{T_e}{T_i}\right)^2$$

The calculations clearly show that the reactor, with the parameters of Case B1, cannot work when trapped-ion instabilities govern the plasma diffusion losses, as the required $\tau_{Ei} \gg \tau_{Ei(TI)}$. Nevertheless a self-sustaining thermonuclear plasma could be obtained for a reactor of the same dimensions (a = 2.25 m, R = 9m), where the diffusion losses are dominated by the trapped-ion instability, only for toroidal magnet fields which would exceed the maximum field allowed by Nb-Ti superconducting material by almost a factor of 2, and the plasma current would be of about 10 MA.

2.3   Model C

To investigate the effect of temperature profiles for a self-sustained reactor state obtained from Model B, a 1-D plasma model is to be used (Model C). Thereby a plasma transport model must be adopted as priority. This was done using the Düchs code [6,7] which is based on a time-dependent, quasi-cylindrical (1-D) multi-fluid model. It includes toroidal effects to lowest significant order in the inverse aspect ratio 1/A, as long as they do not vanish for 1/A → 0, and is in so far equivalent with Model B. For simplicity, a two-fluid plasma model (electrons and one ion species) has been adopted. A complete account of Model C is given in references [1], [5]. As it can be seen from Table I, the results are reasonably close to estimates obtained with Model B.

As a result of the "three models" approach, the main FINTOR 1 Reactor parameters, summarised in Table II, have been obtained.

A first preliminary choice can be made using a strongly simplified plasma-engineering model (A) into which the plasma confinement properties enter only implicitly through the choice

215

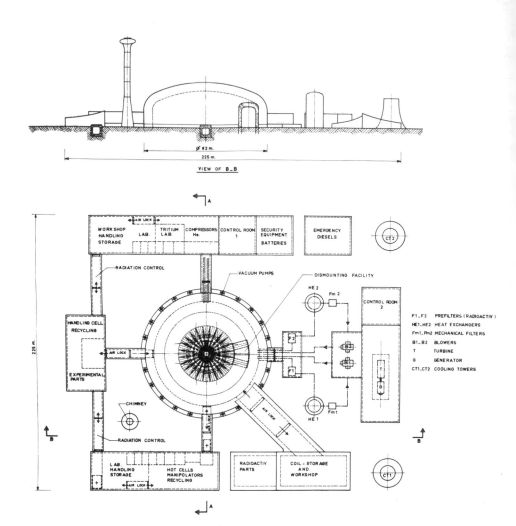

VIEW OF B_B

Ø 82 m.

225 m.

F1_F2   PREFILTERS (RADIOACTIV)
HE1_HE2 HEAT EXCHANGERS
Fm1_Fm2 MECHANICAL FILTERS
B1_B2   BLOWERS
T   TURBINE
G   GENERATOR
CT1.CT2 COOLING TOWERS

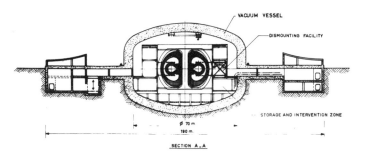

VACUUM VESSEL

DISMOUNTING FACILITY

STORAGE AND INTERVENTION ZONE

Ø 70 m.

190 m.

SECTION A_A

Fig. 5   Overall Layout of FINTOR 1   Reactor Plant

216

of the plasma current.   A rather good assessment of the influence of plasma confinement and of the functional interdepend-
ences between the reactor and plasma parameters can be obtained
adopting a multi-fluid O-D model (B),taking into account the
(space averaged) energy balances.   The influence of the radial
profiles was calculated using the time-dependent multi-fluid
1-D model of Düchs (C).   The next steps along this line towards
further sophistication would be the use of two and three-
dimensional models, which, however, would lead to a degree of
complication , not justifiable, at the present moment, for the
purpose of this work.

2.4   Start-up and Operating Cycle

Among the uncertainties which affect the "credibility"
of fusion reactor designs, two important ones were given some
thought by the Fintor Group: the expected impurity rate due to
Plasma-Wall Interaction which affects the useful pulse lengths
and the Reactor Start-up.

2.4.1   Impurities

A first calculation of the impurity concentration in
FINTOR 1 has been carried out, based on a model suggested by
Scherzer [8].   The results show that, without divertor, in the
case of stainless steel the maximum time interval for which the
FINTOR 1 plasma can stay ignited would be less than 1/10 of the
confinement time.   Pyrolytic graphite and B_4C allow conditions
for about the confinement time.   LiD would allow steady state
operation: however, in this case, the great instability of any
local overheating of the system, combined with an uncontrolled
pressure increase of the fuel and the necessity of operating
below $100^{\circ}C$ would introduce other practical difficulties.

2.4.2   Start-up

An orientative study of the start-up conditions of FINTOR 1
has brought to identify the following operation phases:
- in the first phase, which lasts 0.1 sec, the transformer is

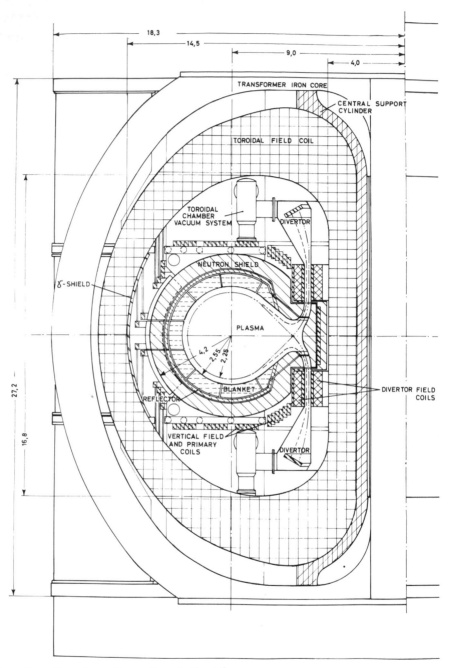

Fig. 6   Reactor Core Cross-Section

218

energised through a condenser bank and produces the breakdown, the rise of the plasma current to 0.5 MA and the plasma
ohmic heating to a temperature of about 0.15 keV.
- in the second phase, the transformer is energised at constant
  voltage and has the   only function of bringing the plasma
  current to 5 MA in 10 sec, while the plasma heating to
  ignition temperature ($\sim$ 12 keV) is provided by the injection
  of fast deuterium atoms.

The value of 0.5 MA for the first phase current was chosen
so that the eddy current losses induced in the toroidal field
coils remain at the same tolerable level during both the first
and the second start-up phase.   The energy of the condenser
bank for the breakdown is 16 MJ.

During the second phase the required transformer energy to
get 5 MA is 1.2 GJ.   With the loss-model assumed, the heating
power to reach ignition by neutral injectors is rather limited
($\sim$ 15 MW) as compared to the transformer power, so it was
considered convenient to begin heating during the start-up
second phase.   Eight injectors with an energy of 80 keV and a
neutral beam current of 1 kA/m^2 would be adequate for this
purpose.

3.   MAIN FEATURE OF THE REACTOR COMPONENTS

An overall FINTOR 1 reactor layout, including the required
ancillary facilities, and a cross-section of the reactor core
are shown in Figs. 5 and 6 respectively.   In the following,
the main Reactor components, alternative design proposals, and
associated problems will be described briefly [1].

3.1   Vacuum Vessel

The whole reactor is housed in a vacuum-tight vessel,
structured to support the external pressure.   The vacuum facing
wall is a stainless liner.   The idea of transferring the vacuum
boundary outside of all the components of the Tokamak reactor
has been considered a reasonable, and at least a provocative

TABLE III

Main Parameters of the Reactor Vacuum Vessel

a) Vacuum-Vessel Parameters	
Liner material	stainless steel
Liner area	10^4 m^2
Liner outgassing rate	10^7 torr.l.sec^{-1}.cm^{-2}
Vessel pressure	10^{-5} torr
Vessel pumping system	
Diffusion pumps:	$20 \times 5 \cdot 10^4$ l.sec^{-1}
Mechanical forepumps:	10×10^4 m^3.h^{-1}

b) Reactor Chamber Vacuum Parameters	
First Wall material	stainless steel
First Wall area	4×10^3 m^2
First Wall outgassing rate	10^{-8} torr.l.sec^{-1}.cm^{-2}
Chamber pressure	5×10^{-7} torr
Chamber pumping speed	10^6 l.sec^{-1}

TABLE IV

Main Parameters of the Iron Core Transformer

	Standard (non-biased)	Biased
Primary windings		
Max m.m.f. $[MA]$	5.52	5.52
Copper weight $[t]$	740	740
Equilibrium field windings		
B_v $[T]$	0.15	0.15
Decay index	0.7	0.7
Copper weight $[t]$	560	560
Plasma		
Plasma current $[MA]$	5	
Iron weight $[t]$	24,500	15,000
Divertor current $[MA]$	4.3	

alternative [9],[10],to the almost impossible task of a conventional solution (high temperature vacuum joints reliability
and,more generally, maintenance and repair operations).   As an
important side effect the proposed arrangements allow the avoidance of cryostats for the superconducting coils.

The technical feasibility of such a large vacuum vessel,
is supported by existing vacuum vessels of similar dimensions
in the USA,   (NASA test facilities).

Assuming an outgassing rate of about 10^{-7} torr.l.sec^{-1}.cm^{-2}
from the walls, a vacuum of about 10^{-5} torr. can be attained
inside the vessel by a system of diffusion pumps, forepumped by
mechanical pumps.

The gaps allowed between adjacent moduli are narrow enough
to ensure a high impedance between the vessel and the toroidal
plasma chamber (a conductance lower than 10^3 l.sec^{-1} is
achievable).

Assuming an outgassing of about 10^{-8} torr. l.sec^{-1}.cm^{-2}
from the reactor walls, a vacuum of about 5×10^{-7} torr can be
attained, before reactor start-up, by a system of diffusion or
cryogenic pumps.   The same system can be used to pump away
from the divertor the exhausting gas, provided that the burning
factor is not lower than 5%.

Although a detailed design of the vessel has not been
performed, and its potentialities not explored in detail, both
the vessel and the pumping system look technically feasible and
engineering sound.

It is felt that the proposed arrangement would improve
the reactor feasibility and somewhat ease the remote handling
problems.

The vessel inner diameter is 70 m , its height 50 m and
its volume about 20.000 m^3 .   Table III gives the vacuum
system main parameters.

3.2   The Poloidal Field System
Extensive MHD equilibrium and stability studies have been

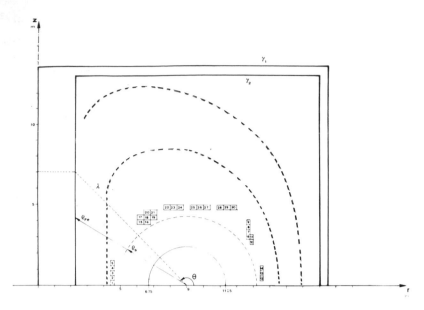

Fig. 7   Poloidal Field Windings Location for Iron Core
          Transformer

TABLE   V

Main Parameters of the Toroidal Magnet

Number of D-shape coils	24
Superconductor	Nb-Ti
Stabilizer	Copper
Structural material	stainless steel
Coil internal dimensions	$10.5 \times 16.8 \ m^2$
Coil cross section	min $1.15 \times 0.6 \ m^2$
	max $3.5 \times 0.6 \ m^2$
Number of turns per coil	661
Conductor shape	hollow square
Conductor cross section	min $(2.0)^2 \ cm^2$
	max $(3.32)^2 \ cm^2$
Average current density	$0.9 \ KA/cm^2$
Temperature	5.2^oK
Coolant	Two-phase helium
Helium cooling power/coil	94 kW
- eddy current losses	0.4 kW
- hysteresis losses	0.2 kW
- nuclear heat	7 kW
Max steel stress	$43 \ kg/mm^2$
Max copper stress	$10.5 \ kg/mm^2$
Energy stored	60 GJ
Weight of one coil	242 tons

222

performed including two alternative axisymmetric poloidal divertors, 2-stagnation points and one-stagnation point respectively. The latter has been eventually chosen for plasma equilibrium reasons although it is more demanding in reactor radial dimensions.

With the divertor coils arrangement inside the toroidal windings, the plasma equilibrium is not affected too much by the divertor. In order to calculate the divertor currents, the magnetic field due to the plasma current and the vertical field for the plasma equilibrium have been evaluated first. As a second step, divertor currents giving an equal and opposite value of the field, in the region where the stagnation point is desired, were calculated. By superposing the calculated magnetic fields, the electromagnetic configuration was obtained.

As the divertor field can be kept constant during the start-up phase, the divertor coils are made by Nb-Ti, whereas the time-varying currents for the vertical field and transformer require copper coils. They are located inside the toroidal coils and they are assembled co-axially to the plasma to improve electromagnetic coupling.

Two transformer solutions [11] involving an air core and an iron core transformer, respectively, have been investigated and particular attention has been paid to the following constraints:

a)   the presence of the divertor channels which limit the possibility to locate the conductors uniformly all around the magnet shield,

b)   the presence of superconducting materials (for toroidal and divertor coils) which imposes strong limits to the value of time-varying fields (a limit of 0.5 T/10 s has been assumed),

c)   the interest in minimizing the magnetic linkage between the vertical field and transformer windings.

A compromise has been reached, in order to obtain the following, for both iron and air core transformers:

1)   To determine a suitable distribution of the toroidal

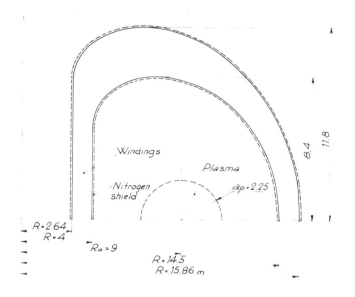

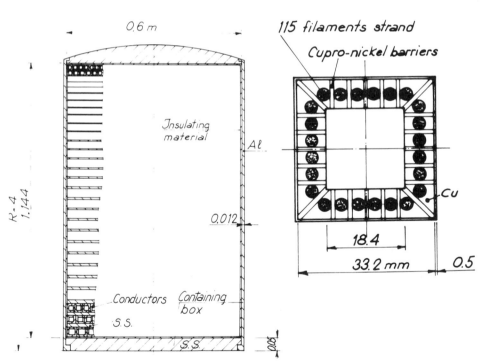

Fig. 8 Toroidal Coil and Conductor Bar Cross-Sections

currents, which give the equilibrium vertical field.

2)   To find such an arrangement of the primary m.m.f., so that
      a negligible field appears in the plasma region,

3)   To calculate the divertor currents values, which lead to
      an equilibrium configuration with a single-null.

The position of the transformer windings (iron core case) is shown in Fig. 7 and the opportunity to pre-polarize the transformer ("biased" solution) was also considered in order to reduce the copper weight in the air core transformer case and the core weight in the iron case.   Although a clear preference for an air core or an iron core transformer could not be fully proven [11] an iron core biased transformer was chosen for the FINTOR 1 design.   The main parameters of this transformer are summarized in Table IV.

3.3   The Toroidal Field System

The technology pattern chosen in FINTOR 1 allows to consider Nb-Ti only, as a mid-term reasonable choice as superconducting material.   The toroidal magnet, made up of 24 coils (so as to allow a ripple at the plasma edge $<$ 2%), has two unique features:

a)   The D-shape, bending moment free pattern, has been given
      to each conductor layer and its supporting stainless steel
      ring.   The overall coil cross-section is therefore wider
      at the top and bottom of the coil.   The centripetal
      pressure in the vertical part of the magnet is taken by an
      inner support cylinder with cradles to accommodate the
      coils.

b)   There are no cryostats, because the whole magnet is con-
      tained in the reactor vacuum vessel.   It is cooled by
      forced circulation of two-phase helium since this system
      allows a more effective and uniform refrigeration as com-
      pared to the case of immersion in liquid helium.

The coil dimensions can be deduced by the cross-sections shown in Fig. 8.   The helium cooling is provided through a

TABLE VI

Main Parameters of the Blanket and Shield

. Number of segments		24
. Number of modules		96
. Number of blanket submodules		432
. Breeding	material	lithium metal
	volume % of total	90%
	thickness	72.5 cm
. Reflector	material	stainless steel
	volume % of total	90%
	thickness	15 cm
. Neutron shield	material	B_4C powder
	volume % of total	90%
	thickness	75 cm
. Gamma shield	material	lead
	volume % of total	90%
	thickness	15 cm
. Coolant	material	helium
	volume % of total	5%
	pressure	30 bars
	inlet temperature	175°C
	outlet temperature	375°C
	flow velocity	10 m/sec
. Structural	material	SS 316
	volume % of total	5%
	max temperature	420°C
	max stress	38 kg/mm^2
. First wall	material	SS 316
	arrangement	welded tubes
	thickness	2.5 cm
. Total weight		4922 tons

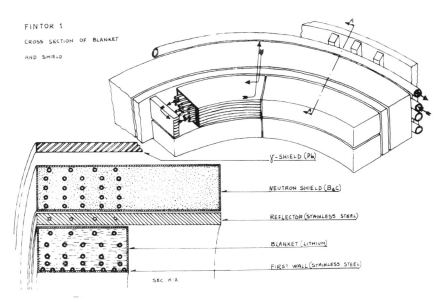

FINTOR 1

CROSS SECTION OF BLANKET AND SHIELD

γ-SHIELD (Pb)

NEUTRON SHIELD (B₄C)

REFLECTOR (STAINLESS STEEL)

BLANKET (LITHIUM)

FIRST WALL (STAINLESS STEEL)

SEC. A-A

Fig. 9   Perspective and Cross-Sectional Views of the
Blanket and Shield Modules

central hole in the conductor.   The number of Nb-Ti filaments
(200 µm in diameter) imbedded in a copper matrix varies considerably from the inner to the outer layer, according to the magnetic field variation through the coil.

The thermal radiation from the external bodies and from the
mechanical supports is avoided by screening the magnet coils
with two copper shields cooled by liquid nitrogen and water,
respectively.

Nuclear heating eddy current and hysteresis losses have
been evaluated and they bring about an additional cooling power
requirement of less than 10%.   This corresponds to a small
variation of the helium coolant quality between inlet and outlet.

Only preliminary mechanical calculations have been performed and therefore a more extensive work is required to assure
a complete mechanical stability of the magnet.

The toroidal magnet main characteristics are summarized in
Table V.

3.4   Blanket System and Magnet Shield

The blanket and shield are divided (as the toroidal magnet)
in 24 equal segments, each one of which can be withdrawn in the
radial direction, between the coils.   The shield (4 modules
per segment, total 96 modules) and the blanket (18 sub-modules
per segment, total 432 sub-modules) are decoupled so as to allow
an independent maintenance of the blanket and the shield.   The
structural material is stainless steel.

In Fig. 9 a perspective and cross-sectional views of a
module are given.   The modules are independently cooled by
helium flowing,in parallel,through stainless steel tubes. This
arrangement allows to realize the first wall by a set of
parallel tubes welded together.

The modular arrangement of the blanket in the reactor is
shown in Fig. 6.   The main data concerning characteristics of
the blanket and the shield are given in Table VI.

The neutron and gamma ray fluxes have been calculated by

TABLE VII

Nuclear Performance Data

Tritium Breeding			
Tritium Breeding over 360° T = 1.41			
Tritium Breeding on 270° (D-shaped blanket) T = 1.24			
Energy Multiplication factor = 1.16			

Radiation Damage in the SS-316 First Wall

Neutron Wall Loading	FINTOR-1 0.07 MW/m^2	Fission Test Reactor EBR	Power Reactor 1 MW/m^2
14 MeV neutron current $\left[\dfrac{n}{cm^2.sec}\right]$	$2.3 \cdot 10^{12}$	-	$4.4 \cdot 10^{13}$
Total neutron flux $\left[\dfrac{n}{cm^2.sec}\right]$	$2.4 \cdot 10^{13}$	-	$3.7 \cdot 10^{14}$
Total yearly fluence $\left[\dfrac{n}{cm^2.year}\right]$	$7.5 \cdot 10^{20}$	-	$1.1 \cdot 10^{22}$
Displacements per atom/year	0.9	44	15
Hydrogen production [appm/year]	30	270	468
Helium production [appm/year]	14	5	219

Radiation Damage in the Magnet Coils

Displacement per atom/year in the copper	$1.5 \cdot 10^{-5}$	-	$2.2 \cdot 10^{-2}$
Max heat deposition [W/cm^3]	$4.76 \cdot 10^{-6}$	-	-

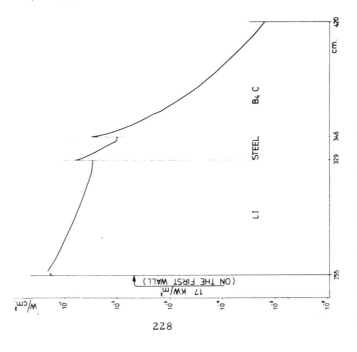

Fig. 10   Total Heat Deposition across a Blanket Module

the one-dimensional transport code ANISN in cylindrical geometry. A detailed engineering design requires, of course, transport or Monte Carlo multi-dimensional calculations [12] for an
accurate treatment of toroidal effects, for a reliable design
of the shield of the divertor coils, etc. Some of the nuclear
characteristics are summarized in Table VII, where the radiation effects on the stainless steel first wall are compared with
those of fission test breeders and fusion power reactors (wall
loading $Q \simeq 1$ MW/m^2). From these figures and considering the
reduced time operation of FINTOR, it can be concluded that no
radiation damage problems or changes of mechanical properties,
are expected in the FINTOR 1 first wall and in the blanket
structural components.

This can be considered an advantage, but also a drawback,
because it means that no relevant nuclear engineering information would be obtained.

The shielding has been designed with the aim of minimizing
the damage to the copper stabilizer. The maximum radiation
induced resistivity in the toroidal field coils at the end of
one year full-time operation is 2.5×10^{-9} Ωcm, to be compared to a maximum magneto-resistivity of $3.6 \times 10^{-8} \Omega$ cm.

In Fig. 10 the total heat deposition across a typical
module is represented. Extensive thermo-mechanical calculations have been carried out on the blanket module by means of
the finite element method.

The stresses and the associated deformations (see Fig. 11)
have been calculated. Being the highest stresses about equal
to the elastic limit of the material, the chosen size of the
modules was considered feasible from the point of view of the
mechanical stability.

3.5 Radio Active Aspects

These aspects play an important role in the viability of a
fusion reactor, and they have been analysed in some detail.

229

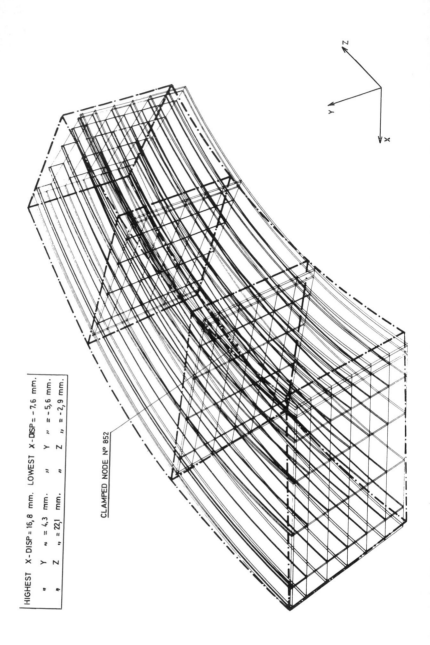

HIGHEST X-DISP = 16,8 mm. LOWEST X-DISP = -7,6 mm.
" Y " = 4,3 mm. " Y " = -5,6 mm.
" Z " = 22,1 mm. " Z " = -2,9 mm.

CLAMPED NODE N° 852

Fig. 11   Stress Associated Deformation Plot for a Blanket
Module

230

3.5.1   Activation and After Heat

The radioactivity at shut-down and the radioactive decay
of the activated structural components have been evaluated with
a computer code prepared at the JRC-Ispra[13].Operation times
from 3 months up to 10 years have been considered.   The
calculations have been carried out on five zones, i.e. the
first wall, the blanket, the reflector, the neutron shield and
the magnet zone (Table VIII).

The time dependence of the radioactive decay in the first
wall and the blanket is shown in Fig. 12 .   The decrease is
relatively slow.   With a 10-years operation, a reduction of a
decade requires more than 3 years of cooling.

As far as the decay heat is concerned, it can be said that
the influence of the operating time on the decay heat is less
than on the activation, and that the decay heat decrease with
the cooling time is faster than the activity decrease.

3.5.2   Dose Rates

The dose rate calculations have been performed in several
points of the FINTOR 1 plant, by means of the code MERCURE-4 .
From the results obtained the following conclusions can be
drawn :

- There are no serious problems for repairs in zones behind
  the gamma shield,
- Repairs in the zone of the neutron shield should require
  employing auxiliary gamma shielding,
- Repairs in the reflector and blanket zone should require
  remote handling systems.

3.5.3   Disassembly in Active Conditions

An attempt has been made to define a proper system to
dismount the activated components of FINTOR 1, which are all
contained in the vacuum vessel.   The system is based on two
principles, i.e. all the dismounting facilities must work
inside the vacuum vessel, and a complete spare segment of the
toroidal zone must be provided.

TABLE VIII

Radioactivity at Shut-Down of FINTOR-I for 1 year and 10 years of Operation, Ci/kW_t

	First Wall		Blanket		Reflector		Neutron Shield		Magnets Zone	
	1 year	10 years	1 year	10 years	1 year	10 years	1 year	10 years	1 year	10 years
Total per zones	$3.721.10^2$	$5.524.10^2$	$4.720.10^2$	$6.511.10^2$	$1.819.10^2$	$2.126.10^2$	1.447	1.793	$3.102.10^{-3}$	$3.533.10^{-3}$
Percentage of the total	36.2%	38.9%	45.9%	45.9%	17.8%	15%	0.14%	0.13%	-	-

Total Activity of 316 SS:

. 1 year operation: $1.027 . 10^3 \; Ci/kW_t$

. 10 years operation: $1.418 . 10^3 \; Ci/kW_t$

B. H. P. of FINTOR-I and of a Fission Reactor Fuel, km^3/kW_t

FINTOR-I		Fission Reactor Fuel	
First wall	110	I-131	330
		I-131 (milk pathway)	230,000
Other 316 SS compo-		Pu-239	1,000
nents	190	Total plutonium isotopes	8,300
Tritium	5.3		

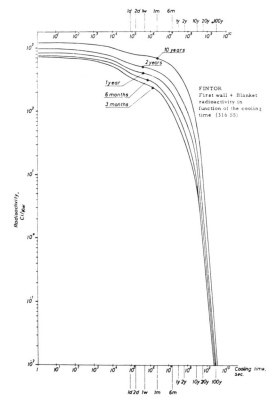

FINTOR
First wall + Blanket
radioactivity in
function of the cooling
time (316 SS)

Fig. 12   Radioactivity Decay Time in the
First Wall and in the Blanket

232

The main dismounting facilities are (<u>Fig. 13</u>), a circular rail
system around the torus, and a movable dismounting facility
which runs on the rail system.   This facility has a shielded
cell with the equipment to recover the damaged segment and to
make partial disassembling for repairing minor damages.   Moreover, at each side of the shielded cells, it has :

- A loading platform which must support the complete spare
  segment or a module; this platform is partially shielded.
- An auxiliary platform which must have the equipment to
  extract and support the outer cover and the gamma shield
  of the damaged segment to be replaced.
- A shielded storage and intervention zone which can recover
  all the 24 segments of blanket and shield.   This zone
  allows important in-site interventions and replacements in
  the non-activated parts of the plant (D-shaped magnets and
  other windings).

3.5.4   Tritium Recovery

The problem of tritium recovery from the blanket has been
assessed by two different approaches, i.e.   "in situ" continuous
recovery and remote recovery by periodically batch-processing
the content of a blanket module set, using yttrium metal as the
extractor.   The latter method is suggested for FINTOR 1,
because :

- The replacement of lithium containing tritium with fresh
  lithium can be accomplished during normal operation and
  running conditions without shutdown of the reactor.
- The lithium batches may be processed away from the radiation field without turning off the reactor.
- Since only 2-4% of the total lithium fluid in the reactor
  is diverted per day, this fluid can be cooled to low temperature, so reducing the inventory of the tritium remaining
  in the lithium after processing.

By assuming a temperature of $200^{\circ}C$ in the recovery system and
an efficiency of 1% for the process, the estimated yttrium

233

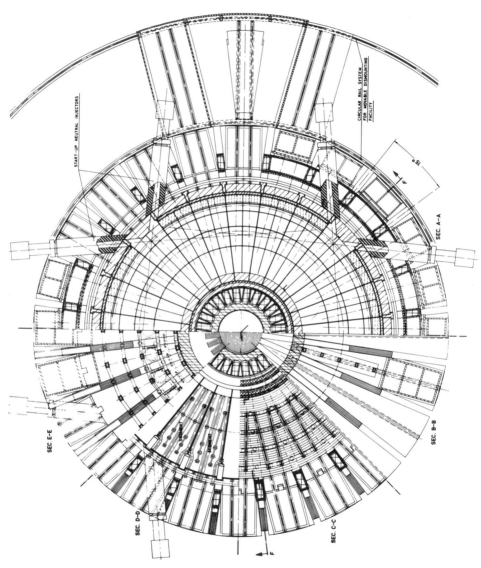

Fig. 13   Top View of FINTOR 1 Reactor with Main Remotely
Controlled Maintenance Facilities

234

required is about 75 kg.

4.   CONCLUSIONS

At the end of the FINTOR 1 design work it seems obvious to note that the overall dimensions of FINTOR 1 look "too big" for a minimum size reactor, while the thermal power output and the wall loading are "too small" to allow meaningful nuclear engineering studies to be performed.   This is a consequence of the assumptions made.

In order to obtain a reactor of reduced dimensions, important changes in these assumptions are necessary, like the removal of the divertor (or consider a two-stagnation point divertor), envisage a superconductor suitable for $B_{TM} > 8T$ (whose availability in the near future is doubtful), etc.

On the other hand, the plasma energy confinement assumed for FINTOR 1 is by far too optimistic; more realistic assumptions would lead to larger plasma (and reactor) dimensions.

The main conclusion is therefore that,
EVEN FOR QUITE AN "OPTIMISTIC" THERMONUCLEAR PLASMA BEHAVIOUR, A "CONVENTIONAL" ENGINEERING WOULD NOT ALLOW THE BUILDING OF A REASONABLY SMALL TECHNOLOGICAL FEASIBILITY REACTOR, WHICH CONTAINS ALL THE COMPONENTS FORESEEN FOR A COMMERCIAL FUSION REACTOR.

The work performed during the FINTOR 1 design exercise has allowed the FINTOR Group to :

a) Acquire knowledge, experience and computational tools covering most of the technical areas of interest in fusion power development.

b) Clarify some of the key interface problems between various reactor components and, more generally, between physics and engineering design constraints.

c) Analyse some novel ideas, like the outer reactor vacuum vessel, the toroidal coil composite structure and shape, etc.

With experience and tools readily available, the FINTOR Group can now proceed toward fusion reactor design work, more

precisely oriented towards specific goals (i.e. detail design
of one of the Technical Feasibility Experiments foreseen
between the large Tokamaks under design at present, and a
Commercial Reactor) with a higher degree of confidence and at
a much faster pace.

ACKNOWLEDGEMENT

This lecture has been based on the work performed by the
FINTOR Group during the past three years and recently published
in the FINTOR 1 Report (Ref. [1]).

My thanks go therefore to all my colleagues of the Group and
in particular to those who gave me very valuable help in preparing this lecture at Erice, J. Bodansky, B. Brunelli,
E. Coccorese, F. Farfalletti-Casali, P. Fenici and P. Rocco.

REFERENCES

[1] The Fintor Group "FINTOR 1 - A minimum Size Tokamak DT
     Experimental Reactor" Edited by F. Farfalletti-Casali,
     J.R.C. EURATOM Ispra - September 1976.

[2] E.R.D.A. "Fusion Power by Magnetic Confinement Program
     Plan". Vol. 1 ERDA 76/110/1 - July 1976.

[3] Conference Report - Dubna 4-7 July 1975. Nuclear Fusion 15
     (1976) p.909.

[4] E.Bertolini, M.A.Hoffman, F. Engelmann"Proc. of V Symposium
     on Engineering Problems of Fusion Research, Princeton, N.J.
     November 1973

[5] E.Bertolini, F. Engelmann,M.A.Hoffman,A. Taroni. "Plasma
     Power Balance Models for Self-Sustained Tokamak Reactors"
     submitted for publication to Nuclear Fusion in September,
     1976.

[6] D.F. Düchs et al.   PPL Technical Memorandum TM-265
     January 1973 - Princeton

[7] D.M. Meade et al Plasma Physics and Controlled Nuclear
     Fusion Research, Tokyo 1974 Vo. I, p.605.

[8] B.M.U. Scherzer to be published in Journal of Vacuum
     Science and Technology.

[9] Fintor Group "Informal Discussion" IAEA Workshop on
     Fusion Technology Culham (GB) February 1974.

[10] Fintor Group   "Conceptual Design Studies of an Experimental
     Power Tokamak Reactor." Proc. of 8th Symposium on Fusion
     Technology Noordwiskerhout, Nederlands June 1974

[11] S. Bobbio, L. Egiziano, G. Lupo, R.Martone "The Transformer
     Design for a Proposed Technical Feasibility Tokamak Reactor"
     Proc. of 9th Symposium on Fusion Technology,Garmisch-
     Partenkirchen 1976.

[12] K. Burn, U. Canali, R. Cuniberti, R.Nicks, C.Ponti, G.
     Realini, R. Van Heusden "Neutron Streaming Problems in the
     'Fintor 1' Design" Proc. of 9th Symposium on Fusion Tech-
     nology, Garmisch-Partenkirchen 1976.

[13] F. Farfaletti-Casali, F. Peter, P. Rocco "Radioactivity and
     Afterhead of Fintor and Related Problems of Maintenance and
     Waste Disposal." Proc. of 9th Symposium on Fusion Technol-
     ogy, Garmisch-Partenkirchen 1976.

PHILOSOPHY AND PHYSICS OF PREDEMONSTRATION FUSION DEVICES[*]

John F. Clarke

Oak Ridge National Laboratory
Oak Ridge, Tennessee 37830

ABSTRACT

A PDFD will operate in the 1980's and must provide the plasma and
plasma support technology information necessary to warrant design,
construction, and operation of succeeding experimental power reactors
and then the demonstration plant. The PDFD must be prototypical of
economic fusion devices to justify its cost. Therefore, development
of the fusion core will be the focus of the PDFD. The physics performance, power production objectives, and characteristics of the PDFD,
and their relationship to the research and development needs to
achieve them are outlined. The design criteria for a PDFD which
satisfied these constraints will be established.

During the past seven years a number of tokamak fusion power
reactor designs have been undertaken with the goal of identifying
major technological obstacles to the realization of practical fusion
power [1]. In addition to these long range studies a number of shorter
range projects have attempted to identify the technological requirements for the construction of an experimental power reactor which
would demonstrate the basic technology necessary to produce net fusion
power [2,3,4]. These latter studies have proved useful in defining
relative priorities within the ongoing tokamak R&D program. However,

---

[*]Research sponsored by the US Energy Research and Development Administration under contract with Union Carbide Corporation.

the devices themselves have proved to be large complex machines with a high cost benefit ratio. In fact, the cost benefit ratio is so high and the uncertainties in the present designs so large that one is forced to conclude that the EPR's as presently conceived are not the next logical step in the fusion program.

The original objective of the EPR program was to advance the science and technology required for commercial fusion power by providing on a timely basis: (1) system operating experience and testing of components and subsystems for a larger demonstration plant to follow; (2) an intermediate focal point for research and development programs; and (3) large scale testing of plasma physics scaling, including deuterium, tritium, and alpha particle effects. The attempt to satisfy all of these objectives in one device while at the same time generating significant amounts of electrical power resulted in the complex designs referred to above. The dilemma in defining the characteristics of the next logical step in the fusion program stems from the desire to eliminate as much of this complexity as possible, while at the same time providing for a major step forward in the fusion program.

If the US fusion program is to achieve its goal of a demonstration reactor by the year 2000, a number of critical scientific and technological problems must be addressed in the mid-1980's. Some of these questions will be addressed in the TFTR device, but many others will require the construction of an additional facility which we shall call a predemonstration fusion device. We shall attempt to define the nature of this device in this paper.

The primary objective of the TFTR experiment is the attainment of $Q \simeq 1$. Calculations have indicated that this objective can be reached in a device with a plasma radius on the order of 90 cm. The next step in the fusion program should be a reasonable extrapolation in performance and size beyond TFTR. On the basis of available scaling laws, to be discussed below, it follows that $Q = 5$ operation can be attained with a reasonable step in size beyond TFTR. It can also be shown that with a Q of approximately 5, ignition can be achieved. Thus, it is reasonable that the PDFD should operate with a near ignition plasma. Operation of such a plasma in hydrogen would require a supplemental heating power of over 200 MW. This prohibitive requirement can be reduced to 50-75 MW when D-T is used and advantage is taken of alpha heating. Therefore, real ignition of the PDFD plasma seems necessary.

An ignition experiment will require sufficient pulse time to investigate the consequences of ignition. Again, simple considerations of fusion reactor rates lead one to conclude that the machine should have a pulse time on the order of tens of seconds. Designs of magnets for $Q = 1$ devices [5,6] indicate that on a purely economic basis superconducting coils would seem to be an essential part of the device. Long pulse operation also implies efficient control of impurity build-up if ignition is to be sustained. Furthermore, since the ignition implies high temperatures and densities, the device must be capable of withstanding large pulses of fusion power. Thus a logical extrapolation beyond the performance of TFTR leads us to a device possessing most of the characteristics of the EPR. The device need not possess the ability to breed tritium, recover the neutron energy, operate with a high duty cycle, or convert heat to electricity. All of these functions are peripheral to operation of the fusion reactor core and can be developed separately from this core. Thus it seems that a PDFD, which is the minimum significant extrapolation beyond TFTR, can be identified by the characteristics of a fusion reactor core.

Since the PDFD is a fusion reactor core, its characteristics should be compatible with the generation of economic fusion power, if tokamaks are to qualify as power systems. In order to define the most relevant version of the PDFD, it would be well to consider some of the factors which determine the economics of fusion power. The first point to be made is that, unlike the fission process in which energy is generated and recovered in the same physical region, fusion energy is generated in a plasma and recovered in an external blanket. Since in building a reactor one pays for materials and construction, not the vacuum region where the plasma resides, the costs associated with fusion systems are those associated with the external structure surrounding the plasma. Let us consider a general idealized tokamak power system. To simplify calculations, the elliptical system shown in Fig. 1 is assumed to include the general features of noncircular tokamaks. It is easy to show that the average power density in the blanket region is

$$\frac{P_B}{V_B} = \frac{2P_W}{\Delta} \sqrt{\frac{1 + \sigma^2}{2}} \left\{ \frac{1}{1 + \sigma + \Delta/a} \right\} \tag{1}$$

where the elongation, $\sigma = b/a$, the structure thickness, $\Delta = \Delta_S + \Delta_C$, includes the coil thickness, and P_W is the wall loading. For a fixed

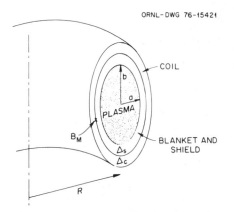

ORNL-DWG 76-15421

Fig. 1. The geometry of an idealized tokamak power reactor is shown.
The elliptic shape is chosen to include the effects of elongation on
reactor economics.

wall loading, plasma elongation
increases the structure power
density by increasing the surface
area; however, it also decreases
the power density by increasing
the volume of structural material.
The thickness of the coil, Δ_C, is
determined by the strength of the
magnetic field. It is easy to
show that

$$\Delta_C = a\sigma \left\{ \frac{B_M^2}{2\mu_0 S} \right\} \left\{ 1 - \frac{(1 + \Delta_s/a)}{A}^2 \right\}^2 \tag{2}$$

where B_M is the maximum value of the toroidal field on the coil as
indicated in Fig. 1, S is the stress in the coil, and $A = R/a$. For
equal stress in the toroidal field coil, elongation is accompanied by
a thickening of the coil and consequently a lowering of the average
power density in the structure.

Since the capital cost of constructing the fusion power reactor
is proportional to the volume of material used, the unit cost of
fusion power is simply proportional to the inverse of the average
power density in the reactor structure. Thus

$$C_P \propto \frac{V_B}{P_B} = \frac{\Delta}{2P_W} \sqrt{\frac{2}{1 + \sigma^2}} \left\{ 1 + \sigma + \frac{\Delta}{a} \right\} \tag{3}$$

It is clear that in order to minimize the unit cost of fusion power,
one must operate with the highest possible wall loading, P_W, and the
thinnest possible structure, Δ. In view of the above, any economic

243

reactor will operate with the highest practicable P_W. The practical limit on P_W is defined by the material damage of the wall material combined with the detailed economics of wall replacement. It is generally accepted that wall replacement times of less than five years would be economically unacceptable.

The scaling of unit power cost is independent of plasma performance if the plasma is capable of loading the wall to the material limit. From the plasma side, the wall loading depends on the plasma power density. This is conventionally written as $P_V \propto \beta^2 B^4 (\langle \sigma v \rangle_F / \overline{T}^2) E_F$ where $\beta = 4\mu_0 \, \overline{nT}/B^2$. Many reactor studies point out the power density advantage to be gained from high beta and magnetic field operation so that the wall loading, $P_W = V \, P_V / A_W$, can be increased.

Elongation of the system has an advantageous effect on the unit power cost because the increase in surface area admitting power to this structure outweighs the increased volume of the blanket structure. However, the increased coil thickness necessary to keep the coil stresses at a constant level adds to the volume of the structure. In terms of unit power cost, elongation of the plasma would be more advantageous if it were necessary to increase the plasma power output in order to raise the wall loading, P_W, to the limit imposed by the materials used in the construction of the first wall.

High magnetic fields affect the unit power cost in a deleterious way through Eq. 2 since the increased structure necessary to contain higher magnetic fields increases the structural volume. High magnetic fields would only be advantageous if they were necessary to raise the level of wall loading to the material limit. Thus in terms of unit power cost, the ideal fusion reactor would be moderately elongated and would operate with the lowest possible magnetic field.

Aside from minimizing unit power cost, it is also desirable to minimize the total cost of a fusion reactor system. The total cost, C, is obtained by multiplying the unit cost by the total power output of the reactor.

$$C \propto A \Delta a^2 \left\{ 1 + \sigma + \frac{\Delta}{a} \right\} \tag{4}$$

Equation 4 shows clearly that to minimize the total cost of a fusion system, one should choose the lowest possible aspect ratio, the smallest plasma radius, and the smallest elongation consistent with reactor operation. There are lower limits on each of these parameters.

One of the difficulties with the EPR designs is the fact that they all employ low aspect ratio construction. This creates problems with respect to the assembly of the reactor and its remote maintenance [2]. Thus, for ease of remote maintenance, there is a practical lower limit of $A \sim 4$.

The minimum size of the plasma radius is determined by the condition that the plasma, $n\tau$, satisfies the ignition condition. The plasma $n\tau$ is determined by the scaling law which applies to the reactor regime. This scaling law is typically a function of magnetic field, elongation, impurity content, and aspect ratio. Consequently, the precise dependence of the cost of a fusion reactor on these parameters depends on the operative scaling law determining the plasma radius.

There are presently two scaling laws which might be thought to apply to reactor operation. The first is the scaling law derived [7]

from the theory of trapped ion modes (TIM). When this theory is
modified to apply to our elliptical plasma model, the plasma nτ
becomes

$$(n\tau)_{TIM} = 3.1 \times 10^{13} \; a^4 \left(\frac{B_T}{5}\right)^6 B^2 Z \left(\frac{A}{3}\right)^{5/2} \frac{1}{T_e^{11/2}} \left(\frac{1 + \sigma^2}{2}\right) \quad (5)^*$$

The linear instability theory on which Eq. 5 is based should apply if
the plasma collisionality is sufficiently low. Collisionality is conveniently defined as the ratio of the effective plasma particle
collision frequency to the so-called bounce frequency, or the frequency with which particles complete their orbits in tokamak magnetic
fields. This is measured by the collisionality parameter, $\nu^* =$
$3.3 \times 10^{-18} \; q(A/3)^{3/2}(an/T^2)$. Thus reactors can have varying degrees
of collisionality depending on their aspect ratio, plasma size,
density, and temperature. Figure 2 shows the range of collisionality
expected for the EPR and some other existing or planned experiments.

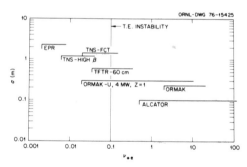

Fig. 2. The collisionality of some existing, planned and contemplated
experiments are shown for comparison. ORMAK-Upgrade with 4 MW neutral
heating, to be operational in 1978, should explore the range of collisionality appropriate to the devices operating in the 1980 time period.

---

*
The units in the practical equations in this paper are centimeters,
teslas, and kiloelectron volts.

For each device a range of ν^* is shown since the collisionality is a
function of radius in the plasma, reaching its minimum value at about
half the plasma radius. The trapped particle modes are expected to
appear clearly below $\nu^* \sim 0.1$. Thus they may not have been observed
in existing experiments and will only be found in the new generation
of tokamaks represented in the figure by ORMAK-Upgrade and TFTR. If
instabilities do occur for collisionalities less than 0.1, they should
appear in reactor plasmas. However, their effect on the transport is
still not well understood and Eq. 5 represents only an approximate
estimate of this effect. It will remain for experiments such as PLT,
ORMAK-Upgrade, and TFTR to verify the effects of these particular
modes as well as their existence.

Present day tokamak experiments obey an empirical scaling law [8],
$n_T \alpha n^2 a^2$. Plasmas obeying this scaling law have collisionalities
ranging from a ν^* of 0.5 to 40.0. Reactors will operate in the range
of $\nu^* \approx 0.001$ to .1. Consequently, there is a gap between the lowest
collisionality observed in present day experiments and the highest
collisionality which we can reasonably expect for a fusion reactor.
If this gap in collisionality could be kept small by operating
reactors at higher collisionality, there might be some hope [9] that
the empirical scaling law observed on present day experiments might
describe the behavior of fusion reactor plasmas. However, it is clear
from Fig. 2 that it remains for machines like ORMAK-Upgrade to bridge
this gap.

Operating reactor plasmas with collisionality comparable to
present experiments in order to preserve the empirical scaling

requires high density or equivalently high beta operation. Written in terms of beta, the empirical scaling law becomes

$$(n\tau)_{emp} = 2.5 \times 10^{15} \, a^2 \left(\frac{B}{5} \right)^4 \frac{\beta^2}{T^2} \tag{6}$$

showing a strong dependence on magnetic field strength [10] and beta.

We can use Eqs. 5 and 6 to determine the size of an ignition experiment for collisionless ($\nu^* < .1$) or collisional ($\nu^* > .1$) reactor operation. From Eq. 5

$$a^2_{TIM} \, \alpha \, \frac{1}{B^3 \beta} \, \sqrt{\frac{2(n\tau)_{IGN}}{1 + \sigma^2}} \tag{7}$$

and from Eq. 6

$$a^2_{EMP} \, \alpha \, \frac{\sqrt{(n\tau)_{IGN}}}{B^4 \beta^2} \tag{8}$$

In both cases, beta can be seen to be advantageous for lowering cost. The effect of elongation appears only in the case of the collisionless reactor and has a weaker effect on the plasma size. Returning to Eq. 4 we see that the total cost or size of the system is roughly independent of elongation for a collisionless reactor and rises with σ for a collisional reactor.

It is known that the minimum ignition $(n\tau)_{IGN}$ is a sensitive function of the plasma impurity content. The rapid increase of $(n\tau)_{IGN}$ with effective plasma charge Z and the direct dependence of a^2 on this increase dictates an extremely effective impurity control method in an economic reactor.

The advantage gained from high magnetic field is real, but limited. Extremely high magnetic fields can be defined as those which cause the coil structure to dominate the blanket and shield structure. Equations 3, 4, 7, and 8 show that although the total cost can be minimized by extremely high field operation, the specific power cost will rise as B^2 for a fixed wall loading.

Let us now turn to an examination of an application of these considerations to the specification of a PDFD which satisfies the constraint of relevance to economic fusion power as well as the advancement of fusion science and technology. A consideration of the EPR designs reveals that these general rules do reflect the economic trends found in those studies.

Table I provides a comparison of the three EPR designs and a candidate for the PDFD. This candidate, the TNS, has been developed by ORNL and Westinghouse to satisfy both the research and development needs of the US fusion program in the 1980's and the need to have a demonstration of the essential components of an economically attractive fusion reactor core. The device was sized using either empirical or trapped particle mode scaling as used in the ANL and ORNL EPR studies. High beta operation allows ignition for either scaling and provides high wall loading. The aspect ratio was enlarged relative to the EPR designs to allow for easier remote maintenance. Moderate D elongation was chosen to optimize trapped particle mode scaling and unit cost. This equilibrium also has a natural divertor property for impurity control. The device shown in Fig. 3 will be described further in the following paper.

Because of the present uncertainties associated with many features of the science and technology of TNS, which will be continually

TABLE I

	EPR			TNS
	ORNL*	ANL[+]	GA[‡]	ORNL/ (W)
Major Radius (M)	6.75	6.25	4.5	5
Minor Radius (M)	2.25	2.1	1.1	1.25
Structure Δ (M)	2.25	2.25	2.25	2.25
Elongation	1	1	3	1.6
Maximum Toroidal Field (T)	11	8.0	7.9	8.0
Plasma Toroidal Field (T)	5	3.5	3.9	4.1
Plasma Beta (%)	3	4.8	10	14
Wall Loading (MW/M^2)	.55	.35	.85	3.7
Thermal Power (MW)	357	261	303	1510
Fusion Core Cost (M$)	249[‖]	249	218	218[#]
Unit Power Cost[§] ($/kW)	695[‖]	923	645	144[#]

*ORNL EPR study, ORNL/TM-5572-77, M. Roberts, Ed.

[+]ANL/CTR-76-3, August 1976, W. Stacey, Ed.

[‡]GA-A14043, July 1976, C. Baker.

[§]Thermal power decided by the cost of the fusion core.

[‖]For purpose of comparison we have used the cost of similar size ANL EPR.

[#]For purpose of comparison we have used the cost of similar size GA EPR rather than an independent cost estimate.

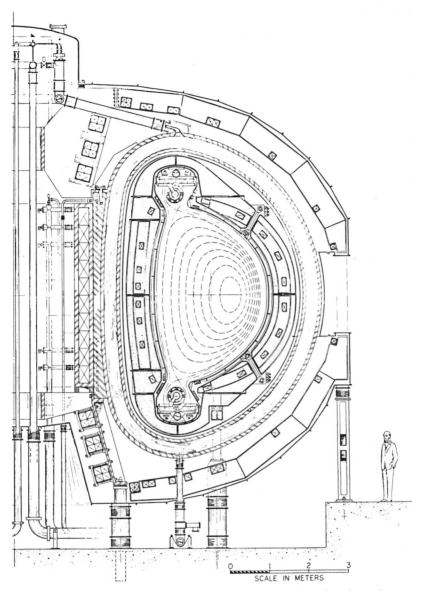

SCALE IN METERS

Fig. 3. The cross section of the initial TNS design is shown to illustrate the attempt at size reduction and economical structure utilization in a high performance tokamak reactor experiment.

251

resolved over the next four years, the design will undoubtedly evolve from its present form in that same time period. However, the reference parameters indicated illustrate the improvement over the performance of the EPR's which can be realized in tokamak systems by a focusing on the key design elements as revealed in Eqs. 3 and 4.

It should be emphasized here that the absolute dollar estimates quoted in the table are not to be taken seriously at this time. We are concerned only with the trends in total and specific power cost as a result of the selection of different design choices. For the sake of consistency, we have estimated the ORNL EPR cost and the TNS cost from the other two EPR's which should have relatively consistent costing since they were performed concurrently. We have also used the nominal structure thickness of 2.5 M for all the devices to simplify comparison. The model in Fig. 1 assumes a compact structure. Reference to a typical EPR cross section (Fig. 4) shows that the structure is greatly extended. By comparison, the TNS design (Fig. 3) attempts to economize on the volume of material to better accord with the prescriptions of Eq. 2 and 3.

As seen from Eq. 3 the dominant factor in determining the specific cost is the wall loading. The General Atomic EPR has the highest wall loading and the lowest specific cost. However, according to Eq. 3 the cost advantage would be 2.75 over the ANL design and 1.7 over the ORNL design. In fact, the actual advantage is 1.4 over the ANL design and the ORNL EPR achieves comparable cost. The reason is unclear. Compared to the ANL EPR, the GA design has larger β and B which should contribute to higher power density in the plasma at the same operating

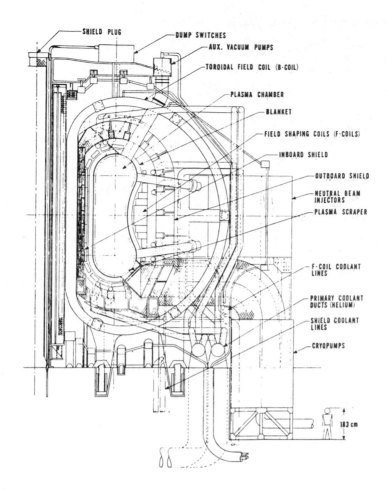

SHIELD PLUG
DUMP SWITCHES
AUX. VACUUM PUMPS
TOROIDAL FIELD COIL (B-COIL)
PLASMA CHAMBER
BLANKET
FIELD SHAPING COILS (F-COILS)
INBOARD SHIELD
OUTBOARD SHIELD
NEUTRAL BEAM INJECTORS
PLASMA SCRAPER
F-COIL COOLANT LINES
PRIMARY COOLANT DUCTS (HELIUM)
SHIELD COOLANT LINES
CRYOPUMPS
183 cm

Fig. 4. The cross section of the best performing EPR design, the
GA EPR is shown. This design possesses many of the desirable characteristics of a PDFD and is a candidate for this function.

temperature. Thus, the average wall loading in the GA EPR,

$\bar{P}_W \propto \beta^2 B^4 (V/A_W)$, should be about five times larger than in the

ANL EPR. However, the wall loading advantage quoted is only 2.4. By

comparison, the wall loadings of the ORNL EPR is larger than the ANL

EPR by the amount expected on the basis of the $\beta^2 B^4$ scaling. Thus, although Eq. 3 indicates a reduction in unit power cost with elongation in the elliptical model, it does not appear that this advantage is fully realized in the GA EPR. Therefore, the nominal performance of the GA EPR does not differ significantly from the others shown.

As mentioned above, the use of higher field on the ORNL EPR compensates for the low beta plasma and high wall loading is achieved. However, the field increase is modest and does not result in significant increased structure because of a novel design utilizing force cooled Nb_3Sn superconductor within stainless steel conduits [2] which support the increased forces. This results in roughly the same coil size as the ANL and GA designs. Larger fields would require increased structure and possibly new types of superconductor. Thus the range for further performance improvement through magnetic field is limited.

The performance of the TNS as indicated in Table 2 is greatly improved over all of the EPR designs. This improvement results from the operation of the device at a high beta and a moderate plasma elongation, in a D configuration which provides for an efficient utilization of available volume within the coils. This results in the wall loading in the TNS design being much larger than that in any of the EPR designs and the power output being five times greater than these devices.

The details of the TNS design will be described in the following paper [11]. However, it is important here to consider the reason for assuming that high beta can be achieved in such a device and that the resulting high wall loadings have practical application for fusion

TABLE II

REQUIREMENT FOR PDFD	CHARACTERISTICS OF TNS
High wall load	3.7 MW/M^2
High beta	14 %
Z control	Natural divertor
Moderate elongation	1.6
Moderate magnetic field	4.1 T
Minimum structure Δ	2.5 M
High performance	1500 MW_{TH}
Moderate size	a) Coils half EPR size
	b) $a_{TNS}/a_{TFTR} = 1.4$

power systems. The possibility of high beta operation is based on the flux conserving tokamak concept [12,13]. This concept utilizes the fact that heating of large reactor plasmas from their initial low beta state to the desired high beta ignition state must occur on time scales which are much faster than magnetic diffusion times. Consequently this heating will occur through a sequence of equilibria containing the same amount of magnetic flux. As a result the low beta q profile is preserved during the heating process and a unique final equilibrium is obtained. This final equilibrium has been shown to possess many interesting stability properties such as good shear and an absolute magnetic well [14]. The ORMAK-Upgrade experiment which will be operational in 1978 is aimed at verifying the existence of such equilibria. The belt pinch [15] experiment seems to possess such an equilibrium at betas of 30 to 70% and to exhibit stable operation for 20 MHD growth times.

Operation of tokamak plasmas at high beta with efficient use of the plasma volume inevitably results in high wall loading such as shown in Table II. Equation 3 indicates that this high wall loading has economic advantages with regard to reducing the unit cost of fusion power. However, if this high wall loading results in such severe wall damage that the reactor wall must be replaced on short time intervals, this advantage would be more than offset.

Recent results on the simulation of CTR radiation damage in stainless steel [16] has shown that wall lifetimes in excess of 20 megawatt years/m^2 could be expected from a reasonable materials development program. Consequently, the wall loading indicated for the TNS design would result in a wall lifetime in excess of five years.

An examination of the costs of TNS components resulting from our initial design shows that the fusion core of TNS has about the same cost per unit thermal output as the fission core of the Clinch River Breeder Reactor. Thus, we have some indication that a device with TNS parameters has characteristics of an economic power system. Therefore, the TNS parameters satisfy the major requirement of a PDFD. Elements of these requirements are indicated in Table II. We conclude that the plasma size is a reasonable step beyond the TFTR and the ignition condition seems to be attainable in a reactor core which is directly on the path leading to an economic fusion system. The key assumptions about the attainability of the beta and confinement necessary for the performance indicated must be verified by experiment over the next few years; however, the logic for TNS is clear. Results and plans from the ongoing fusion program lead naturally to a DT fueled ignition device following the TFTR. The objective of ignition in turn sets definitive criteria for numerous of the key physics and technological systems. With these criteria set, the goals of the research and development program necessary to insure that the criteria can be met on a time scale which will support TNS construction can be established. In the following paper the criteria for the key physics and technological systems will be established. In the last paper of this series, the research and development programs for some of the necessary technology supportive of the TNS design will be described.

REFERENCES

[1] BADGER, B. et al., Reports UWFDM68 1, 1974; 2, 1975, University of Wisconsin, Madison.

[2] ROBERTS, M., Ed., "Oak Ridge Tokamak EPR Study", ORNL/TM-5572-77, 1976.

[3] STACEY, W., Ed., "Tokamak EPR Conceptual Design" ANL/CTR-76-3, August 1976.

[4] BAKER, C., "Review of the Conceptual Design of a Doublet Fusion EPR", GA-A14043, July 1976.

[5] HAUBENREICH, P. N., AND ROBERTS, M., Eds., "ORMAK F/BX, A Tokamak Fusion Test Reactor", ORNL-TM-4634, June 1974.

[6] "Two Component Torus Joint Conceptual Design Study", Plasma Physics Laboratory, Princeton, New Jersey and Westinghouse Electric Corporation, Pittsburgh, Pennsylvania, September 1974.

[7] KADOMTSEV, B., and POGUTSE, O. P., Nuclear Fusion 11 (1971) 67.

[8] MURAKAMI, M., Private communication.

[9] KADOMTSEV, B., Fizika Plazmy 1 (4) (1975) 53.

[10] COPPI, B., 1976 Erice School on Tokamak Reactors for Breakeven, Sicily, Italy, September 20 - October 1, 1976.

[11] McALEES, D. G., 1976 Erice School on Tokamak Reactors for Breakeven, Sicily, Italy, September 20 - October 1, 1976.

[12] CLARKE, J. F., "High Beta Flux Conserving Tokamaks", ORNL/TM-5429, June 1976.

[13] CALLEN, J. D., et al., "Tokamak Plasma Magnetics", International Conf. on Plasma Physics and Controlled Nuclear Fusion Research, Berchtesgaden, FRG, October 6-13, 1976, Paper CN35/B10.

[14] DORY, R. A., and PENG, Y-K. M., "High Pressure, Flux-Conserving Tokamak Equilibria", ORNL/TM-5555, August 1976.

[15] GRUBER, O., and WILHEIM, R., Nuclear Fusion 16 (1976) 243.

[16] BLOOM, E. E., et al., "Temperature and Fluences Limits for a Type 316 Stainless Steel CTR First Wall", to be published in Nuclear Technology.

CHARACTERISTICS OF A PREDEMONSTRATION
FUSION DEVICE*

D. G. McAlees[+]

Oak Ridge National Laboratory
Oak Ridge, Tennessee   37830

ABSTRACT

Low beta reactor system studies have shown that criteria necessitated by the physics lead to difficult engineering problems. The low power density produced in these systems also results in large reactors and makes the question of economic viability a serious one. It is clear that if higher pressure plasmas could be confined and operated, then many of these problems could be alleviated. The incentives for high plasma density as a means of achieving higher pressures are discussed. The possible methods of achieving high density are outlined with particular attention to the high beta approaches. The requirements which are imposed on the reactor by using the flux-conserving tokamak concept are described.

The characteristics of a predemonstration fusion device which can test the necessary aspects of such a system are developed. The plasma size required to attain ignition is found to be relatively small provided other criteria can be satisfied. These criteria are described and the technology developments and operating procedures required by them are given. The dynamic behavior and parameters of the reference system during the operating cycle are also outlined.

---

*Research sponsored by the Energy Research and Development Administration under contract with Union Carbide Corporation.

[+]Exxon Nuclear Co., Inc.

I. INTRODUCTION

Several of the features which are desirable for a tokamak reactor are discussed in this lecture. The considerations which lead to the parameters of a predemonstration fusion device (PDFD) which can test the necessary aspects of such a system are then developed.

The power density produced in a reactor plasma depends on the density and temperature of the fuel ions burned. Tokamak reactors will have to operate under ignition conditions to attain economic viability. This requirement imposes parametric constraints on the containment system since ignition depends on density, temperature, and energy confinement time. The energy confinement time in turn depends on other variables, but at present this dependence is not known. There are indications, however, that confinement time will scale favorably with an increase in density. This result is suggested both by experiments and by theoretical analyses. Thus from several points of view there are incentives to operate the reactor plasma at high density.

Several approaches for achieving high density are envisioned. A method which requires a moderate magnetic field (or high beta) is desirable for economic reasons. It may be attainable if, as expected, flux is conserved during heating and if the resulting magnetic configuration can be maintained during the burn phase. This concept is that of a flux-conserving tokamak.[1] Noncircular plasmas may offer a means to high density.[2] Finally, high magnetic fields (or low beta) may be necessary.[3] In this paper a reference system which includes a moderate magnetic field, a moderately elongated plasma, and high beta will be developed. Since the operating temperature of the plasma will be in the range of reactor interest, i.e., $\sim$ 10 — 15 keV, high beta implies high density and vice versa in this discussion.

High beta (density) imposes constraints on the design and operation of the system. For example, stable high beta plasma equilibria must be created and maintained and a compatible poloidal field system must be developed. High power density means that the first wall will be subjected to high neutron fluxes (> 1 MW/m^2) as well as a high thermal flux. The breakdown and current rise phases of start-up may require higher voltages when increased filling pressures are used. Supplemental heating using neutral beams may be difficult because of the beam energies required to achieve adequate beam particle penetration. Finally, impurity and burn product effects in a dense plasma must be examined. In some cases technological developments may be needed to meet design criteria, while in other cases it may be possible to develop operating procedures which provide the desired solutions. Specific features of this type are discussed below.

II. SYSTEM MODELING

There are many ways to simulate the detailed energy balance processes in a tokamak. In the present work the same model used previously in the ORNL Experimental Power Reactor Design Study[4] has been adopted, and the impact of high density has been assessed. The time dependent model includes pseudoclassical, neoclassical, and dissipative trapped particle mode scaling laws as well as radiation effects and cold fueling. Conduction and convection times are functions of plasma collisionality.

Figure 1 shows the plasma radius required by this model for ignition as a function of temperature; the axis magnetic field for three cases is shown, where the density is 2×10^{20} m^{-3} and the plasma elongation (height to width ratio) is 1.6. The minimum size for ignition occurs at about T = 13 keV; below this temperature, the Nτ required for ignition increases sharply and an increase in size is necessary to attain it. Above 13 keV, trapped ion mode transport (N$\tau \propto \beta^2 T^{-11/2}$) dominates the scaling and an increase in plasma size again is required. The neutron wall loadings associated with various operating temperatures are also given. Note that at N = 2×10^{20} m^{-3}, T = 13 keV, and B$_T$ = 4.5 T, the wall loading exceeds 3 MW/m^2, and the plasma radius necessary for ignition is 1.25 m. With a field strength of 6.5 T, the size required is reduced to $\sim$ 1 m and the wall loading is $\sim$ 2.5 MW/m^2.

ORNL-DWG 76-15687

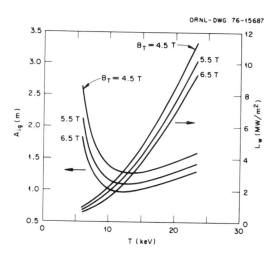

FIGURE 1. Plasma radius for ignition (A$_{ig}$) and corresponding neutron wall loading as a function of temperature.

Figure 2 shows the plasma radius required for ignition as a function of density. In these cases T = 13 keV. At lower densities the collisionality of the plasma is such that transport is dominated by the trapped ion mode which requires an increased plasma size. As density is increased the size required for ignition decreases, since the impact of the trapped ion mode becomes less pronounced.

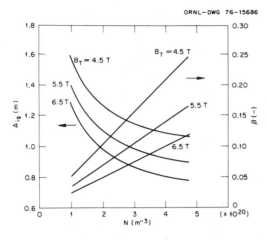

FIGURE 2. Plasma radius for ignition and β as a function of density.

There is no quantitative answer to the question of what maximum density is acceptable. Also, theoretical work[5] suggests that β ∿ 20% or higher may be attainable. For this analysis an operating density of about 2×10^{20} m^{-3} is assumed, which results in a β ∿ 0.14 at 13 keV, a wall loading less than 4 MW/m^2, and an ignition size of 1.25 m. Table 1 summarizes the reference system parameters computed from a time dependent, self-consistent simulation model.

TABLE 1. Reference Parameters

Major radius, R_o(m)	5
Minor radius, a(m)	1.25
Elongation, b/a (-)	1.6
Toroidal field at winding, B_{max}(T)	8
Toroidal field on axis, B_T(T)	4.32

TABLE 1 (cont'd)

Plasma current, I(MA)	4
Electron density, $N_e (m^{-3})$	2.2×10^{20}
Ion density, $N_i (m^{-3})$	2.0×10^{20}
Ion temperature, T_i(keV)	13.5
Energy confinement time, τ_E(s)	1.2
Toroidal beta, $\beta(-)$	0.14
Average neutron wall loading, $L_w (MW/m^2)$	3.7
Thermal fusion power, P_{th}(MW)	1510

III.   PARAMETER SENSITIVITIES AND SYSTEM DYNAMICS

The operating parameters shown in Table 1 represent steady state conditions. Since the model used includes the trapped ion mode where $N\tau_E \propto \beta^2 T^{-11/2}$, the temperature increases to the point where the energy loss rate equals the heating rate due to thermonuclear alpha particles and a steady state is achieved. Figure 3 shows $N\tau_E$ vs T_e for three different values of magnetic field. The system size is that given in the table and $Z_{EFF} = 1$ is assumed. As the temperature increases, $N\tau_E$ is dominated first by pseudoclassical, then by trapped electron mode, and finally by trapped ion mode diffusion losses.[6] $N\tau_E$ required for ignition is also shown. For example, when $B_T = 4.3$ T, ignition is attained at 10 keV. The temperature continues to rise after ignition since P(alpha) exceeds P(loss) and steady state conditions are reached at $\sim$ 15 keV, i.e., $N\tau$(operation) = $N\tau$(ignition). There is a minimum field for this size device which can provide ignition; it is about 4.0 T and results in an $N\tau$(operation) curve which is tangent to the $N\tau$(ignition) curve. Note that this would occur at about 13 keV and 2 x 10^{20} m^{-3}, which in turn implies a fusion power output of 1200 MW. This is an important point since it means (assuming this scaling model) that the minimum power output associated with an ignited device of this size is greater than 1000 MW. A heat removal capability compatible with this power level would have to be installed in such a device.

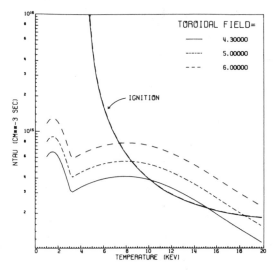

FIGURE 3. Nτ required for ignition and system Nτ operating characteristics as a function of temperature and various field strengths.

The system could be operated at a higher field. In the 6.0 T case, ignition occurs at 7 keV, but the equilibrium temperature is greater than 20 keV. This is the result of improved confinement at the higher field. Under these conditions the system could achieve ignition at a density less than 2 x 10^{20} m^{-3}. The power produced is proportional to N^2 so that ignition conditions at the lower density would be compatible with a power output less than 500 MW. This, of course, represents a low beta system and is less attractive economically. Alternatively, a smaller size system could be used to achieve ignition under high field conditions. Figure 1 showed that a 1 m plasma radius may be sufficient when B_T = 6.5 T. In this case the power output would be 630 MW when $N = 2 \times 10^{20}$ m^{-3} and T_i = 12 keV.

The ignition calculations above assumed a pure (Z_{EFF} = 1) plasma. When impurities are included, both the ignition requirement and the system operating characteristics change. Figure 4 shows this for the reference (Table 1) system. Iron is assumed to be the impurity. Due to enhanced radiation losses, ignition cannot be achieved above $Z_{EFF} \sim 1.5$. This establishes the impurity control criteria which will have to be met during the burn cycle of the device. Note also that in the second trapped electron mode $N\tau \propto 1/Z_{EFF}$ and in the trapped ion mode $N\tau \propto Z_{EFF}$ so that the confinement characteristics shown in Fig. 4 cross over after the transition into the trapped ion mode.

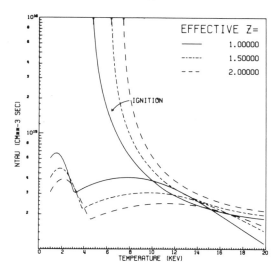

FIGURE 4. $N\tau$ required for ignition and system $N\tau$ operating characteristics as a function of temperature and various impurity (iron) levels.

The desired reactor plasma performance can be attained at B_T = 4.3 T, $N = 2 \times 10^{20}$ m^{-3}, and T_i = 13 keV provided that a plasma with a beta of about 0.15 can be stably confined. The equilibria of such plasmas are discussed in the next section.

IV. MHD EQUILIBRIUM AND STABILITY

The maximum attainable value of beta in a stable toroidal plasma is influenced by the plasma cross sectional shape,[7,8] safety factor profile, and toroidal current profile. The usual approach to and results of MHD equilibria calculations are discussed first and compared to those of the high beta flux-conserving tokamak (FCT) equilibria.[1,5]

Under the assumptions of large aspect ratio and nearly circular flux surfaces, it has been shown that $\beta_p = A$ is an upper limit in an axisymmetric equilibrium.[9-11] If β_p exceeds A, a poloidal field separatrix would intersect the plasma boundary. For nearly circular flux surfaces, the local values of poloidal field strength can be used to relate q, A, β_p and β as follows:

$$q = \frac{B_T}{AB_p} , \qquad \beta = \left(\frac{B_p}{B_T}\right)^2 \beta_p = \frac{\beta_p}{q^2 A^2} .$$

This then implies that beta is limited to $1/q^2 A$, which yields $\beta \sim 1 - 5\%$, using typical values for q and A.

It has been shown numerically that when these assumptions are not made, equilibria with $\beta_p \sim A^2$ can exist in an ideal, circular shell.[12] These equilibria were computed in the usual way, i.e., the pressure function, $p(\psi)$, and the toroidal diamagnetism function, $F(\psi)$, were chosen arbitrarily so that reasonable, bell shaped toroidal current distributions resulted when $\beta_p \sim 1$. These conditions are consistent with those observed in present experiments.

The high beta profiles obtained in this way for $\beta_p \sim A$, typically include reversed currents at small values of R, that is, on the inside of

267

the plasma cross section. Also, to prevent q < 1 at the plasma center so that local (Mercier[13]) stability criteria are satisfied, as β_p is increased, it is necessary to increase q at the plasma boundary. This is accomplished by increasing B_T or decreasing the total plasma current. Because of this, as β_p is increased, β first increases but ultimately decreases. Figure 5 shows this behavior for an equilibria in an optimized D-shaped cross section. The maximum β(2.5 − 5.6%) depends on the shapes of the current and pressure profiles assumed, but it occurs at $\beta_p \sim 2.4$.

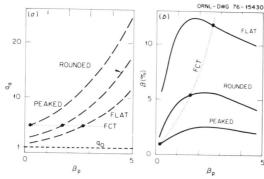

ORNL-DWG 76-15430

FIGURE 5. Local stability limits for non-FCT MHD equilibria in a tokamak with an optimum D-shaped cross section. (A = 3); (a) behavior of a(limiter) and q(center) and (b) maximum β as a function of β_p. Flat, rounded, and peaked refer to the pressure profile.

In the same beta optimization calculations,[4] it was found that with bell shaped profiles there is a limit to increasing beta by elongating the plasma. As the elongation is increased, successively more peaked current profiles are required to satisfy local stability criteria. The increase in beta obtained from the slight noncircularity is nullfied as the current profile becomes more peaked. The peaked profile enhances the importance of the plasma center where the flux surfaces remain approximately circular. Thus, in the highly elongated cases the peripheral regions of the plasma are found to contribute little to the average beta. A moderate elongation of $\sim$ 1.6 is found to provide the maximum benefit with respect to beta within the assumptions used here.

Next the concept of flux conservation and its impact on attaining
high beta equilibria will be outlined. As described above, high beta
equilibrium calculations previously have used relatively arbitrary
profiles for $p(\psi)$ and $F(\psi)$. When heating leads to a pressure excursion
on a time scale which is short compared to the magnetic diffusion time,
however, the toroidal and poloidal fluxes are frozen into the plasma. The
$p(\psi)$ and $F(\psi)$ functions which describe FCT equilibria are then not
arbitrary. Rather $p(\psi)$ is determined by the heating and transport processes
and $F(\psi)$ by the requirement that flux be conserved. Also, since

$$q(\psi) \propto \delta\phi/\delta\psi ,$$

the q profile established at low beta will persist during the pressure
excursion due to flux conservation. The beta limitations observed in Fig.
5 due to the necessity to increase q(limiter) to prevent q < 1 at the
plasma center no longer occur. Numerical procedures have been developed[5]
to preserve $q(\psi)$ as $p(\psi)$ is increased.

A typical FCT equilibrium sequence where q(0) = 1, q(a) = 4.8, and
$\beta \leq 0.18$ is shown in Fig. 6. In these equilibria, it is not necessary to
decrease the total plasma current or increase B_T to maintain local stability as β_p is increased. Thus, unlike the previously calculated high beta
equilibria, the FCT approach permits β to increase monotonically with β_p.

The FCT equilibrium configurations do not show reversed currents; however, the total plasma current does increase. It has been estimated analytically[14] that the total plasma current scales as $p^{1/3}$. Using the same
analytical procedure but performing the integrals numerically, it is found
that the current scales as $p^{1/4}$ in the range of pressure increases of

interest.[15] This has been observed numerically[5] as well. The increased current in a high beta FCT equilibrium is produced by diamagnetic effects and does not have to be driven by the ohmic system primary. However, the poloidal field system must be designed specifically to accommodate the changes in the plasma current distribution and magnitude during the heating phase. A low impedance poloidal coil configuration which can provide automatically most of the needed current change during the FCT pulse is desirable.

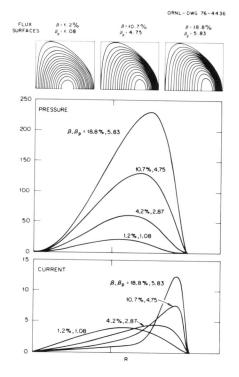

FIGURE 6. Character of (a) flux surfaces, (b) mid-plane pressure profiles, and (c) current profiles as they evolve while β increases. q(limiter) = 4.8.

Only local stability criteria which are satisfied by $q(0) \sim 1$ have been addressed. There is experimental evidence from the belt-pinch[16] and SPICA[17] that high beta plasmas can operate in a stable way. There is, nonetheless, considerable effort required to resolve the nonlocal and more general stability questions associated with FCT equilibria. This work is under way.

V. POLOIDAL FIELD SYSTEM

The high beta equilibria computed in the previous section were obtained using a fixed boundary MHD code. In this model the eddy currents induced in the shell which surrounds the plasma respond to changes in the plasma. The eddy currents provide the proper equilibrium shaping field, cancel the poloidal field which otherwise would exist outside of the shell, and have a sum equal to the plasma current but flow in the opposite direction.[18] The shell can be replaced by a set of discrete coils in the vicinity of the plasma which produce the ideal equilibrium field pattern. When the coils are properly located they will respond in the same way as the ideal shell.

The shape of the current induced in the shell on either side of the "tip" of the D-shaped plasma remains unchanged for large changes in β and I_p. Thus, the coils on either side of the tip can be located[18] to carry equal currents, which reduces the number of power supplies and the degree of control required. Also, since the coils must be placed a finite distance from the plasma boundary, it is found that a divertor-like pair of coils must be located near the tips of the D to reproduce the proper field pattern.[19] Such a poloidal field system retains the features of the STATIC system[18,20] and in addition it is compatible with the equilibrium requirements of the rapid changes of the D-shaped plasma during heating.

The divertor-like coils described above carry currents which flow in the same direction as the plasma current. The separatrix formed as a result of this can be shifted to the edge of the plasma by making modest changes in the poloidal field system. This divertor configuration is "natural" for D-shaped plasmas. The coil locations in this case are determined as described above except that the shell used is D-shaped and has a sharp tip similar to the geometry of a separatrix. The resulting coil system is shown in Fig. 7. Using these coil locations, a typical FCT equilibrium ($\overline{\beta}$ = 0.15) which includes the natural divertor has been

computed using the free boundary MHD code; it also is shown in Fig. 7.
The current required in a divertor coil is 1.54 I_p compared with 1.14 I_p
for an equivalent equilibrium without the natural divertor feature. The
current level is small compared with conventional divertor coil requirements. Since the coils in this system simultaneously maintain the equilibrium and the separatrix configuration, the location of the separatrix is
fixed when the plasma is centered and the difficulty of controlling the
separatrix location should be minimized.

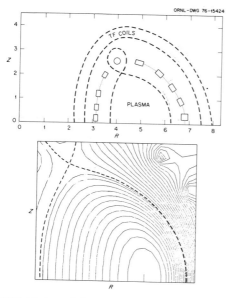

Finally, the current wave forms
in the coils during the pulse are
of interest. The pulse is composed
of start-up, ohmic heating, FCT
heating, ignition excursion, burn,
and shutdown phases. Start-up
$(0 < t \leq 1s)$ is defined as the time
required for breakdown and for the
plasma current to reach about 0.5
MA. During the ohmic heating phase
$(1 \text{ s} \leq t \leq 2 \text{ s})$ the plasma current
rises to its final value and a low

FIGURE 7. (a) Optimized coil locations for the D-shaped FCT with
natural divertor, and (b) a high $\overline{\beta}$
(~ 0.15), free boundary, FCT equilibrium confined by the coils.

beta discharge is established. The
FCT heating phase $(2 \text{ s} \leq t \leq 6 \text{ s})$ is
when rapid heating occurs and the

equilibrium evolves in a way dictated by flux conservation. Ignition
occurs and the temperature continues to rise until the burn phase is
established $(10 \text{ s} \leq t \leq 30 \text{ s})$. Finally, the shutdown phase $(30 \text{ s} \leq t \leq 40 \text{ s})$

is entered and results in termination of the discharge. Figure 8 shows the current time histories in the plasma and the equilibrium field and ohmic coil systems which are required to maintain plasma equilibrium during the cycle. The average beta, which is also consistent with the cycle for the predemonstration device (see Table 1), is also shown as a function of time. This typical case does not include a natural divertor. Table 2 summarizes the results.

ORNL-DWG 76-15431

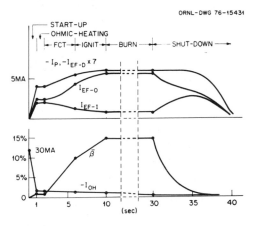

FIGURE 8. β and current time histories compatible with the operating cycle of a high density system.

TABLE 2. Parameters for PDFD Reference Cycle

$\overline{\beta}(\%)$	$\beta_{axis}(\%)$	$\overline{\beta}_p(\%)$	$I_p,7 \times I_{EF-D}(MA)$	$I_{EF-0}(MA)$	$I_{EF-I}(MA)$
0.8	2.7	0.5	4.1	2.43	1.94
1.8	6.4	1.1	4.2	2.68	1.85
5.7	17.3	2.6	4.8	3.60	1.54
11.7	33.6	4.0	5.7	5.02	1.06
14.7	41.2	4.4	6.1	5.75	0.82

The conclusion of these analyses is that the high density desired may well be accessible at moderate magnetic field strengths (high beta) by

operating in the flux-conserving mode. The poloidal field system required
to permit such operation is reasonable and, in fact, may facilitate impurity control as a result of its natural divertor configuration.

VI.   PLASMA START-UP

The gas breakdown and current rise phase in a large tokamak will
establish certain design criteria for the system. The parameters of special
interest are voltage and current wave forms, maximum filling pressure and
plasma density, volt-second consumption, tolerable impurity level and
line radiation, wall interactions, and runaway electron flux. Innovative
start-up schemes may be required for large tokamaks ($R_o \gtrless 5$ meters) because
high density operation may need extremely large voltages to ionize the gas,
high runaway electron fluxes may occur, and large tokamaks are very sensitive to impurities.

Previous research in this area[21-23] has been concerned with both
gas discharge physics, and atomic and molecular physical processes. The
model used in this analysis[4] includes these processes as well as wall interactions, runaway electrons, and time dependent impurities. These are
incorporated into a set of coupled rate equations. The equations are
used to compute the time dependent average values of the density and
temperature of electrons, H_1^+, H_2^+, H_1^o, H_2^o, the non-equilibrium ionization
states of oxygen, plasma current from a circuit equation, and the runaway
electron flux.

The model has been applied to the PDFD reference design (cf. Table
1). For an applied voltage of 50V/turn, which is typical of present
tokamaks, and an initial H_2^o density of 0.2×10^{20} m^{-3} which is five times

less than is desirable for high density operation, the density, temperature, and plasma current stagnate at low levels. This indicates that much higher voltages are required in larger tokamaks to achieve start-up, even at relatively low filling pressures. With $N_{20} = 1.0 \times 10^{20}$ m^{-3} and 450 V/turn the plasma is fully ionized after 14 msec. However, even a small amount of impurities would cause excessive line radiation at these very high densities; thus the plasma must be kept extremely pure.

The ionization and plasma current initiation in large tokamaks requires very large applied voltages and low impurity levels. Runaway electron creation may be an important effect. Research is required on small radius and low density start-up methods.

VII. PLASMA HEATING

The fundamental aspects of neutral beam heating during the plasma start-up are discussed in this section. The supplemental heating power level necessary to achieve ignition in the reference system in about 6 s is found to be 75 MW. This calculation is based on energy balance considerations using the final density ($\sim 2 \times 10^{20}$ m^{-3}) and does not include the problems of beam penetration into a dense plasma. High density and high impurity content make penetration into a large plasma increasingly difficult. Beam focusing, perpendicular injection, and high beam energies may be necessary to overcome the penetration problem. Small radius start-up and low density start-up may provide other means to solve the problem. Here the concept of low density start-up, which has previously been suggested,[24,25] is outlined.

Beam deposition can be characterized by the function H(r) which was first introduced by Rome et al.[26] H(r) is a geometric measure of the deposited fast ion number density profile which results from injection. It is desirable to have the beam deposition peaked in the central region of the plasma for maximum heating efficiency. In these calculations a deuteron beam, 30 cm in diameter, is assumed to be injected into a $Z_{EFF} = 1$ plasma.

Figure 9 shows the deposition profiles for several different beam energies. In these cases injection is nearly perpendicular (16^{0} from perpendicular to prevent ripple trapping of the fast ions), and the density profile is assumed to be $N = 3 \times 10^{20} (1 - (r/a)^2)m^{-3}$. Adequate penetration can be achieved in this way if a beam energy of ~ 250 keV can be made available. Impurity effects dictate an even higher energy.

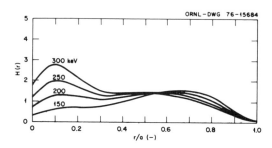

ORNL-DWG 76-15684

FIGURE 9. Neutral injection fast ion deposition profiles for various injection energies (perpendicular injection).

High density and temperature are required at the ignition point where the beams are no longer needed to sustain the plasma energetically. It may be possible to establish the temperature conditions at low density, increase the fueling rate to build up the necessary density while continuing injection, and tolerate the less favorable injection profiles which will occur near the end of the heating phase. Figure 10 shows H(r) for 150 keV tangential injection into average densities of 7.5, 10, and 20 x 10^{19} m^{-3}. The injection deposition profile is favorably

peaked when N = 7.5 x 10^{19} cm^{-3}, is relatively flat for N = 1 x 10^{20} m^{-3},

and becomes peaked at r/a = 0.75 when N = 2 x 10^{20} m^{-3}. Therefore,

heating efficiency at this beam energy will decrease when N > 1 x 10^{20} m^{-3},

but if this part of the heating cycle occurs over a time on the order of

an energy confinement time, the less efficient heating process may well

be tolerable.

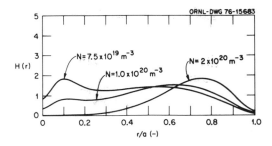

FIGURE 10. Neutral injection fast ion deposition profiles for various plasma densities and a beam energy of 150 keV (tangential injection).

Figures 11 and 12 show the time dependent density and power histories

during heating. Initially N = 7.5 x 10^{19} m^{-3}, T = 100 eV, and P_{beam} = 75

MW. At t = 2 s, T_i = 12 keV, and the fueling rate is increased to begin

the density buildup phase. The plasma ignites and beam injection is

terminated after t = 4.4 s. Density buildup occurs over a period of 2.4 s,

and the average confinement time during this phase is $\tau_E \approx 1.5$ s. Energeti-

cally this scenario is feasible since 75 MW is sufficient to heat the

plasma to ignition at full density.

In fact, more detailed studies of the low density approach using a

coupled beam deposition-spatially dependent transport calculation are

necessary to optimize the start-up procedure. Also, as the flux surfaces

shift when high beta is attained, beam penetration should improve.

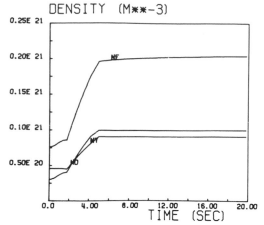

FIGURE 11. Plasma density buildup during low density start-up procedure.

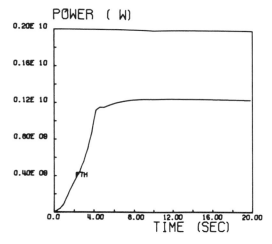

FIGURE 12. Fusion power output during low density start-up and burn cycle.

VIII. CONCLUSIONS

There are incentives for high density operation in a tokamak reactor. Several features of a predemonstration fusion device (PDFD) which will demonstrate the feasibility of attaining such operation in flux-conserving tokamaks were described. Other key aspects such as fueling, impurity control, and the effects of magnetic field ripple were not included here.

278

Suitable high beta equilibria were found and a poloidal field system design compatible with the equilibria was developed. The natural divertor configuration inherent in the system led to the inclusion of a poloidal divertor in the PDFD. The start-up analyses showed that further innovations are required unless very large voltages are applied to achieve breakdown and to establish the plasma current. Neutral beam energies of $\sim$ 300 keV were found to be necessary when the plasma is heated under full bore and full density conditions. A low density start-up procedure was outlined which, if feasible, suggests that injectors with an energy capability similar to that being developed for TFTR may be adequate to achieve ignition in the PDFD. Successful operation of a PDFD of this type means that fusion reactors can be designed with high power density and high beta. These features lead to reduced size and enhance the possibility of attaining economic viability in the tokamak.

ACKNOWLEDGMENTS

Direct contributions to this paper from Drs. S. E. Attenberger, J. D. Callen, F. B. Marcus, A. T. Mense, Y-K. M. Peng, J. A. Rome, and N. A. Uckan and discussions of the text with G. G. Kelley were appreciated.

REFERENCES

1. J. F. Clarke, "High Beta Flux-Conserving Tokamaks," ORNL/TM-5429, June 1976.

2. C. C. Baker et al., "Experimental Power Reactor Design Study," GAC Report GA-A14043, July 1976.

3. D. R. Cohn et al., Nucl. Fusion 16 (1976) 31.

4. D. G. McAlees et al., "Plasma Engineering in a Deuterium-Tritium Fueled Tokamak," ORNL/TM-5573, September 1976.

5.   R. A. Dory and Y-K. M. Peng, "High-Pressure, Flux-Conserving Tokamak Equilibria," ORNL/TM-5555, August 1976.

6.   S. O. Dean et al., "Status and Objectives of Tokamak Systems for Fusion Research," USAEC Report WASH-1295, 1974.

7.   H. P. Furth, Nucl. Fusion 15 (1975) 487.

8.   Y-K. M. Peng et al., "Magnetohydrodynamic Equilibria and Local Stability of Axisymmetric Tokamak Plasmas," ORNL/TM-5267, July 1976.

9.   V. D. Shafranov, Soviet Phys. — JETP 6 (1958) 545.

10. S. Yoshikawa, Phys. Fluids 7 (1964) 278.

11. F. A. Haas and C. Ll. Thomas, Phys. Fluids 16 (1973) 152.

12. J. D. Callen and R. A. Dory, Phys. Fluids 15 (1971) 1523.

13. C. Mercier, Proc. Conf. on Plasma Physics and Controlled Nucl. Fusion, Vol. II, (1962) 801; L. S. Solovev, Soviet Phys. — JETP 26 (1968) 400.

14. J. F. Clarke and D. J. Sigmar, "Global Properties of Flux-Conserving Tokamak Equilibria," ORNL/TM-5599, August 1976.

15. G. G. Kelley (ORNL), private communication, 1976.

16. O. Gruber et al., "The Belt-Pinch — A High-β Tokamak with Noncircular Cross Section," Max-Planck Institut für Plasmaphysik, Report IPP 1/156, October 1975.

17. C. Bolbeldyk et al., "Current Decay and Stability in SPICA," IAEA-CN-35/E-6, Berchtesgaden, FRG, October 1976 (to be published).

18. Y-K. M. Peng et al., Proc. of Sixth Symp. on Engineering Problems of Fusion Research, (IEEE Pub. No. 75CH1097-5-NPS, New York, 1975) p. 1114.

19. Y-K. M. Peng et al., "Poloidal Field Considerations for Tokamaks," ORNL/TM-5648 (to be published).

20. F. B. Marcus et al., Proc. of Sixth Symp. on Engineering Problems of Fusion Research, (IEEE Pub. No. 75CH1097-5-NPS, New York, 1975) p. 1110.

21. R. Papoular, EUR-CEA-FC-769, Fontenay-aux-Roses, France (April 1975).

22. R. J. Hawryluk and J. A. Schmidt, MATT-1201, Princeton Plasma Physics Lab, New Jersey, USA (January, 1976).

23. E. Hinnov, A. S. Bishop, H. Fallon, Jr., Plasma Phys. 10 (1968) 291.

24. J. P. Girard et al., Nucl. Fusion $\underline{13}$ (1973) 685.

25. D. G. McAlees, "Alpha Particle Energetics and Neutral Beam Heating in Tokamak Plasmas," ORNL/TM-4661, November 1974.

26. J. A. Rome et al., Nucl. Fusion $\underline{14}$ (1974) 141.

PREDEMONSTRATION FUSION DEVICES
RESEARCH AND DEVELOPMENT NEEDS*

John F. Clarke

Oak Ridge National Laboratory
Oak Ridge, Tennessee 37830

ABSTRACT

In this concluding lecture, the criteria and reference design
information for the PDFD will be examined with respect to the presently
planned R&D programs, and the major research and development needs will
be described. Emphasis will remain on the fusion core rather than the
nuclear shell. Scaling, heating, fueling and other needs will be outlined. In the area of technology, magnet systems, and tritium handling
compatible with PDFD operation will be discussed. Presently planned
research and development programs will provide the information needed
to support the TNS version of the PDFD by 1980.

Since the design of the PDFD must be based on our knowledge of
present experiments and theoretical predictions, it is not surprising
that the tokamak experimental program planned on the basis of this same
knowledge and theory should be addressing itself to the key uncertainties in the physical requirements of the PDFD plasma. Table I outlines
the key physical problems and the experiments which will provide
answers to these questions. These experiments utilize neutral beams
for plasma heating and with the exception of TFTR are hydrogen fueled
by gas recycle. The exception, TFTR, has a DT handling system which

---

* Research sponsored by the US Energy Research and Development Administration under contract with Union Carbide Corporation.

283

TABLE I

US EXPERIMENTS ADDRESSING KEY PDFD QUESTIONS

Problem	Experiment	Start
Impurity evolution and control	ISX	1977
	PDX	1978
Scaling	PLT	1976
	PLT + beams	1977
	ORMAK-Upgrade	1978
	Doublet-III	1978
	TFTR	1981
High beta equilibrium and stability	ORMAK-Upgrade	1978
Non-circular plasma operation	Doublet-III	1978
	PDX	1978
Neutral heating	PLT	1977
	ORMAK-Upgrade	1978
High field effects	Alcator	1978
D-T operation	TFTR	1981

is not designed to be extrapolated to a fusion reactor system.
None of these devices are long pulse or operate with superconducting
magnet systems. Since PDFD will have to address the ignition of a DT
plasma and the maintenance of the ignited plasma for tens of seconds,
it will require additional development in two areas not now receiving
much support—fueling and tritium handling, and one area now being
heavily supported because of its near-term impact—neutral injection.

In the latter area, the emphasis in development beyond that
required for TFTR must be on efficiency. There may also be a need to
raise the maximum beam energy, possibly to as high as 300 keV which
would require development beyond the TFTR beams. In this paper we will
discuss these three technologies in the context of the requirements
imposed by the TNS design.

Neutral Beams in TNS

Penetration of the injected fast neutrals must be adequate to heat the plasma core under all imaginable experimental operating conditions. If we examine the problem of adequate penetration in the high-density mode of operation ($\bar{n} = 2.10^{14}$ cm^{-3}) and allow some impurity accumulation during startup ($Z_{eff} = 1.5$) when charge exchange induced wall sputtering may be expected, a fast neutral energy of 300 keV results in slightly center-peaked deposition profiles for ± 20° off-radial injection [1] into a 125 cm radius plasma. This injection scheme minimizes perturbation of the "normal" tokamak startup and operation physics, but requires a difficult technology to be developed.

Three alternative startup schemes which are presently being investigated theoretically, and which might lower the beam energy especially during the startup phase, are low density startup at full plasma radius, low radius startup at full density by programming the divertor separatrix, and injection of fast ions into ripple "windows" [2]. The possibly reduced complexity of neutral beam technology makes these schemes attractive, but further analysis of their effect on the tokamak physics is needed.

The requirement of 75 MW of power delivered to the plasma at energies up to 300 keV makes the problem of beam efficiency one of critical importance. Possible efficiencies range from about 12% in a $D^+ \rightarrow D^0$ "brute force" system to about 70% for a $D^- \rightarrow D^0$ system with an energy recovery system. The corresponding electric power required for the beam injectors would be about 625 MW and 107 MW, respectively. If we take a startup time for TNS of 4 seconds, after which the α heating dominates and the beams may be turned off, the energy required for each TNS pulse in the two extremes above is 2500 MJ and 428 MJ, respectively. The former case raises large problems with respect to pulsed energy storage. The existing schemes for generating intense beams of negative ions, 1) direct extraction and acceleration of D$^-$ ions from a D$^-$ plasma generator and, 2) double charge exchange conversion of low-energy D$^+$ into a D$^-$ beam followed by post-acceleration, may both have their inherently high neutralization efficiency further enhanced by the direct recovery of the un-neutralized ions' energy. This may result in a net efficiency approaching 70%.

All of these three high-efficiency schemes have their own sets of difficult physics and technology problems, the solution of which requires an aggressive beam research and development program. Table II shows the possible schemes being explored to enhance high energy efficiency. All of these options will be examined in the next three years.

TABLE II

POSSIBLE APPROACHES TO EFFICIENT NEUTRAL BEAMS AT 300 keV

Approach	Efficiency
1. D^+ (300 keV) $\rightarrow$ D^o (300 keV)	12%
2. D^+ (300 keV) $\rightarrow$ D^o (300 keV) + direct recovery	35%
3. D^+ (1 keV) $\rightarrow$ D^- (1 keV) $\rightarrow$ D^- (300 keV) $\rightarrow$ D^o (300 keV)	50%
4. D^- (300 keV) $\rightarrow$ D^o (300 keV)	50%
5. Approach 3 or 4 + direct recovery	70%

Superconducting Magnets in TNS

The TNS machine is primarily aimed at achieving a deuterium-tritium fueled ignition. Plasma engineering studies on energy containment and ignition conditions indicate the need for a plasma major radius of about 5 m and minor radius of about 1 m with moderate elongation. This plus the requirement of the space for plasma shaping coils, magnetic ripple criteria [3], a divertor and shield dictates a TF coil bore of about 4 x 6 m.

The maximum field required for the machine is not completely fixed yet. A maximum field of 8 T will give an ignition plasma density at $\beta \approx 10\%$. Higher fields will allow operation at a lower value of β and improve the confinement. Therefore, there is still a good incentive for increased field. A 12 T or even higher field might be required as a fall-back position if high β cannot be achieved.

The ignited plasma is expected to deliver about 1500 MW of thermal power during the burn cycle. If a radiation shield (about 60 cm) capable of three-decade attenuation is used and we assume 1/3 of the radiation hits the TF coils, the peak radiation load on the TF coils would be about 500 kW. This is 30 times higher than the EPR design point. However, it is expected that the TNS machine will not have a duty cycle any higher than 10% so this load can be removed by the refrigeration system.

The poloidal field coil system of the TNS will have similar features to that used in the ORNL-EPR conceptual design study [4]. The EF coil inside the TF coils can serve to reduce the pulse field on the TF coils. However, in the flux-conserving mode operation, the plasma current is raised from 4 MA to 8 MA in about 4 sec [5]. In this time period the shielding function of the EF coils may be degraded. Therefore, the average $\dot{B}$ on TF coil may be somewhat higher than 0.05 T/sec (the value used in the EPR study), but possibly lower than 0.5 T/sec (the value to be employed in the coil fabricating program).

The poloidal field coils required for the TNS consists of two types of coils: the Ohmic Heating (OH) coils and the Equilibrium Field (EF) coils. The main part of the OH coil will be inside the center column of about 4 m in diameter. It serves to supply a flux swing of about 80 V-sec for the plasma current start-up and burn period. The EF coils should be as close to the plasma surface as possible in order to shape the plasma. On the other hand, if the EF coils are placed inside the shield, the radiation load would be very high. If we assume 10% interception of the peak load, then the value is about 150 MW. This is probably too high even for a water-cooled copper coil and consequently optimization of the position and shielding of these coils is an important area of development.

Toroidal Field Coils

Superconducting, resistive, and hybrid magnet designs were all considered for the TF magnet system. At first thought, one might assume that a smaller TF coil would suffice if it is resistive. One needs less shielding for high temperature resistive coils operating at 350-600 K than for superconducting coils at 4 K. However, because of

288

the need for space for the EF coils and some shield and the consideration of field ripple, the size of resistive coils would not be less
than superconducting ones. The minimum space between the plasma edge
and the TF coil necessary to meet the field ripple and access requirements might just be enough for shielding the superconducting coils.
Furthermore, a resistive coil should be operated in pulsed mode. The
capital cost of a resistive system and the operating costs (power
consumption) are both probably higher than a comparable superconducting coil and the latter allows dc operation. The pulse nature also
introduces a serious mechanical fatigue problem. An all-resistive TF
magnet is most probably not scaleable to the larger devices such as an
EPR. The technique of using resistive insert coils in a superconducting coil (hybrid design) for the size and environment required in TNS
is still uncertain. A much closer look at this approach both from a
mechanical and thermal point of view is necessary. The TNS design em-
ploys superconducting TF coils because they are feasible and can be
applicable to larger machines. However, their design is not fixed
and the options are many. A direct conclusion of the design project
to date is the requirement of fields in excess of 80 kG at the coil
to provide a margin for ignition in the presence of uncertain plasma
behavior.

To select the best set of design choices, the US has a research,
development, and demonstration program for superconducting toroidal
magnets centered at ORNL. The program includes both in-house
research, industrial design and fabrication and international cooperation. The industrial and international cooperation is centered

around the Large Coil Project (LCP). This project will let industrial subcontracts for building 6 coils with 2.5 x 3.5 m bore covering the range of viable possibilities above. These coils will be assembled in a compact torus which will allow full field testing by 1979. Co-operation with the International Energy Agency (IEA) on LCP will result in 1 to 3 coils being contributed to the compact torus demonstration by the IEA members.

Fueling Requirements for TNS

The principal problem to be addressed in fueling a TNS-like plasma is that of injecting D-T material deeply into the central portion of a hot, dense plasma against the confining effects of the magnetic fields. It is generally acknowledged that fueling by neutral beams would require prohibitively large power expenditures and that the most practical method of introducing fuel into such a device is by injection of solid D-T pellets accelerated to hypervelocities. The work of Gralnick [6] suggests that millimeter size pellets moving into the plasma at speeds of the order of 10^4 m/sec would be required to refuel a reactor grade plasma. Velocities of this magnitude have been reached in the laboratory using ballistic techniques, but the technology for accelerating hydrogen ice to hypervelocities has not been developed. In recent experiments, small frozen hydrogen pellets (about 0.1 mm) were successfully formed and injected into ORMAK [7] but only at velocities of 100 m/sec. Acceleration was accomplished by gas dynamic drag and it is believed that this technique can be extended with relative ease to reach velocities of about 1000 m/sec. The magnitude of the problem associated with attaining higher speed is made clear by recognizing that at 10^4 m/sec each molecule in the pellet acquires about 1 eV of directed kinetic energy. Since the binding energy of the solid hydrogen phase is only about 0.01 eV, it is apparent that any acceleration scheme must be highly efficient in the sense that only a very small fraction of the energy involved could be permitted to appear as internal energy or heat. This criterion alone would ostensibly eliminate many of the known hypervelocity techniques that rely on strong impulsive loading of the pellet for acceleration.

Pellet Ablation Scaling Law

In addition to the work of Gralnick, a recent pellet ablation theory was proposed by Parks, Foster and Turnbull [8] to explain the observed behavior of pellets injected into the research tokamak, ORMAK [7]. The model accurately predicts the observed pellet life-times under the conditions investigated (ORMAK edge plasma parameters). The essential difference between this latest theory and that proposed by Gralnick is in the treatment of the interaction of the hot plasma with the pellet surface. Gralnick assumes that the incident plasma electrons evaporate ion pairs from the surface and that the resultant warm plasma cloud partially shields the pellet from the hot plasma by Coulomb collisions. The energy input required for the solid-to-plasma phase transition at the surface is the 32 eV corresponding to dissociative ionization of molecular hydrogen. In the model of Parks, et al., the assumption is made that a cold neutral gas is created at the surface and that the expanding neutral cloud shields the pellet by providing a dense medium for inelastic collisions with incoming

291

electrons. The energy requirement for this neutral ablation process is the 0.01 eV corresponding to the latent heat of evaporation of hydrogen. The key elements of the two theories differ widely but the predictions agree surprisingly well for a plasma representative of TNS. The neutral ablation model lends itself more easily to numerical calculations and a pellet lifetime scaling law can be extracted from it easily.

Larger pellet radii contribute to longer lifetimes and this effect tends to relieve the requirements for high injection velocities. However, pellets cannot be made arbitrarily large, but must be limited to sizes that contain no more than a small fraction of the total plasma ion content. In terms of this fractional value f and the plasma volume V, the expression for the pellet lifetime τ reduces to

$$\tau \propto 7.1 \times 10^{-3} \ (fV)^{5/9} \ n_e^{2/9} \ T_e^{-11/6} \tag{1}$$

One important result is that pellet lifetimes are <u>increased</u> with plasma density and volume. This is due to the fact that large dense plasmas will accommodate bigger pellets which provide a more effective gas dynamic shield. High β operation affords a slight but definite advantage in this respect.

The most important parameter in this scaling law is the plasma energy and it is this fact that may make hotter plasmas difficult to fuel. To demonstrate this we take as typical average parameters for TNS the values $n_e = 2 \times 10^{20}/m^3$, $T_e = 13$ keV, $V = 250 \ m^3$. Assuming that pellets contain 10% of the plasma ion content (f = 0.1) a characteristic lifetime of 40 μsec results. If the pellet must penetrate 0.60 m to the center of the TNS shifted plasma, character-istic velocities in the 10^4 m/sec range are required. Non-circular shifted plasmas might offer significant advantages with respect to injection velocity requirements since pellets can be injected along the plasma minor axis from the outer wall which is closest to the plasma center. Because of the projected high velocity requirements, perpendicular injections seems wise.

Calculated fueling parameters for various experimental tokamaks are presented in Table Table III based upon the complete neutral ablation model and accounting for the variation of plasma parameters along the pellet trajectory. Pellet radii were limited to values that would contribute only 10% of the plasma ion content. It is clear that velocity requirements become more demanding as plasma parameters progress toward breakeven values. The 6000 m/sec velocity projected for TNS is high but as will be shown in the following section, at least one acceleration scheme appears to be capable of providing this performance.

It should be noted that none of the various pellet ablation models have had sufficient experimental verification and, in view of this, the results presented here should be viewed only as approximate guidelines for future hardware development programs.

Pellet Acceleration Methods

Some of the methods that may be applicable to acceleration of the needed millimeter size D-T pellets to hypervelocities are listed in Table IV, along with estimates of velocity capabilities. Other schemes will undoubtedly be proposed in the future and so this

293

TABLE III

ESTIMATES OF EXPERIMENTAL PELLET FUELING REQUIREMENTS

For Pellets Containing 10% of Device Total Ion Content and for Perpendicular Injection. Parameters are Plasma Center Values.

| Device | Device Characteristics | | | Pellet Radius $(10^{-6}m)$ | Velocity (m/sec) |
	$a(m)$	$Te(keV)$	$ne(m^{-3})$		
ORMAK	0.23	1.0	7×10^{19}	220	2000
ISX	0.25	1.0	7×10^{19}	240	2000
ORMAK Upgrade	0.30	0.9 3.0^+	1×10^{20} 1×10^{20}	300 300	1500 5000
PDX	0.47/0.57	1.0 3.0^+	3×10^{19} 1×10^{20}	300 475	1500 3000
PLT	0.45	1.7 4.0^+	5×10^{19} 1×10^{20}	360 450	3000 6000
TNS$^+$	1.25/2.0	20.0^+	3×10^{20}	3000	6000

$^+$with neutral beams.

$^+$parameters chosen to preserve line average values.

TABLE IV

CATALOGUE OF ACCELERATING DEVICES FOR 1 mm RADIUS D-T PELLETS

Device	Speed (m/sec)	Comments
Gas dynamic Light gas gun	1000	Requires moderate working pressures and temperatures.
Electrostatic	600	Charging limit requires acceleration voltage = 13×10^6 V.
Laser rocket	500	Requires higher performance.
Electron beam rocket	2000	90% of pellet mass consummed as propellant.
Mechanical centrifuge	2500 5400	Aluminum rotor. Fiber composite rotor.

discussion is not intended to exhaust all possibilities.

No distinction is made between low pressure gas dynamic acceleration and the light gas gun because hydrogen pellets are not expected to survive the extremely high pressure atmosphere at which present design gas guns operate. The limit of 10^3 m/sec corresponds to the approximate speed of sound of hydrogen gas at ordinary working pressures and temperatures. Because of the proven performance of light gas guns, this device should not be excluded from consideration. Its exact capabilities with hydrogen pellets should be determined experimentally.

Electrostatic acceleration techniques are limited by the inability to attain sufficiently large charge-to-mass ratios on pellets above 1 μ size. The low tensile strength of solid hydrogen limits the amount of charge that a 1 mm pellet can accommodate (surface space charge fields $< 2.5 \times 10^8$ V/m).

It has been suggested that a giant laser pulse could be used to ablate a small amount of material from one side of a pellet to provide a large impulse fo acceleration. The strong shock wave that is transmitted to the pellet bulk imparts half of its energy as directed kinetic energy with the remaining half appearing as internal heat. The low binding energy of the condensed phase limits the strength of the shock wave and hence the velocity. A limit of about 500 m/sec has been estimated for this concept; but it is recognized that it needs to be explored. The use of a focused electron beam to provide an ablation driven acceleration is also possible. In a properly designed device, the velocity of the expanding material would not be as high as that for the laser ablation concept and the impulse would be less severe.

The final acceleration concept considered is a purely mechanical device that would utilize centrifugal forces to accelerate pellets outward along the arms of a rotating arbor designed for optimum peripheral, or tip, speeds. In the present concept, pellets would be introduced at the arbor hub after being cut from a continuously extruded filament of solid hydrogen and would be guided by channels formed in the arbor arms to the tip where they leave with equal components of radial and tangential velocity. Present designs attain tip speeds of 1800 m/sec using high strength aluminum alloys as the material of construction in the arbor. This should provide pellet velocities of the order of 2500 m/sec. For this device, performance varies as the square root of the strength-to-weight ratio of the material used. Several fiber composite materials exhibit strength-to-weight ratios several times larger than that of aluminum. Highest performance would result from a choice of either Kevlar 49 or Thornel 300 fibers embedded in an epoxy resin matrix. Pellet velocities as high as 5400 m/sec are indicated with 1 m diameter rotors operating at 1000 Hz rotational speed. Development of this device might lead to a slight increase in performance, but velocities higher than 10,000 m/sec do not seem possible.

Of the concepts considered here, the centrifugal accelerator appears to be the most practical solution for a TNS fueling device with 6000 m/sec capabilities. The gas gun is a likely back-up device although it has not been proven with hydrogen pellets. If the

296

theoretical predictions probe to be correct, it is clear that the most advanced type of accelerator will need to be developed.

Tritium Recycle in TNS

The tritium recycle system for TNS will remove unburned fuel, ash (helium), and impurities from the divertor chamber; purify the unburned fuel by removing helium and other impurities; separate high purity deuterium for recycle to the neutral beam injectors; prepare a small hydrogen purge stream for discharge; supply suitable fuel surge and storage capacity; and provide containment sufficient to hold tritium exposures and releases to acceptable levels. The ORNL design for a TNS tritium recycle system is expected to resemble that proposed for EPR. This would be a system with essentially the same process components expected to be used in EPR, and they could be easily applied to even larger devices.

Removal of fuel from the divertor chamber will require a vacuum pump with high pumping speeds, complete containment of tritium, complete recovery of fuel, minimum probability for adding contaminants to the plasma, and ability to operate under the space restrictions, radiation fields, and magnetic fields around the torus. A compound cryosorption pump is recommended with separate sorption panels for pumping the unburned D-T fuel and the helium ash. The fuel can be pumped by a molecular sieve panel cooled to approximately 15°K; helium can be pumped by a similar panel cooled to 4.2°K (liquid helium temperature). The 4.2°K panel would be located "behind" the 15°K panel so only a minimum amount of fuel would reach the helium pump. The most likely alternative to this would be substitution of a 4.2°K cryocondensation panel for the 15°K cryosorption panel in the fuel pump. All other pump systems considered failed to meet one or more of the TNS requirements.

In the compound pump system just proposed, considerable separation of helium from the fuel will be accomplished in the vacuum pump itself. The final purification step would take place in beds of hydride-forming materials, probably uranium. These beds can be operated to remove all likely impurities (He, oxygen, carbon, nitrogen, etc.), and they will serve other purposes as well. They will provide necessary surge capacity at low pressures and fuel accumulation capacity in the event that the pumps or other portions of the fuel recycle system loses refrigeration. Other purification systems have been considered, particularly permeable Pd membranes, but these require compressors, or at least transfer pumps, which are undesirable and do not appear to be necessary.

To supply a high purity deuterium stream for the injectors, an isotope separation system will be required. The most promising isotope separation technique for TNS (and later fusion devices) is cryogenic distillation. This is a well established technique, and it appears to be clearly the most economic technique, principally because of the low tritium inventory and energy costs. One column would produce the high purity deuterium stream, and a second smaller column would process a portion of the deuterium to produce a hydrogen stream for discharge.

Tritium containment is usually associated with the fuel recycle system partly because most of the TNS tritium inventory (hazard) will

298

be associated with the recycle system. Since tems of millions of curies of tritium will be used and stored in TNS, a most effective and reliable containment system will be required. We suggest a multi-level containment system with active atmosphere process systems to remove tritium at each level. The fuel recycle system will be constructed as compactly as possible once allowances are made for convenient maintenance of all components. All portions containing large quantities of tritium or having significant probabilities for tritium release will be contained in a tight enclosure with continuous processing of the atmosphere to hold the tritium concentration low enough to prevent unacceptable releases of tritium from diffusion and leaks through the enclosure and from the necessary exhaust to the atmosphere. In addition, an emergency tritium cleanup system is required for the much larger reactor room.

Although the tritium recycle system may not be either the most expensive system to build or the most difficult system to develop for TNS, it certainly will be one of the most critical systems because of the crucial role it plays in containing the relatively hazardous tritium containing fuel and preventing excessive contamination of either the TNS equipment or the environment. The TNS can serve as a test stand for many components of tokamak fusion reactors, but it will be necessary to have a high reliability tritium system for TNS. Most of the development and testing of the tritium system should precede TNS, and we have proposed a relatively rapid development and testing program to assure that the components we have selected can fit into a working tritium recycle system. This assurance will be needed by the time the design for TNS is finalized, perhaps as early as 1980. This requires detailed design of a major tritium test facility to begin almost immediately, and construction should begin within a year.

Conclusion

The physical uncertainties in the TNS version of the PDFD will be addressed by an array of experiments over the next five years. The beams and superconducting magnets are also under development on a large scale. These programs expect to have minimally acceptable products for TNS by 1980. The pellet fueling and the tritium handling systems require the largest increase in effort in the immediate future but promising techniques exist in both areas. We conclude that with a reasonably aggressive development program, firm answers to the questions existing in the TNS design can be obtained by 1980. This would allow this PDFD to be in operation by 1987.

Acknowledgement

The material discussed in this paper was furnished by M. Lubell, W. Lul, S. Shilling, L. Stewart, J. Watson and S. Milora, all members of the ORNL Fusion Program.

300

REFERENCES

[1]   CALLEN, J. D., CLARKE, J. F., ROME, J. A., "Theory of Neutral Injection into Tokamaks, Paper E14, Third Internation Symp. on Toroidal Plasma Confinement, Garching (1973).

[2]   CALLEN, J. D., "Magnetic Field Ripple Effects in Tokamaks", Seminar at DCTR, ERDA, Wash., D. C., Jan. 9, 1976; JASSBY, D. L., and GOLDSTON, R. J., "Enhanced Penetration of Neutral-Beam-Injected Ions by Vertically Asymmetric Toroidal Field Ripple,"Princeton Report MATT-1206, Feb. 5, 1976; MATT-1244, May 1976.

[3]   UCKAN, N. W., TSANG, K. T., CALLEN, J. D., "Effects of the Poloidal Variation of the Magnetic Field Ripple on Enhanced Heat Transport Tokamaks", ORNL/TM-5438, Oak Ridge (1976).

[4]   ROBERTS, M., Ed. "Oak Ridge Tokamak, EPR Study", ORNL/TM-5572-77, Oak Ridge (1976).

[5]   CLARKE, J. F., "High Beta Flux Conserving Tokamaks", ORNL/TM-5429, Oak Ridge (1976).

[6]   GRALNICK, S. L., Nuclear Fusion 13 (1973) 703.

[7]   FOSTER, C. A., et al., Bull. Am. Phys. Soc. 10 (1975) 1300.

[8]   PARKS, T., FOSTER, C. A., TURNBULL, R. J., "A Model for the Ablation Rate of a Solid Hydrogen Pellet in a Plasma", submitted for publication in Nuclear Fusion.

COMPACT EXPERIMENTS FOR α-PARTICLE HEATING

B. Coppi

I. Introduction

Recent experiments, and scalings deduced from them, have indicated that relatively high density plasmas can be produced and well confined in toroidal configurations capable of sustaining high current densities without inducing macroscopic instabilities. Here we propose to develop a line of compact devices sustaining sufficiently high plasma currents to confine the 3.5 MeV α-particles that are produced in D-T reactions. This line is proposed as a parallel program to the development of large volume Tokamaks which is being undertaken on a worldwide basis.

The experimental apparatuses we propose have the following features:

a)   a relatively high Ohmic heating rate so that given the high particle densities at which they can operate, they can produce a substantial amount of D-T fusion power even for relatively modest temperatures.

b)   the central core of the machine is reduced to a basic minimum of elements with little concession to diagnostic access ports. Thus the complexity and expenditure in remote handling systems are greatly reduced in comparison to those foreseen for other types of Tritium burning experiments.[1]

c)   the core can be produced in duplicates, one of which can be tested with D-D plasmas in existing laboratories which do not have Tritium handling facilities. Another possibility, that obviously cannot be considered for the largest size devices

303

which are presently under development, is to test the core in a laboratory and then transport it to a site having the necessary power sources and Tritium handling facilities.

d)   their main scientific goal is to investigate the effects of α-particle heating and all the problems (diffusion, instabilities, etc.) connected with it. In particular, they can lead to realize a relatively fast paced and low cost program to demonstrate ignition experimentally.

e)   they can be developed into 14 MeV neutron and 3.5 MeV α-particle sources for a variety of applications.

f)   they can be employed as material testing reactors[3]. In this case the plasma chamber itself would serve as a test surface and be subjected to mechanical loadings to simulate operating conditions of a power reactor.

g)   auxiliary heating systems involving relatively modest power levels can be adopted. In particular, heating processes that are specially suitable for the compact devices (to be referred to as Ignitors) under consideration, in the sense that they would not require major modifications of the basic magnet structure, have been proposed recently (see Section Vl).

II. Tunnelling of Bremsstrahlung Limit

The procedure for bremmsstrahlung tunnelling during heating of a confined DT plasma toward ignition temperatures with relatively high densities is based on:

a) Choosing the plasma density during the preheating phase to be sufficiently high that the ion temperature remains relatively close to the electron temperature and the effective average charge number (Z_{eff}) is near unity but below the limit for which the electron temperature would be bounded by radiation emission (bremsstrahlung)

b) maintaining a total plasma current in excess of 2.5 MA so that a large fraction of the produced 3.5 MeV α-particle can be contained.

c) raising the plasma density after an average ion temperature in excess of 4 keV, over the central portion of the plasma column, is reached and the α-particle heating can compensate the bremsstrahlung radiation emission.

A heating cycle with these characteristics can be realized on the basis of classical resistivity, collisional energy transfer from electrons to ions, and ion thermal conductivity when the poloidal magnetic is larger than a critical value that is given in Section III. This depends on the assumption the empirical scaling laws which are presently known for the electron energy replacement time, hold for higher temperature regimes than attained so far and it is one of the reasons why the adoption of auxiliary heating systems is considered.

An alternative possibility may be offered by realizing densitites that are sufficiently low as to set up a "slide-away" regime[4] where the transverse ion temperature, to the magnetic field, is maintained at values close to those of the transverse electron temperature by the excitation of current driven lower-hybr frequency modes. In this case the ion distribution tends to develop an energetic tail which can increase considerably the total rate of fusion reactions with respect to the case of a pure ion Maxwelli with equal transverse temperatures.

Thus we can distinguish three phases of evolution for temperature and density as indicated of Fig. 1.

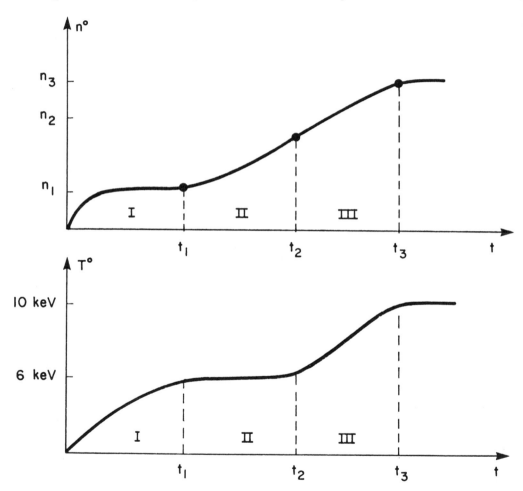

Phase I ($0 < t < t_1$) is that of preheating of the plasma to the temperature range $T_i^0 \approx 5 - 6$ keV where α-particle heating begins to overtake bremsstrahlung emission and ion collisional thermal conductivity losses (see next Section). Here T_i^0 indicates the central ion temperature.

Phase II ($t_1 < t < t_2$) is the heating phase of the cold plasma that is produced by bleeding neutral gas into the plasma chamber. During this phase the peak ion temperature is assumed to remain nearly constant.

Phase III ($t > t_2$) begins when the plasma density has become sufficiently high that the α-particle heating can over-come the difference between electron thermal conductivity losses and external heating. The end of this phase corresponds to a typical temperature of 10 keV and the maximum density that is compatible with the adopted confinement configuration (see Section V).

While we discuss a more complete form of the thermal energy balance equation in the next section, here we notice that, if the only source of heating in the center of the plasma column is Ohmic, with classical longitudinal electrical resistivity, and the major energy loss is by bremmsstrahlung we have

$$(\eta_{cl} J_\parallel^2)^0 \approx (n^0{}_{14})^2 (T^0{}_{e4})^{1/2} \times 1.22 \times 10^{-2} \text{ W/cm}^3$$

Here $J_\parallel$ is the current density in A/cm^2, η_{cl} the classical resistivity, $n_{14}^0 = n^0 / (10^{14} \text{cm}^{-3})$, and $T_{e4}^0 = T_e^0 / (4$ keV$)$, n^0 and T_e^0 being the electron central density and temperature. Then we normalize $J_\parallel^0$ to 5 kA/cm^2, that we consider as a desirable value for an "Ignitor-Oh" device (See Section VI) in which Ohmic heating plays the major role, and obtain

$$n^0 \approx 2.5 \times 10^{15} \times \frac{J_5^0}{T_{e4}^0} \, (\hat{\lambda})^{1/2} \, cm^{-3} \, .$$

Here $J_5^0 = J_{\parallel}^0/(5 \, kA/cm^3)$, $\hat{\lambda} = \ln \Lambda/(15)$, and $\ln \Lambda$ is the well known Coulomb logarithm.

III. Simplified Energy Balance Equation

In order to evaluate approximately the density values for which ignition temperatures can be achieved, we write a simplified form of the average thermal energy balance equation. We assume that the electron thermal energy is lost by anomalous processes while for the ions only the collisional thermal conductivity is taken into account. Thus we write

$$\frac{1}{\tau_E} < n(T_e + T_i) > = \frac{1}{\tau_{an}} <n \, T_e> + \frac{1}{\tau_{ic}} <nT_i> , \tag{1}$$

where τ_E is the energy replacement time,

$$\tau_{an} = 16 \, n_{14}^0 \, \hat{a}^2 \alpha_\tau^e \, msec , \tag{2}$$

$\hat{a} = a/(20 \, cm)$, a is the plasma radius, τ_{ic} the collisional ion energy containment time, and α_τ^e is a coefficient of order unity. Notice that

$$\frac{\tau_{an}}{\tau_{ic}} \propto \frac{n^2}{T_i^{1/2}} \, . \tag{3}$$

Besides these losses we take into account the average bremsstrahlung emission as

$$<P_B> \approx 0.24 \, (n_{15}^0)^2 \, (T_{e6}^0)^{1/2} \times \alpha_B \, watt/cm^3 , \tag{4}$$

where $\hat{T}_{e5}^0 = T_e^0/(6 \, keV)$, $n_{15}^0 = n^0/(10^{15} \, cm^{-3})$ and $\alpha_B \equiv 5<n^2T_e>/(n^0 T_e^0$ takes into account the effects of density and temperature profile

The cyclotron emission loss is expressed by

$$<P_S> \approx 0.6 \, \varepsilon_S \left(\frac{<n_{15}>}{\hat{a}} \right)^{1/2} \times <T_{e6}>^{2.1} \, (B_{15})^{5/2} \, watt/cm^3 , \tag{5}$$

308

where $B_{15} = B_T/(150 kG)$, B_T is the toroidal magnetic field and ϵ_s is a reduction factor that takes into account the effects of geometry, wall reflectivity, and finite β.

The α-particle heating is expressed by

$$<P_\alpha> \simeq (n_{15}^0)^2 (T_{i6}^0)^{-2/3} \exp\left(11[1 - (1/T_{i6}^0)^{1/3}]\right)$$
$$\times 0.7 \times \alpha_h \text{ watt/cm}^3, \tag{6}$$

where $T_{i6}^0 = T_i^0/(6 keV)$ and $\alpha_h \equiv 5<P_\alpha>/P_\alpha (r=0)$ includes both the effects of temperature and density profiles and the α-particle losses due to diffusion.

The average Ohmic heating power is written as

$$<P_\Omega> = E_{\|} \bar{J}_{\|} = B_{p3}^2 \frac{1}{\hat{a}^2 (T_{e6}^0)^{3/2}} \times \alpha_J \times 2.56 \text{ watt/cm}^3, \tag{7}$$

where $\hat{B}_{p3} = B_p/(30 \text{ kG})$, $B_p = I_p/(5a)$, I_p is the total plasma current,

$$\alpha_J = \frac{q_s \times 1.5}{4 q_0} \frac{\eta_0}{\eta_{c1}^0} \hat{\lambda},$$

J^0 is the central current density, $\bar{J}_{\|}$ the current density average, η_0 the central resistivity, $\eta_{c1}^0 \approx 10^{-6} \hat{\lambda}/(6T_{e6}^{3/2}) \Omega$ cm the classical expression for it, q_s is the so called safety factor against MHD instabilities, as discussed in Section V, $q(r) = 2\pi/\iota(r)$, $\iota(r)$ is the rotational transform and $q_0 \equiv q(r = 0)$.

The thermal power loss due to classical and anomalous thermal conduction and particle transport can be estimated as

$$<P_L> = \frac{1.8}{\hat{a}^2} \times \left[T_{e6}^0 \frac{\alpha_p}{\alpha_\tau} + 0.1 \times (n_{15}^0)^2 (T_{e6}^0)^{1/2} \frac{\hat{A}^{3/2} (q_s/2)^2}{B_{15}^2} \alpha_\tau^i \right]$$
$$\times \text{ watt/cm}^3, \tag{8}$$

where $\alpha_p = 5 <nT_e>/(n^0 T_e^0)$, $\alpha_\tau^i = [5<nT_i>/(n^0 T_i^0)] < (n/n^0)(T_i/T^0)^{1/2} (q/q_s)^2 >$, and $\hat{A} = R_0/(2.5a)$

It is interesting to observe that if $\tau_{an} < \tau_{ic}$, $T_i \simeq T_e$ and $<P_L>$ and $<P_\Omega>$ are the largest terms in the thermal energy balance equation, the maximum temperature that can be reached is simply a function of the <u>poloidal magnetic field</u>.

We indicate the average power input of the auxiliary heating system that may be adopted by $<P_A> = 2.5\hat{P}_A$ watt/cm^3. Then the corresponding total auxiliary power is

$$W_A \simeq \hat{a}^2 \, \hat{R} \, \hat{P}_A \text{ MW,} \tag{9}$$

where $\hat{R} = R/(50\text{cm})$, and we can write the total energy balance equation as

$$\hat{a}^2 \hat{P}_A + \frac{\alpha_J}{(\hat{T}^0_{e6})^{3/2}} B^2_{P3} = 0.72 T^0_{e6} \frac{\alpha_p}{\alpha_\tau^e}$$

$$+ (n^0_{15})^2 \left\{ 0.1 \, (T^0_{e6})^{1/2} \left[\hat{a}^2 \alpha_B + 0.72 \left(\frac{T^0_i}{T_e}\right)^{1/2} \times \frac{\hat{A}^{3/2}}{B^2_{15}} \, \hat{q}^2 \, \alpha_\tau^i \right] \right.$$

$$\left. - 0.28 (T^0_{i6})^{-2/3} \exp \left[11 - 11/(T^0_{i6})^{1/3} \right] \times \alpha_h \hat{a}^2 \right\}$$

$$+ 0.05 \times \alpha_s \, \hat{a}^{3/2} \, (n^0_{15})^{1/2} (T^0_{e6})^{2.1} (B_{15})^{2.5} \tag{10}$$

where $\hat{q} = q_s/2$. and $\alpha_B = 5 <n/n^0>^{1/2} <T_e/T^0_e>^{2.1} \epsilon_s$. Here the numerical coefficients in front of each term indicate, roughly, the corresponding total power in a configuration with a major radius of 50 cm when an appropriate choice for a and B_p is made. From this equation it appears that a poloidal field of about 20 kG can be sufficient to reach peak temperatures of about 5 keV even when $\hat{P}_A$ is negligible. Notice that the toroidal magnetic field value can be related to that of the poloidal field by

$$B_T = \hat{A}\,\hat{q}\,B_{P3} \times 150 \text{ kG}.$$

Assuming that poloidal fields in the range of 20 - 30 kG can be produced, and that the estimate of the thermal energy loss by conduction as represented by Eq. (8) remains valid, the function of an auxiliary heating system as represented by the term $\hat{P}_A$ in Eq. (10) is mainly that of decreasing the duration of the preheating and heating phases. To this end an optimal choice of the density value n to be maintained during the preheating phase should also be made following the discussion given in the next section.

IV. Regimes of Interest

We notice that, assuming the ion thermal energy transfer as purely collisional, the thermal energy balance for the ions can be reduced simply to

$$3\,\frac{m_e}{m_i}\,n\nu_{ei}\,(T_e - T_i) = \frac{1}{r}\frac{\partial}{\partial r}\left(D_{ic}\,nT_i\,\frac{d\ln T_i}{d\ln r}\right), \tag{11}$$

where

$$D_{ci} \approx 1.6 \times 10^3\,\hat{\lambda}\,\hat{q}^2\,\frac{n_{15}}{(T_{i5})^{1/2}B_{15}^2} \times \hat{A}^{3/2} \times \left(\frac{a}{2r}\right)^{3/2}\,\text{cm}^2/\text{sec},$$

and the term on the left hand side can be written as $nT_i \times \nu_{e \to i}$ with

$$\nu_{e \to i} \approx \left(\frac{T_e}{T_i} - 1\right)\frac{n_{15}}{(T_{e5})^{3/2}} \times \frac{2}{A_i} \times 10^2\,\text{sec}^{-1},$$

A_i being the ion mass number. Then, if we take $(1/r)\partial/\partial r \sim 1/a^2 \approx (400\text{ cm}^2)^{-1}$, it is easy to see that $(T_e/T_i) - 1 \ll 1$.

We recall that for $T_i \approx T_e$

$$\frac{1}{\tau_E} \approx \frac{1}{\tau_e} + \frac{1}{\tau_i} ,$$

where τ_e and τ_i are the electron and ion energy confinement times. Then taking into account bremsstrahlung emission we may write

$$\frac{1}{n_{15}\tau_E} \approx \frac{1}{T_5^{1/2}} \left(\frac{1}{\tau_{bo}} + \frac{1}{\tau_{i_0}\hat{a}^2} \right) + \frac{1}{n_{15}^2 \hat{a}^2 \tau_{an}^0} , \tag{12}$$

where the last term corresponds to the anomalous electron thermal energy loss. At the highest densities, where radiation and ion thermal conductivity losses prevail, we have

$$n\,\tau_E \approx 10^{15} \, (T_5/\Delta)^{1/2} \, \text{sec/cm}^3 ,$$

where Δ is the quantity contained in parenthesis in Eq. (12), that i related to the term in the first set of square brackets, multiplying $(T_{e6}^0)^{1/2}$, in the right hand side of Eq. (10).

The maximum value of τ_E corresponds to the value of $\hat{n}_{15}^0$ for which the two terms in the right hand side of Eq. (12) are equal. Correspondingly, by using the numerical estimates indicated in Eq. (10) we obtain

$$n_{15}^0 \approx 2.5 \, (\hat{T}_{e5}^0)^{1/2} \, (\alpha_p/\Delta)^{1/2} ,$$

and

$$\tau_E \approx 200 \, \hat{a}^2 \alpha_\tau^e \, (T_{e5}^0)^{1/2} \, (\alpha_p/\Delta)^{1/2} \times \text{msec.}$$

In order to have an idea of the maximum energy density that can be confined we evaluate the "poloidal beta" as

$$\beta_p = \frac{3}{2} \, \langle n(T_e + T_i) \rangle \times \frac{8\pi}{B_p^2} \approx \frac{10^{-1}}{1.5} \times \frac{n_{15}^0 T_{e5}^0}{B_{p3}^2}$$

$$\times \, \alpha_p \times [1 + \frac{\langle nT_i \rangle}{\langle nT_e \rangle}], \tag{13}$$

312

For $T_i \simeq T_e$, $T^0 \simeq 10$ keV and $B_p \simeq 30$ kG we have $\beta_p \simeq 2.5$ for $n \simeq 0.94 \times 10^{16}/\alpha_p$ cm^{-3}. The power density produced by fusion reactions in the center of the plasma column for $T^0 \simeq 10$ keV is about

$$P_F^0 \simeq (n_{16}^0)^2 \left(\frac{\overline{\sigma v} \times 10^{16}}{cm^3/sec}\right) \times 7 \times 10^3 \text{ watt/cm}^3, \tag{14}$$

and for $n_{16}^0 \simeq 0.95$ we have $P_F^0 \simeq 6.2$ kW/cm^3. Then, considering a plasma volume $V_0 \simeq 4 \times 10^5$ cm^3, as we shall do for a reference device in Section V, the total power produced by D-T fusion reactions can be measured by

$$W_F \simeq \frac{V_0}{4} P_F^0 \simeq 620 \text{ MW.}$$

The corresponding wall loading would be of the order of 155MW/m^2 for a wall surface of 4m^2 as we shall consider, during the power producing part of the pulse. Thus a duty cycle of about 5×10^{-2} would be sufficient to make a device of this type interesting as a Material Testing Reactor.

In this regard we point out that the D-T fusion reaction rate in reality may be larger than evaluated on the basis of the $\overline{\sigma v}$, used in Eq. (14), which results from averaging over an ion Maxwellian distribution. The reason for this is that there are microinstabilities such as drift modes which are affected by ion Landau damping and tend to decrease the slope of the ion distribution at superthermal velocities. Another contribution which is to be added to the reaction rate is that of the ion runaways that may be produced.

Finally we recall that the value of the streaming parameter to be considered for the onset of the slideaway regime[4], that was mentioned in the context of Section II, is

$$\xi_0 \simeq \left.\frac{v_D}{v_{the}}\right\}_0 = 1.6 \times 10^{-1} \frac{B_{15}}{q_0 \hat{R}} (\hat{T}_{e\perp}^0)^{-1/2} (n_{14}^0)^{-1},$$

where $\hat{T}_{e\perp}^0 = T_{e\perp}^0/(1 \text{ keV})$ and $T_{e\perp}$ is the transverse electron temperature.

On the basis of experiments carried out in Alcator we may consider $\xi_c \simeq 0.1 - 0.2$ as the relevant critical value, and the density threshold is then

$$n_c^0 < (\frac{\xi_c}{0.16}) \times \frac{B_{15}}{q_0 \hat{R}} (\hat{T}_{e\perp}^0)^{-1/2} \times 10^{14} \text{ cm}^{-3} .$$

The advantage of exciting the slideaway regime is that the tail of the ion distribution tends to be inflated by the Landau damping of the lower hybrid mode driven by the current carrying electrons and this increases the rate of α-particle production as was indicated earlier. On the other hand, this advantage can be offset by the enhanced energy loss by radiation emission due to the superthermal electrons that characterize a slideaway electron distribution and to the fact that Z_{eff} is likely to be considerably larger than unity.

314

V. Ideal Ignitor Parameters

The configuration that we envision for an Ignitor-Oh device is one with a tight aspect ratio, such as $A = R_0/a = 2.5$ For this we indicate the rate of utilization of the externally applied toroidal field by the parameter

$$q_s = \frac{B_T(0) \ a}{B_p(a) R_0} \ , \ \text{where} \ B_p(a) \equiv \frac{I_p}{5a} \ ,$$

a being the plasma radius, and recall that in a tight aspect ratio configuration $q(a) > q_s$, where $q(r) = 2\pi/\iota(r)$ and ι is the rotational transform, as indicated earlier.

We also notice that the effects of ion-ion collisions and finite ion gyro-radius have been found theoretically to have a stabilizing influence on the onset of internal resistive modes[5], when plasmas with a combination of relatively high densities and temperatures are realized. Therefore, we consider it realistic to assume values of q_s as low as 2 in evaluating the expected best performance.

We notice that the plasma current I_p is related to the total current (ampere-turn) flowing in the toroidal coil by the relationship

$$I_p = I_c \frac{1}{q_s} \ (\frac{a}{R_0})^2 .$$

Then we have

$$I_c = \hat{I} \ \hat{q} \ \hat{A}^2 \times 37.5 \ \text{MA} \times \text{turn} \ ,$$

where

$\hat{q} \equiv q_s/2, \quad \hat{I} \equiv I_p/(3MA), \quad \hat{A} \equiv R_0/(a \times 2.5).$

In addition, for a plasma with $\beta_p = 2.5 \hat{A}\hat{\beta}_p$ and $q_s = 2\hat{q}$, we have

$$\beta = \frac{0.1}{\hat{q}^2\hat{A}} \hat{\beta}_p.$$

We consider a reference device with the following parameters:

$R_0 = 50$ cm $V_0 \simeq 4 \times 10^5$ cm^3

$a = 20$ cm $S_0 = 4 \times 10^4$ cm^2

$B_p = 24 - 30$ kG $\bar{J} \simeq 1.9 - 2.4$ kA/cm^2

$I_p = 2.4 - 3$ MA $q_s \simeq 2 - 2.5$

$B_{T_0} = 150$ kG

We observe that, the ratio of the magnetic field at the inner edge of the plasma column to the one at the outer edge is more than twice, and this is also the case for the applied electric field, whi $B(R = R_0 - a) \simeq 250$ kG. Here S_0 is the plasma column surface and V_0 its volume. We shall analyze the main characteristics of the equilibrium configuration that is associated with a choice of geometrical parameters and total plasma current somewhat larger than indicated here, in a different report and discuss there the relevant requirements for the transverse magnetic field and the air core transformer systems.

The considered value of the plasma current is sufficient to contain a large fraction of the banana orbits of α-particles with

3.5 MeV energy. The corresponding value of the streaming parameter, in the Tritium burning regime, can be estimated as

$$\xi = 0.5 \times 10^{-4} \frac{\hat{B}_{15}}{q_0 \hat{R}} (n_{16}^0)^{-1} (T_{10}^0)^{-1/2}.$$

where $n_{16}^0 = n^0/(10^{16} \text{ cm}^{-3})$, $T_{10}^0 = T^0/(10 \text{ keV})$. Thus the distorsion of the electron distribution resulting from the induced current is extremely small and, in particular, it ensures the validity of the Spitzer-Härm approximation for the current carrying electron distribution. The collisionality parameter indicating the extent to which the particle distribution is affected by magnetic trapping is estimated as $(\nu_{*i})^{1/4}$ where

$$\nu_{*i} = \frac{\hat{q} \hat{\lambda} n_{16}}{(T_{i5})^2} \hat{A}_0^{3/2} \times 0.7$$

and $\hat{A}_0 = (2R_0/a)/5$.

We notice that the operation of a device with $q(r \simeq a) \simeq 2.5$ that corresponds roughly to $q_s \simeq 2$ for a configuration with $A \simeq 2.5$, up to ignition temperatures is consistent with results obtained from the Alcator experiments. The reason is that, the density n^0 which can be adopted in order to achieve these temperatures is in the range $1 - 2 \times 10^{15}$ cm^{-3}. The corresponding value for β_p as indicated by Eq. (13) is not larger than 0.15. For these low^5 values of β_p, Alcator has produced stable discharges with $q(r = a_L)$ as low as 2.4, a_L being the limiter radius. In addition, as indicated in the next Section, we consider the adoption of auxiliary heating systems which tend to heat the plasma surface preferentially and therefore lead to a relatively wide current density profile. Thus the corresponding values of $q(r = a)$ that can be expected, tend to be lowered. A brief discussion of the heating strategy adopted in order to ensure MHD stability is given in Appendix A.

We also notice that, it may be possible to achieve the same goal, without the aid of an auxiliary heating system, by a relatively fast raise of the current and, therefore, of the plasma temperature to such a level that the current skin penetration time becomes larger than the rise time of the current adopeted in the following phase. In this situation, that appears to be realized in the Alcator experiment with a proper programming of the plasma current, the current density profile is the resultant of the superposition of a centrally peaked profile and of one that to be peaked at the outside. We also recall that the stability of resistive modes, that can be excited near the rational surface where $q(r) = 2,3$, ect., is strongly enhanced by raising the electron temperature near the edge of the plasma column where the surfaces tend to be located if q_s is relatively low.

Finally we observe that it can be advantageous to start the formation of the plasma column in a configuration with major and minor radius smaller than the one of the final configuration. Thus if $R_1 = R_0 - \Delta, a_1 = a - \Delta$, we have $B_{T1} = B_{T_0} R_0/(R_0 - \Delta)$ the peak current density can be expected to increase by the facto

318

$[R_0/(R_0 - \Delta)]^2$. In this case the peak density limit imposed by bremsstrahlung emission is also raised according to the discussion given in Section II. For instance, referring to the ideal Ignitor parameters given above we may choose to start with $R_1 \simeq 40$ cm, so that $a_1 \simeq 10$ cm, $B_{T1} \simeq 187.5$ kG and $J_{o1} \simeq (0.75/q_0) \times 10$ kA/cm^2. The corresponding density limit at 4 keV is about 5×10^{15} cm^{-3}.

VI. Auxiliary Heating Systems and the High Density Approach

The types of auxiliary heating systems that we have considered and that we analyze separately in more detail are:

- transverse neutral beam injection combined with the effects of magnetic ripple,

- lower hybrid frequency heating with or without combination of the effects of magnetic ripple,[7]

- adiabatic compression,

- low frequency modulation of the vertical magnetic field (e.g. axisymmetric pumping[8]),

- low frequency modulated induction of skin currents,[9,10]

Both of the last two heating methods involve the modulation of currents flowing inside a set of horizontal coils at frequencies between 1 and 10 kHz and do not require any major alteration of the basic magnet structure that is foreseen for an Ignitor type of device. We also notice that, given the proposed program of plasma heating, an auxiliary system that provides heating of a surface plasma layer may be more an asset than a liability during phase II where a relatively large amount of cold plasma is produced at the edges of the current carrying column. Thus we envision two possible series of Ignitor devices. One (Ignitor-Oh) in which the Ohmic heating system plays a major role and peak densities of about 10^{15} cm^{-3} can be adopted during the preheating phase, as has been prevalently considered in this report. Another (Ignitor-Au), in which an auxiliary heating system plays a major role, such as for the case discussed in Ref. 3, the relevant magnet systems are chosen in order to accomodate the needed access ports and/or plasma volume, and densities lower than for the former type of devices have to be adopted during the preheating phase.

Finally, when comparing the high plasma density approach as represented by the Ignitor line of devices with that of large volume, relatively low density experiments, we may make the following remarks:

1) Since in the former case the ion temperature is kept very close to the electron temperature, Ignitors can equally prof from heating systems that couple directly to the electrons or to the ions. Either system can, in fact, be used to control the current density distribution and, if possible, to improve the conditions for macroscopic stability of the plasma column.

2) The ratio of the torus major radius to the collision mean free path, indicating the degree of collisionality of the plasmas that can be obtained, is closer to that of presently operating experiments, for equal values of $n\tau_E$ and T_i . Under the same conditions, the numerical values of the energy replaceme time are lower and make high density plasmas more immune to the excitation of slowly growing modes for which there has not been adequate investigation yet.

3) While the typical values of the streaming parameter ξ are not higher than those for the large volume, relatively low density experiments, the linear particle density $\bar{n}a$ is considerab higher. Thus the neutral mean free path for ionization, charge exchange etc., is considerably shorter than the radius of the plasma column. As indicated by the high density Alcator experime the impurity influx should be relatively low.

320

Acknowledgements

It is a pleasure to thank T. Consoli and H. Skargard, for pointing out the appealing characteristics of the low frequency heating systems mentioned in Section VI, and L. DeMenna and A. Taroni for their continuing collaboration and their suggestions. The work reported herein was performed in part under the sponsorship of C.N.E.N. of Italy, in part privately, and in part with the support of the U.S. Energy Research and Development Administration.

References

1. B. Coppi, R.L.E. Quarterly Progress Report No. 97, pg. 50 (Massachusetts Institute of Technology, Cambridge, Mass., April 1970).

2. E. Apgar, B. Coppi, A. Gondalekhar, H. Helava, D. Komm, F. Martin, B. Montgomery, D. Pappas, and R. Parker, Paper IAEA-CN-35/A-5 presented at the VI International Conference on Plasma Physics and Controlled Nuclear Fusion Research (Berchtesgaden, W. Germany, 1976) and Massachusetts Institute of Technology, R.L.E. Report PRR76/36 (Cambridge, Mass., September, 1976).

3. B. Coppi, Massachusetts Institute of Technology, R.L.E. Report, PRR-7518 (Cambridge, Mass., August, 1975).

4. B. Coppi, F. Pegoraro, R. Pozzoli, and G. Rewoldt, Nucl. Fusion 16, 309 (1976).

5. B. Coppi, G. Lampis, F. Pegoraro, L. Pieroni and S. Segre, Massachusetts Institute of Technology, R.L.E. Report, PRR-7524 (Cambridge, Mass., October, 1975).

321

6. B. Basu and B. Coppi in Paper IAEA-CN-35/B13 presented at the VI International Conference on Plasma Physics and Controlled Nuclear Fusion Research (Berchtesgaden, W. German-, 1976).

7. Following a suggestion of M. Porkolab (1976) the injection of microwave power at the lower hybrid frequency would have the effect of producing energetic particles at the plasma surface. These could then be taken to the center of the plasma column, when trapped in a magnetic field ripple, by th toroidal curvature drift.

8. E. Canobbio, Paper IAEA-CN-35/G6 presented at the VI International Conference on Plasma Physics and Controlled Nuclear Fusion Research (Berchtesgaden, W. Germany, 1976).

9. Y. Amagishi, A. Hirose, H. W. Piekaar and H1M. Skargard, Paper IAEA-CN-35/G5-2 presented at the VI International Conference on Plasma Physics and Nuclear Fusion Research (Berchtesgaden, W. Germany, 1976).

10. R. L. Freeman and Doublet IIA Group, Bull. Am. Phys. Soc. 21, 1053 (1976).

11. L.S. Soloviev in Reviews of Plasma Physics, Vol. 6, (Ed. M. A. Leontovich; Publ. Consultant Bureau New York, 1975).

12. G. Cenacchi, A. Taroni and A. Sestero, Nuovo Cimento 25 B, 279 (1975).

13. R. Englade and R. Gajewski, (to be published, 1977).

APPENDIX

STRATEGY FOR MACROSCOPIC STABILITY

In order to preserve the macroscopic stability of the considered configuration we propose to proceed as follows:

a) we start the heating cycle with a well distributed current density profile of the type indicated in Fig. A-1. The corresponding profile for $q(r)$ is such that the region affected by possible localized MHD instabilities is only a very small fraction (such as 1/16) of the total plasma cross section area. This is, in fact, the case illustrated in Fig. 1A and corresponding to $q_s \simeq 2$.

b) ignition is reached with $\beta_p \leq 0.3$, so that throughout the heating phase the conditions indicated in a) hold.

c) as ignition is attained, the plasma density is raised to the point where β_p becomes considerably larger than unity, such as 2.5 - 4. The increase in β_p is accomplished in a time considerably shorter than the characteristic time for magnetic field diffusion. Thus the profile $q = q\,(\psi)$, ψ being the magnetic surface label, remains frozen and the stability conditions indicated in a) persist.

We notice that in a fat configuration, with an aspect ratio of about 2.5, values of β_p around 2.5 can be achieved by magnetic flux conservation with current density profiles that are relatively well spread in comparison with those that would be realized in a configuration with equal values of β but larger aspect ratio.

In Fig. A-2 we indicate the profile of $q(R)$ and the region affected by localized MHD unstable modes for the case where $\beta_p \simeq 2.5$ is reached without magnetic flux conservation. We see that the minimum value of q is about 0.3 and we may argue that in order to bring this close to unity we need have a safety factor q_s close to 6. This is consistent with the values for q_s that have been required in the Alcator experiment ($q_s \gtrsim 6.5$) in order to attain values of $q(R = R_0)$ larger than unity. The instability criterion adopted in the cases described by Figs. A-1, and 2, has been taken from Ref. 11 and incorporated in the equilibrium code[12] EQUISA by R. Englade and R. Gajewski.[13]

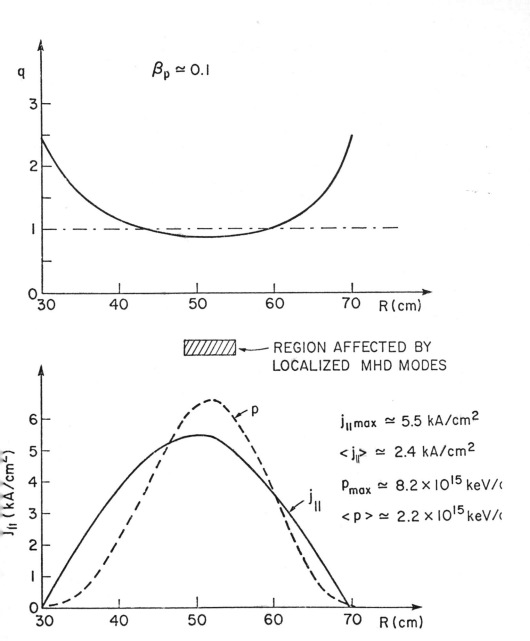

Fig. A-1 Profiles of q(R), J_{II} (R), and p (R) for a configuration
with $\beta_p \simeq 0.1$, aspect ratio 2.5, and safety factor
$q_s \approx 2$ (B_T = 150 kG, I_p = 2 MA, R_o = 50 cm, and
a = 20 cm).

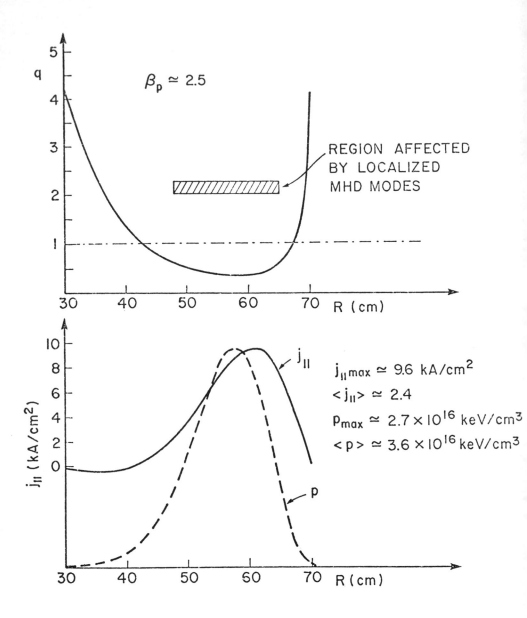

Fig. A-2 Profiles of the same quantities as in Fig. 1-A for

$\beta_p \approx 2.5$ and without assuming magnetic flux conservation.

during the heating from $\beta_p \approx 0.3$ to $\beta_p \approx 2.5$.

JET AND THE LARGE TOKAMAK EXPERIMENTS

P.H. Rebut (EUR-CEA)

JET Design Team, Culham Laboratory

1. Definition of the parameters of an apparatus: JET
2. Alternative solutions and comparison with the other large Tokamaks
3. The solutions of two main problems in JET
   a) the mechanical structure
   b) assembly and remote handling

ABSTRACT

The JET Design was the consequence of the following principles: simplicity of the apparatus, minimum forces and momentum on the components (e.g. D-shape coils, aspect ratio, free thermal expansion), equivalent stress loading in different parts of the apparatus, modular design, accessibility.

In a general presentation of JET the reasons for the choices of particular solutions will be discussed. A comparison with other large Tokamaks will show the different options.

A more detailed analysis of the mechanical structure and the remote handling technique on assembly will show some of the real difficulties of this large apparatus working under active operation. This presentation will include a review of the forces acting on the structure, the problem of insulation, access, general assembly and remote assembly.

The automatic welding and cutting equipment will be presented.

A general conclusion is that these problems of remote handling and accessibility will be the basic problems which must be overcome for a real DT burner.

"The paper was not available at the time of publishing the book".

TECHNOLOGY

FUEL TECHNOLOGY AND THE ENVIRONMENT

J. Darvas

Association EURATOM/KFA
Kernforschungsanlage Jülich
Postfach 1913, D 517o Jülich

ABSTRACT

The environmental problems related to the use of large quantities
of tritium are reviewed. Particular attention is given to the health
physics aspects arising from chronic and acute exposures to tritium,
and to permissible release rates from large fusion devices. It is
concluded that damage to man, including mutagenic effects, resulting
from tritium intake is sufficiently known for maximum permissible
dose rates to be defined, and that routine release rates from a
stack of the order of 1ooo Ci/y would not lead to dose rates to the
public in excess of permissible limits. The technologies required
in large fusion devices, like the experimental power reactor, are
commented with a view on the future European fusion programme, em-
phasizing the need for research and development in the areas of
tritium recovery from the exhaust and from the blanket, of tritium
containment and waste disposal. Finally, licensing problems are
discussed, suggesting a few supplementary points, insufficiently
covered in the present or in the forthcoming regulations, especially
those related to the transport and to the disposal of tritium wastes.

----

1. INTRODUCTION

The discussion of fusion break-even involves three aspects
related to tritium. First, a large fusion device, like an experimental power reactor (EPR) will require well developed
techniques for the handling of large amounts of tritium, of
the order of 1o MCi and perhaps more. Second, the working personnel in the power station and the general public will have
to be protected against tritium exposure above the permissible
level. This includes efficient containment, hazard analysis for
routine and accidental releases and waste disposal techniques.
Last not least, the very desirability of controlled fusion relies
much on our present belief that fusion power on a large scale
is potentially safer than other nuclear power systems. Arguments
in support to this belief will span from a thorough knowlegde
of the deleterious effects of tritium on the human body, to the

331

demonstration of safe containment techniques and to the difficulty, if not impossibility, of diverted uses of tritium.

In this lecture we shall put more emphasis on the environmental
aspects of tritium rather than on the technologies required for
the operation of a large fusion device. In section 2 the relevant
properties of tritium and its present sources in the atmosphere
are reviewed. Section 3 deals with health physics aspects: damage
to man, incorporation mechanisms, chronic and acute exposures.
In the next section the relation between emission rates and dose
to thepublic is outlined and rejects to the atmosphere and to
surface water is discussed, including two illustrative examples
of accidental releases. In section 5 short comments are given
on the state of the art of tritium technologies required for an
EPR and areas are indicated where, to the authors opinion, research and development is needed. Finally, in section 6 licensing
problems are discussed in view of the forthcoming regulations.

For the reader's concenience, some definitions used througout
this lecture are given below.

$$1 \text{ rad} = 10^{-2} \text{ j/kg} \qquad \text{absorbed dose}$$

$$1 \text{ Ci} = 3.7 \times 10^{10} \text{decays/s} \quad \text{activity}$$

$$1 \text{ rem} = 1 \text{ rad} \times \text{RBE}$$

biological dose equivalent
RBE = relative biological
effectiveness

$$1 \text{ TU} = 10^{-18} \frac{^{3}\text{H-Atoms}}{^{1}\text{H-Atoms}}$$

tritium concentration
TU = tritium unit
$$(1 \text{ TU} = 0.33 \times 10^{-8} \text{Ci/m}^3$$
$$\text{in water})$$

2. RELEVANT PROPERTIES OF TRITIUM AND ITS OCCURENCE IN NATURE

Tritium decays to ^{3}He by pure β^- radiation:

$$^{3}\text{H} \xrightarrow{\beta^-} {}^{3}\text{He}$$

The decay electron has a maximum energy of 18.5 keV and an average energy of 5.7 keV. Hence 1 Ci of tritium radiates about 1.82×10^{16} keV/d or 3×10^{-5} w. The half life of tritium is 12.26 years, i.e. the decay constant is $\lambda = 1.79 \times 10^{-9}$ 1/s. Hence 1 Ci is equivalent to 2.07×10^{19} tritium atoms or 0.103 mg of T_2. The range of the β^- particle in air is 0.5 mm average and 6 mm maximum, the stopping distance in unit-density material, like water, is of 6 μm. (For further data see ref. /1 /).

Tritium occurs in nature as a product of nuclear reactions induced by cosmic radiation (fast neutrons, protons, deuterons) in the upper atmosphere, e.g.

$$^{14}N + n \rightarrow \, ^{3}H + \, ^{12}C$$
$$^{14}N + p \rightarrow \, ^{3}H + \text{fragments}$$
$$^{2}H + \, ^{2}H \rightarrow \, ^{3}H + p$$

The world inventory prior to thermonuclear weapons testing (1954) was calculated to be about 9 MCi /1 /, but other estimates based on a production rate of 4 - 8 MCi/y give a steady inventory of 70 - 140 MCi / 2/.
Most of the tritium is in form of tritiated water (HTO) as a result of oxidation and exchange of T_2 with ordinary hydrogen. The inventory distributes roughly as follows:

90 % in the hydrosphere, HTO   ($\leq$ 10 TU)
10 % in the stratosphere, HTO   ($\sim 6 \times 10^5$ TU)
0.1 % in the troposphere, mostly as water vapor
        (HTO), but also small amounts of T_2 and
        methane

The natural inventory has considerably increased after the atmospheric tests of thermonuclear weapons, with an estimated release of about $\geq$ 10 MCi/Megatonne explosion / 2/.

The concentration of tritium in the atmosphere at ground level is about 10 TU from natural sources; by 1972 the atmospheric concentration was about 10 times higher as a result of weapons testing /3 /. Other sources quote 0.5 - 5 TU in rainwater prior to, and 500 TU immediately after the tests in 1954 /1 /.

Fission Reactors and Reprocessing plants will be an important source of tritium in the future. A recent estimate for the Federal Republic of Germany gives a long term forecast of about 1o MCi/y produced and o.1 MCi/y released to the atmosphere in that country / 4/.

3. HEALTH PHYSICS OF TRITIUM

3.1 Damage to Man

There are 2 damage mechanisms to be considered :
- energy deposition (radiation effect)
- transmutation   (genetic effect)

Because of the extremely small range of the β-particles in water (o.8μ average, 6 μ maximum) toxic effects by energy deposition are likely to occur within subcellular dimensions in the vicinity of the decaying particle. Transmutation damage is only important in genetic material and its probability is very small compared to the radiation effect.

Radiation effect

The energy released by 1 mCi of tritium is

$$3.7 \times 10^7 (1/s) \times 5.7 \text{ (keV)} \times 1.6 \times 10^{-16} \text{j/keV} = 3.37 \times 10^{-8} w$$

1 mCi/kg -body tissue will then give an absorbed dose of
$$D = 3.37 \times 10^{-8} (j/s/kg) \times 3.15 \times 10^7 (s/y) \times 10^2 \text{ rad/j/kg}$$
$$= 1.06 \times 10^2 \text{ rad/y}$$

In the case of chronic intake of tritium, when we may consider the body hydrogen uniformaly labeled, 1 mCi body burden to the "standard man" (7o kg weight, 8o% water concentration in soft tissue) results in D = 1.9 rad/y or 37 mrad/week. Assuming a water half-time in the body of 12 days, a body burden of 1 mCi will be maintained by chronic intake of o.o57 mCi/day tritiated water. 1 mCi is the maximum permissible burden for body tissues according to the recommendations of the International Commission

on Radiological Protection (ICRP). This limiting value is valid
for occupational exposure, with the premises of regular radiological and medical control. For the general public, a value
1o times less was recommended, but forthcoming regulations are
going to be more restrictive.

The relative biological effectiveness (RBE) for the β-decay
of tritium has been extensively discussed in the past, the actual
recommendation of the ICRP is to set the RBE equal to one and hence
numerically equate the biological dose to the physical dose (1 rem/
1 rad).

With 1 mCi corresponding to 2.07×10^{16} T-atoms and the number
of H-atoms in the soft tissue of the "standard man" of

$$7o \times 0.8 \times \frac{6 \times 10^{26}}{18} \times 2 = 3.73 \times 10^{27}$$

we find a dose equivalent rate of

$$0.34 \times 10^{-6} \text{ rem/y/TU}$$

in the body. Hence the tritium concentration of 1oo TU in the
northern hemisphere as observed by 1972 was equivalent to a burden
of the public of 3.4×10^{-5} rem/y, orders of magnitude lower than
the burden limit and even small compared to the dose rate of o.1rem/y
from natural radiation.

On these grounds, according to Osborne /3/ a maintained worldwide
inventory of tritium released from reactors (fission or fusion)
of 2oo MCi appears acceptable. This would result in about 1o - 5o TU
(allowing for disequilibrium in the oceans) and hence a radiation
dose rate of $3 - 15 \times 10^{-6}$ rem/y. We may note that a steady inventory
of 2oo MCi corresponds to a global release rate of

$$F = \frac{\ln 2}{12.3} \times 200 \approx 1o \text{ MCi/y.}$$

Genetic effects

According to our introductory remarks, the radiation effect
consists in local energy transfer leading to β radiolysis. When
tritium is incorporated into the cell nucleus, one decay leads

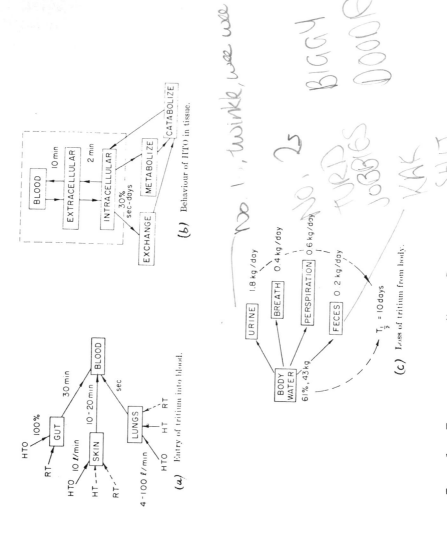

(a) Entry of tritium into blood.

(b) Behaviour of HTO in tissue.

(c) Loss of tritium from body.

Fig. 1   Tritium in the Human Body
(from R.V. Osborne, Ref. 3)

336

to an average deposition of 5.7 keV corresponding to a dose to the
nucleus (diameter = 8 μ) of o.3 rad. The most sensitive material
in the nucleus is DNA and the main deleterious effect is DNA-strand
breaks. According to Feinendegen / 5/, about 1 single strand break
is produced per decay, but this is very efficiently repaired
(nearly 1oo %). Double strand break is much less probable, about
o.o5 - o.1 per decay, but also less efficiently repaired (about
5o %)

The decay of tritium in most positions of the DNA molecule leads
with somewhat smaller probability to similar effects, about o.3
single breaks and less than o.o1 double breaks. Hence from
both transmutation and radiation effects a total of about 1.3
single strand breaks and less than o.11 double strand breaks
should be expected.

There has been concern about tritium decaying in very particular positions of DNA, the so-called mutagenic positions, which
lead not to a destruction of the molecule but to a permanent
alteration of the genetic substance. According to Feinendegen / 5/
only 2 % of the DNA hydrogen is located at this mutagenic positon
and furthermore an average mammalian cell nucleus contains only
2 % DNA. Hence only o.o4 % of the total hydrogen contained in
the cell nucleus is placed in this particular position and assuming equilibrium with tritiated water in the nucleus, only
o.o4 % of the tritium will be located in this position. The conclusion is that a mutagenic effect is orders of magnitude less
probable than strand breaks.

3.2   Incorporation Routes

There are three main incorporation routes (Fig. 1 (a) from ref. 3/

- ingestion in food or water (HTO, other tritiated compounds,
  (RT);this is 1oo % efficient and in about 3o minutes tritium
  is absorbed to the blood;

337

- inhalation through the lungs (HTO,HT,RT); inhalation of HTO is nearly 1oo % efficient because of the rapid exchange of water vapor molecules in the lungs. Absorption to the blood is almost immediate;

- diffusion through the skin (mainly HTO), at a rate corresponding to 1o l - air/min and absorption to the blood in about 1o - 2o minutes.

From the blood, HTO is distributed within minutes to extracellular, and from there to intracellular water spaces (Fig.1 (b)) About 3o % of molecularly bound tissue hydrogen may be labeled by tritium from the body water in a time ranging from seconds to days. Some tritium will be returned by catabolic processes to the body water.

Elimination of tritium occurs mainly in urine, but also with exhaled air, by transpiration and with feces (Fig. 1 (c)). The half life depends on the mass of body water and on the excretion rate and has an average value of 1o - 12 days; ICRP has adopted 12 d for the "standard man". / 3/.

3.3   Chronic Intake by Special Routes

From the above figures, together with the limiting body tissue burden for the general public of o.1 mCi we can deduce the maximum permissible concentrations of tritium in air and water.

Consider first chronic intake of tritium by water ingestion. Assuming all the body hydrogen (1o % of the body weight) labeled, we have

$$\text{MPC in water} = \frac{\text{o.1}}{7\text{o} \times \text{o.1} \times \frac{18}{2}} = 1.6 \times 10^{-3} \text{ mCi/kg}$$

This value recommended by ICRP, can be shown to be conservative by about a factor of 3 /3/, because only 4o % of man's water throughput is taken in as water and the highest concentration of water in the critical organ (soft tissue) is 8o %; hence the tritium concentration in soft tissue is about 32 % that of ingeste water.

Similarly, for inhalation (14 l/min) and skin diffusion (10l/min) of HTO in air we have , recalling that the turnover time is 12 days,

$$\text{MPC in air} = \frac{\ln 2}{12 \times 24 \times 60} \times \frac{0.1}{14 + 10} = 1.6 \times 10^{-7} \text{mCi/l}$$

To obtain this result, we have used the breath and skin diffusion rates given by Osborne / 3/; with a more conservative choice of parameters, ICRP recommends

$$\text{MPC in air} = 0.8 \times 10^{-7} \text{ mCi/l.}$$

Ingestion of tritium in food may either occur in the chemical form of water (e.g. in milk) - in which case the above discussion may be applied again - or in the form of organic compounds. For this latter case there is no experimental evidence that the toxic effects would be different from water ingestion.

Finally, consider inhalation of tritium gas. T_2 or HT is soluble to about 2 % in body fluids, but conversion to oxide is extremely slow and HTO accumulation is 10^4 times less than that resulting from HTO ingestion.

3.4 Acute Exposure

Osborne calculates the total dose resulting from the acute intake of tritium to be 78 mrem/mCi / 3/. This number relies on 10 days turnover time of HTO in the body. There is an additional problem, not well investigated yet, of bound tritium after an acute exposure. In spite of some (not very precise) experimental data there is no real experience on this question. Osborne's conclusion is that the additional "long term dose" from this is not more than 23 mrem/mCi.

4. RELEASE OF TRITIUM FROM NUCLEAR POWER STATIONS

4.1 Release Data from Fission Reactors in Europe

Before going into the discussion of how much tritium (and in which form) may be released in normal operation from nuclear (fission or fusion) power stations within the permissible radio-

TABLE 1

TRITIUM RELEASE DATA FROM NUCLEAR POWER STATIONS

IN THE E.C. (FROM REF. /6/)

(A) GASEOUS EFFLUENTS

Facility	Activity released (Ci/year)				
	1970	1971	1972	1973	1974
Germany					
MZFR	1 190	1 130	542	1 091	1 098
KRB	∼50	∼50	∼50	∼50	∼200
KWO	29.3	20.2	11.46	20.25	11.46
KKS	n. a.	n. a.	< 20	< 20	11.1
France					
Monts d'Arrée			83	696	1 756
Italy					
Trino					7.3
Netherlands					
Borssele	n. a.	n. a.	n. a.	n. a.	9
Great Britain					
Oldbury				30	12
Winfrith	135	155	232	300	283
Wylfa	n. a.		194		

(B) LIQUID EFFLUENTS (AVERAGE 1970 - 1974)

BWR : 36 mCi/GWh;
PWR : 330 mCi/GWh;
GCR : 31 mCi/GWh.

340

active burden limits to the public, it will be illustrative to review measured release rates from existing power stations. The following table contains a compilation of such data, measured in the years 1970 - 1974 in the countries of the European Community / 6/.

4.2 Behaviour in the Atmosphere of Released Tritium

Tritium or HT exchanges with water according to the reaction

$$H_2O + HT \rightleftharpoons H_2 + HTO$$

The equilibrium constant is about 6.5 at $20^{\circ}C$, hence HTO formation is strongly favoured at ambient temperature. However, this information does not tell us what the actual exchange rate with atmospheric water will be.

Some informative data on atmospheric exchange rates have been obtained recently by Friedrich et al. / 7/ by measurements in the vicinity of the tritium stack at Jülich, through which the rejects of a commercial neutron generator are passing. The composition of the rejects war 68 % T_2, 17 % DT, 14 % HT, 1 % CH_3T. First measurements were downwind and under recorded meteorological conditions, giving an exchange rate of about 1 % (gross average). The experiments were repeated in a climate chamber of 14 m^3 volume with definite humidity and temperature. The results are shown in fig. 2. The influence of temperature (fig.2a) and of tritium concentration (fig.2c) are negligible in the range of interest. Humidity also has very little influence, except for very dry climate (fig.2b). Time is the only important factor (fig. 2d). The result of this experiment together with other atmospheric measurements after thermonuclear test explosions / 8/, strongly indicates that ultimately all tritium gas will exchange to HTO, although this is a rather slow process. Hence HTO is relevant to the safety of the public, whereas tritium gas is of concern for the personnel handling tritium experiments or fusion power stations.

341

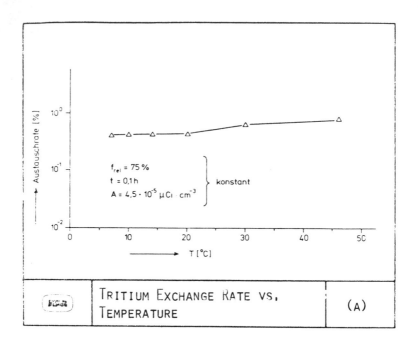

| | TRITIUM EXCHANGE RATE VS. TEMPERATURE | (A) |

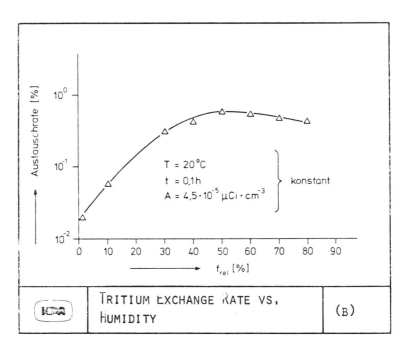

| | TRITIUM EXCHANGE RATE VS. HUMIDITY | (B) |

FIG. 2: TRITIUM EXCHANGE RATE IN AIR

(FROM W. FRIEDRICH ET AL., REF. /7/)

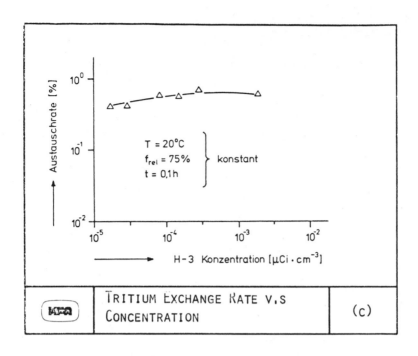

TRITIUM EXCHANGE RATE V,S CONCENTRATION   (c)

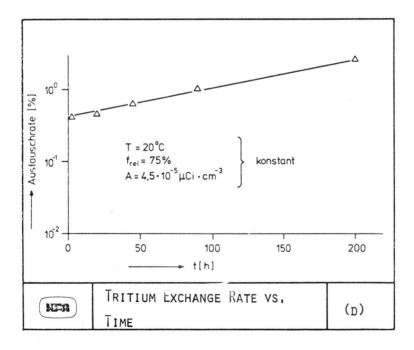

TRITIUM EXCHANGE RATE VS, TIME   (D)

FIG, 2: TRITIUM EXCHANGE RATE IN AIR
(FROM W, FRIEDRICH ET AL,, REF, /7/)

4.3 Emission from a Stack

Routine emission rates from a stack should be such that the resulting dose rate to the public, even under worst conditions, does not exceed the permissible limit. The dose rate related to the emission rate strongly depends on meteorological conditions and therefore individual evaluation is always necessary when referring to a particular site of the tritium emission source.

The general formula relating dose rates from submersion, ingestion and inhalation can be written as follows

$$D = g \cdot \bar{\chi}(x) \cdot Q$$

where   D is the dose rate (rem/y);

   g is the dose rate constant (rem x m^3/Ci x s) which has different numerical values for submersion, ingestion and inhalation;

   $\bar{\chi}$ is the average dispersion factor (s/m^3) depending on the distance x from the stack, and

   Q is the emission rate from the stack (Ci/y).

In the case of ingestion, the formula is only valid for x $\geq$ 2000 m, where the "sediment factors" for fallout and washout are proportio nal to the dispersion factor $\bar{\chi}$:

$$\bar{F} = v_F \cdot \bar{\chi}(x)$$
$$\bar{W} = \bar{v}_w \cdot \bar{\chi}(x) \qquad \text{for } x \geq 2000 \, m$$

Otherwise we have for the ingestion dose rate

$$D = K \cdot (\bar{F} + \bar{W}) \cdot Q$$

with the sediment factors $\bar{F}$ and $\bar{W}$ given in m^{-2} and K is again a constant (rem x m^2/Ci) (see e.g. / 4/).

The dispersion factor $\bar{\chi}(x)$ depends on the distance to the source, on the speed, stability and direction of the wind and on the heigh

344

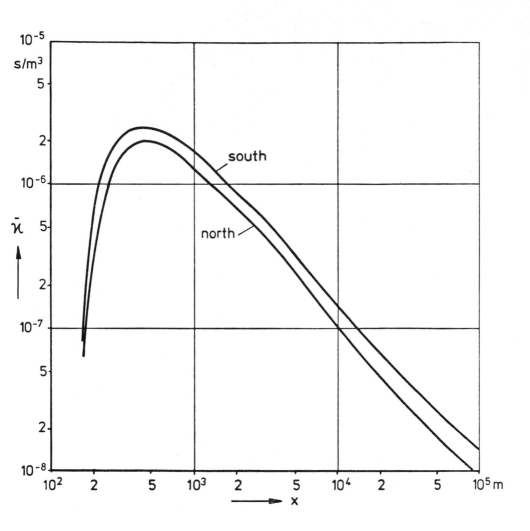

FIG. 3: AVERAGE DISPERSION FACTOR FOR THE FRG
FROM STATISTICAL EVALUATION OF METEOROLOGICAL
DATA (FROM BONKA ET AL., REF. /4/)

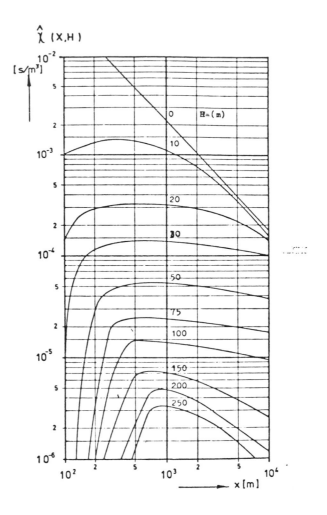

FIG. 4: SHORT TERM DISPERSION FACTORS FOR
UNFAVORABLE METEOROLOGICAL CONDITIONS:
CONSTANT SPEED, DIRECTION AND STABILITY
OF WIND DURING EMISSION (FROM REF. /9/)

of the stack. Without going into details, we note that the dependence on the height of the stack H is of the type

$$exp\left(-\frac{H^2}{2\,\sigma_z^2(x)}\right)$$

(where $\sigma_z(x)$ is the so-called vertical dispersion parameter). The values of $\bar{\chi}$ are usually averaged over statistical wind data. Fig. 3 shows the dispersion factor (broad average) for the Federal Republic of Germany, according to Bonka et al. / 4/. With a 100 m high stack, the maximum occurs at 2000 m distance, $\bar{\chi}_{max} \approx 10^{-6} s/m^3$. Local variations can be as high as 100 % (55 % at Jülich).

Very important for the evaluation of maximum permissible emission rates (as required in licensing procedures) is the dispersion factor for the worst situation, i.e. short-term dispersion under most unfavorable meteorological conditions with wind speed, direction and stability constant during the emission. Such dispersion factors have been computed (see e.g. ref. / 9/) for different stack heights, as shown in fig. 4. A 100 m stack would thus give a maximum dispersion factor of $\sim 10^{-5} s/m^3$.

The dose rate constants for submersion, inhalation and ingestion are given in the following table, as evaluated using ICRP methods and basic data (see e.g. ref. / 4/).

Table 2   Dose Rate Constants (rem x m^3/Ci x s)

Critical organ	submersion	inhalation	ingestion
skin	1.5 x 10^{-3}	–	–
body	–	0.06	0.09
tissue	–	0.07	0.10

To give an illustrative example, we calculate the long term average dose rate to the public by emission of 1000 Ci/y of tritium from a 100 m stack at Jülich. Carefully recorded meteorological data for this particular site give a maximum dispersion factor of

347

$\bar{\chi}^{max} = 4.25 \times 10^{-7}$ s/m^3. Hence the dose rate to the public via air is

$$\text{skin:} \quad 6 \times 10^{-7} \text{ rem/y} \quad \text{(submersion)}$$
$$\text{body:} \quad 6 \times 10^{-5} \text{ rem/y} \quad \text{(ingestion + inhalation)}$$

Taking worst conditions, $\bar{\chi}^{max} = 10^{-5}$ s/m^3 and the resulting body dose rate would be 1.4×10^{-3} rem/y. Anticipating forthcoming regulations (3o mrem/y permissible dose to the public), such an emiss would be permissible, even from a 3o m stack, which would add another orden of magnitude to the dose rate (see fig. 4).

4.4 Release to Surface Waters

The release of proper dilutions of tritiated water to rivers is current practice in most countries. However, with the growth of installed nuclear power there is a concern about the capacity for tritium in rivers. In analogy to dipersion in air, statistical hydrological data enter in the calculation of the dose rate to th public arising from tritium released to surface waters, and parti ular attention is due to the accumulation of tritium by sedimenta at particular locations. According to Bonka et al. / 4/, the capacity of large European rivers is a few 1o.ooo Ci/y and careful computations with expected liquid tritium effluents in the FRG of 2×10^4 Ci/y by the year 199o (2oo GW-el nuclear power installed) lead to 10^{-6} Ci/m^3 tritium concentration in German rivers.

New regulations presently under discussion will be very restrictive on release to surface waters, and the best philosophy would be not to release tritiated water at all. HTO-rejects from large fusion devices (e.g. as a result of catalytic oxidation of T_2 in filters prior to release through the stack) should be solidified for ultimate waste disposal.

Also Easterly and Jacobs /1o/ have shown that continuous release of tritium into the atmosphere leads to a global collective dose to man at least 5 times less (with most probable assumptions on the transfer mechanisms to man about 1o times less) than release to surface waters.

4.5 Experience from Accidental Releases

The forecast of dose to the public from large single releases as
a result of a failure is rather speculative at the moment. For
example, if we assume that such a release may take place in fault
conditions from the base of the stack, then, looking at the short-
term dispersion factors of fig. 4, it will be extremely difficult to
fullfill licensing conditions. Rather than speculate about the probability of such a fault condition in a fusion device, we shall
summarize the findings from two accidental releases reported:
one with potential hazard to occupational individuals, another
with potential hazard to the general public.

The first case, a small accidental release in a research reactor
at Jülich reported by Jahn /11/, illustrates how a well-organised
monitoring and emergency system can prevent damage to the personnel.
During routine work at the reactor, about 78 kg of heavy water
containing 265 Ci of tritium,was quite unexpectedly released to
the reactor hall. The tritium concentration in the hall was raised
immediately and reached a maximum of 1.8×10^{-4} Ci/m^3 90 minutes
later. In the venting system (stack) an increasing tritium emission
was registered, peaking at 330 mCi/h, 5 hours after the incident.
23 mCi were emitted in total. The working personnel was immediately evacuated from the reactor hall. The decontamination crew, with
heavy masks, collected the D_2O in 3 days work. 20 persons underwent
tritium incorporation examination, the activity found in urine
ranged from 1 % to 18 % of the maximum permissible value. Hence the
accident can be classified as minor to insignificant, affecting only
the availability of the reactor.

The second case is a large accidental release of 289 kCi of T_2
through a 36 m stack at Livermore in August 1970, as reported
by Myers, Tinney and Gudikson /12/. The response of the emergency
crew was sampling of air, water, vegetation, milk and urine at
different locations of the environment, selected after checking
the meteorological data at the time the release occured. The
findings are summarized in table 3 .

Table 3: Tritium Activities of Samples Before and After
the Livermore Accidental Release of 289 kCi.
Values are given in nCi/l. (From Myers et al.,
ref. /12/).

Sample	Max-accident levels	Pre-accident levels	Off-site MPC	Detection limit
Water	14	<5	3000 b	5
Air (Water vapor)	$1.4X10^{-3}$ a	$<4X10^{-5}$ a	$2X10^{-1}$ a	$4X10^{-5}$ a
Milk	8.0	<5	3000 b	5
Vegetation (water)	1000	10(av) 250(max)	3000 b	5
Urine	<5	<5	3000 b	5

[a] The minimum sensitivity would depend on the relative humidity, which was 50 percent at the time of the release.

[b] Assuming that tritium was equilibrated in the environment. This would be true only for a chronic release situation, but the values are useful as reference points.

The maximum detected tritium concentrations were well below the
off-site maximum permissible concentration values. A model computation of the maximum possible dose to an individual, with the
prevailing meteorological data and assuming worst conditions
(standing on the centerline of the plume at the point of maximum
ground-level concentration during the entire passage of the tritium
cloud) resulted in about 3 mrem. The maximum dose effectively found
by urine analysis was o.025 mrem. Similarly, theoretical maximum
dose value from ingestion of an infant via milk was about 7o mrem.
However, no tritium activity above background was found in any
of the milk samples collected. This is a clear indication that
the hazard analysis had very safe margins and that the probability
of "worst conditions" was very small in this particular case.

5. TRITIUM-EQUIPMENT FOR LARGE FUSION DEVICES (EPR)

We may assume here that the purpose of an EPR is to develop and
to test the basic technologies required for the fusion reactor.
Tritium recovery and handling is certainly among the basic technologies and in particular we need to consider three major areas:

i)   tritium handling components of the reactor:
      vacuum pumps
      fuelling system
      equipment for separation from gas
      equipment for isotope separation
      equipment for recovery from blanket
      storage (supply and waste)

All this equipment will be maintained under secondary containment,
perhaps some under inert atmosphere.

ii)   equipment required for the reactor building (venting system,
      including tritium filters or trapping), and for monitoring
      professional personnel

iii) Monitors for the environment (air, water, plants, meteorolog-
      ical data)

The state of the art and proposed solutions have been reviewed
on several accasions (see e.g./13/), hence we shall limit ourselves
to short comments highlighting the state of the art in Europe.

Vacuum pumps

No major developments are required, the state of the art is close
to adequate as long a some basic rules are not forgotten:

          - no moving shaft seales
          - no organic fluids
          - no elastomer seals (see e.g./14/)

The preferred type of pump for high vacuum in most studies is
cryopumps (using 4.2 K surface to freeze tritium) or cryosorption
pumps (with surface active material permitting operation at
15 - 25 K for the same pressures). Hg diffusion pumps allowing
for high pumping speed are also suitable up to 10^{-6} torr, Hg
is inert to tritium.

Fuelling system

This is a major uncertainty, because we still do not know how
a reactor will be refuelled. It seems premature to discuss e.g.
the technology required for pellet refuelling because no sound
proposal for a pellet acceleration mechanism (to the $10^6 - 10^7$ m/s
required) exists at the moment. In the US discussions on EPR /15/
there was concern about the possible complexity of the pellet production system.

Separation from the exhaust gas

Again no major developments are required. There is agreement that
U-bed and/or Pd-membranes can be used to purify tritium and that
the techniques are state of the art.

Isotope separation

Cryogenic distillation is the most suitable method according to
most authors. While the US has many years of experience with this
method, the situation in Europe is different. Hence, although the
method is available in principle, EPR development in Europe should
include acquisition of the expertise in this field.

Recovery from the blanket

A major development is required in this field. More than a dozen
methods have been proposed so far, and the viability of these in
a reactor is still open. For sure, the tritium recovery method
will strongly interact with the design of the blanket (choice of
breeding and structural material), and the equipment is likely
to be an important cost item of the EPR.

Storage for supply and for waste

Tritium can be stored either as a gas or dissolved in an appropriate solid (currently Uranium), the latter being considered
safer against accidental release. In industrial applications
tritium waste is usually stored as tritiated concrete in sealed
containers. Improved techniques with extremely small bulk leach

rates have been reported recently /16/. However, these storage techniques have to be demonstrated on a quasi-industrial scale before the waste will be accepted for disposal.

Venting system for the reactor building

The state of the art seems sufficient for EPR purposes. As a ground rule, the reactor building will be kept under slightly negative pressure to avoid outleaks. The laboratory decontamination system described in Ref. /17/ is capable of reducing tritium concentrations to the ppb level (and even lower) in a few minutes, by use of a catalyst to convert H to water (reduction factor of about 10^5 has been reported). Again, these techniques need demonstration on a technical scale prior to licensing.

Monitoring the environment

Most nuclear laboratories and nuclear power stations are currently monitoring tritium in the environment. In order to assess the statistical average dispersion factor from the stack and dispersion by surface waters, it is also necessary to record meteorological and hydrological data.

6. FORTHCOMING REGULATIONS AND LICENSING

In the Federal Republic of Germany new rules are under way, going to be more restrictive than the previous ones based on the values quoted in section 3.3. For the first time there is an explicit statement, that "plasma devices producing more than 10^{12} neutron/second" are subject to a license. One main difference between the old rules and the new ones is that the old rules fixed MPC's in water and in air, while now there is a limit on the yearly incorporation in man.

There is a recent recommendation by the Commission of the European Communities /18/, but national rules can be different. Such is the case of the German Federal Republic with significantly lower

353

incorporation limits for the general public than foreseen by
the EC recommendation. (Table 4).

Table 4: Recent Limiting Values for Chronic Tritium
          Incorporation (Ci/y)

	Occupational	General Public	
	EC and FRG	EC	FRG
Inhalation (air)	1.2×10^{-2}	1.2×10^{-3}	0.7×10^{-4}
ingestion (water, food)	2.6×10^{-2}	2.6×10^{-3}	1.6×10^{-4}

Permissible emission rates have to be deduced from these values.
In the case of the more restictive German regulation, the underlying assumptions for such calculations are still in discussion:
there is always a "worst distance" downwind from the stack, where
the plume touches the soil - should we consider incorporation at
this particular place, even if the possibility of coincidence with
a populated area is virtually zero? If yes, then there is no distinction between a village and the desert and it will be extremely
hard to fullfill the licensing conditions for a large fusion device

There may be a further potentially controversial point. There
is a fundamental legal difference between nuclear fuels and other
radioactive materials: nuclear fuels, according to vigent definitions, are Uranium, Plutonium and Thorium, and these are subject
to very strict legal control arising from national legislations
and international agreements. In a Tritium Burner and in all subsequent fusion devices tritium will undoubtedly be a "nuclear fuel
although, with respect to potential diverted use, very different
from U, Pu, and Th. However, it will be necessary to formulate pre
cise arguments underlying this latter statement; discussions to cla
this point should be initiated at an appropriate time, not too far
in the future.

354

Finally, also the rules for transport and waste disposal of tritium need further clarification. According to the new regulations in the FRG, a dilution of tritium to the MPC, e.g. in water, to get rid of tritium-wastes will not be allowed in the future. On the other hand, none of the present underground waste disposal sites in the FRG is prepared to accept significant amounts of tritiated waste. Current practice of the OECD countries is to sink tritium in coastal waters, but this proves increasingly difficult after the protest of the coastal countries (Portugal, Republic of Ireland). Hence there is a need for clear-cut regulations on tritium-transport and waste disposal, based on well founded recommendations.

7. SUMMARY AND CONCLUSIONS

The health physics of tritium is reasonably well explored to provide safe limits for the chronic incorporation of tritium by man. In particular, mutagenic effects by tritium decay in DNA molecules - which has been a controversial subject for some time - have been shown to be much less probable than DNA strand breakage.

Routine release rates from a power station can be related to the permissible dose rate from chronic exposure, by dispersion calculations using local meteorological data. A yearly emission of about 1ooo Ci from a large fusion device should be compatible with safety standards and would also be comparable to measured emission rates from fission reactors. The calculation is not as straightforward for unintentional single releases, because of the uncertainly in the mode of release. This could prove to be a major point in the hazard analysis, as required to obtain a license for a large fusion device.

Future regulations are expected to preclude the release of reject tritium into surface waters. Hence, apart from the tritium emitted through the stack, rejects like T_2O from the filters of the stack will have to be solidified and enclosed in containers for waste disposal.

In addition to the technologies required for the internal handling and recovery of tritium in a fusion power plant, special attention should be given to the development of reactor containment and of w disposal techniques. All fusion devices using tritium will necessa rily have an extended monitoring system, both internal and environ mental, and also radiological protection, decontamination and medical services.

In order to have clearly defined boundary conditions for the safe ty analysis and for the disposal of tritium waste, the regulations on radiological protection should be extended to the needs of fusion in due time.

In Europe there is a need for more experience in tritium technologies. For the development and construction of a European EPR, a research and development program for these technologies is required.

ACKNOWLEDGEMENTS

Discussions with Mr. W. Anger on tritum waste disposal, Mr. L.E. Feinendegen on radiation effects in the human body, and with Messr H. Jacobs and P.F. Sauermann on licensing problems are gratefully acknowledged.

REFERENCES

/1/   E.A. Evans, Tritium and its compounds,
      Butterworths, London, 1966

/2/   D.G. Jacobs, Sources of Tritium and its Behaviour upon
      Release to the Environment. TID-24635 (1968)

/3/   R.V. Osborne, Permissible Levels of Tritium in Man and
      the Environment, Rad. Res. 5o (1977), p. 197

/4/   H. Bonka et al., Zukünftige radioaktive Umweltbelastung
      in der Bundesrepublik Deutschland durch Radionuklide aus
      kerntechnischen Anlagen im Normalbetrieb. JÜL-122o (1975)

356

SUPERCONDUCTING MAGNETS

A. SOME FUNDAMENTALS AND THEIR STATE OF THE ART

W. HEINZ

Kernforschungszentrum und Universität Karlsruhe
Institut für Experimentelle Kernphysik
7500 Karlsruhe, Postfach 3640
Federal Republic of Germany

Abstract:

An introduction to the basis of superconductivity is given.
Materials suitable for use in superconducting magnets and
their essential properties are presented followed by a discussion on the status of conductor development. A survey
of stabilization methods, behaviour of the superconductor
under normal operating conditions and special requirements
of superconductors for fusions magnets are given.

Magnet technology and its status is briefly discussed.
Existing magnet systems are introduced. Finally some of the
most important problems encountered in developing big
toroidal magnet systems are mentioned.

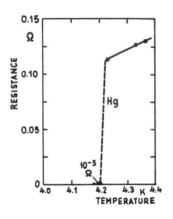

Fig. 1: Superconductivity of mercury as discovered by K. Onnes (after[1])

1	2	3	4	5	6	7	8	9	10	11	12	13	14	15	16	17	18
H																	He
Li (0,08)	Be 0,03											B	C	N	O	F	Ne
Na (0,03)	Mg (0,05)											Al 1,19	Si 6,7	P 4,6-6,1	S	Cl	Ar
K (0,08)	Ca (0,3)	Sc (0,01)	Ti 0,39	V 5,3	Cr	Mn	Fe	Co	Ni	Cu (0,01)	Zn 0,9	Ga 1,09	Ge 5,4	As 0,5	Se 6,9	Br	Kr
Rb (0,01)	Sr (0,3)	Y 1,2	Zr 0,55	Nb 9,2	Mo 0,92	Tc 7,8	Ru 0,5	Rh (0,09)	Pd (0,01)	Ag (0,01)	Cd 0,55	In 3,4	Sn 37;5,3	Sb 3,6	Te 4,5	J	Xe
Cs	Ba	La 4,8;5,9	Hf (0,08)	Ta 4,4	W 0,01	Re 1,7	Os 0,65	Ir 0,14	Pt (0,01)	Au (0,01)	Hg 4,15 3,95	Tl 2,39 1,45	Pb 7,2	Bi 3,9 7,2;8,5	Po	At	Rn
Fr	Ra	Ac															

Ce 1,7	Pr	Nd	Pm	Sm	Eu	Gd	Tb	Dy	Ho	Er	Tm	Yb	Lu 0,1-0,7
Th 1,37	Pa 1,3	U 0,2	Np	Pu	Am	Cm	Bk	Cf	Es	Fm	Md	No	Lw

Fig. 2: Distribution of superconductors in the periodical system. The figures give the transition temperature, figures in brackets the lowest temperature without finding superconductivity. Dark coloured elements are getting superconducting only under high pressure (after[1]).

360

Introduction

The energy demand of industrial societies will grow in the future
despite all energy saving efforts. Primary energy resources are
limited and most of them present serious pollution problems.
Nuclear fusion may offer an attractive possibility to overcome
some of the constraints due to fuels used today. The objective
of the European fusion programme is therefore "... the construction, if possible, of a fusion reactor producing useful energy
at a competitive cost". As far as a reactor based on magnetic
plasma confinement is considered superconducting magnets for
producing the confining fields are vital.

1. FUNDAMENTALS OF SUPERCONDUCTIVITY

1.1 What is superconductivity?

1.1.1 General characterisation

Superconductivity was first observed by Professor Heike Kamerlingh
Onnes of the University of Leiden in 1911. It was observed that
the resistivity of mercury disappeared sharply at 4.15 K and
remained zero below this *transition temperature* (fig. 1). The
existence of a new state of mercury characterized by the absence
of electrical resistivity came to be recognized, and to be known
as the *superconducting state*. Many pure metals become superconducting. Below a certain temperature, the transition or critical
temperature T_c, they show *perfect conductivity*. Their transition
temperatures range from 10 mK (the lower limit of detection) to
9.15 K (niobium). About 40 elements are known to become superconducting some of these in high pressure phases only (fig. 2).

Until 1933 it was assumed that infinite conductivity was the
essential property of a superconductor, but then Meissner and
Ochsenfeld showed that the superconducting state possessed the
additional entirely independent property of excluding magnetic
flux completely, except in a thin surface layer. A metal in the

superconducting state never allows a magnetic flux density to exist in its interior. This effect is to be known as the *Meissner-(Ochsenfeld-)effect*. A sample in which there is no net flux density when a magnetic field is applied is said to exhibit *perfect diamagnetism*. Perfect conductors showing perfect diamagnetism are *superconductors*.

A superconductor applied to a magnetic field will develop *screening currents* which flow within a thin surface layer. Due to Lenz's law the induced currents circulate in such a manner as to create a magnetic flux density which everywhere inside the specimen is exactly equal and opposite to the flux density of the applied magnetic field thus creating zero net field. These screening currents will not die away but persist because of the perfect conductivity of a superconductor. Now a resistanceless circuit operating in a closed loop, e.g. a superconducting ring or a superconducting magnet together with a superconducting switch, will never change the enclosed total magnetic flux as long as the circuit remains resistanceless. A superconducting magnet operating in this way is said to be in the *persistent mode*.

1.1.2 The superconducting state

The appearence of the superconducting state is the consequence of a new interacting mechanism. What is observed are differences between the specific heat values in the superconducting and normal states. The specific heat of a metal, however, is made up of two contributions, a lattice and an electronic one. The properties of the lattice do not change when a metal becomes superconducting, hence the difference between the specific heat values mentioned can only arise from a change in the electronic system.

Since the specific heat is greater in the superconducting state than in the normal one the entropy of the conduction electrons decreases more rapidly with temperature than if they were in the normal state. That means a new form of electron order must begin

to set in below transition temperature. This ordering is now
well understood: Two electrons interact via a polarization of
the lattice and form a so called *Cooper pair*. A Cooper pair
consists of two electrons having equal but opposite momentum
and spin:

$$\text{Cooper pair:} \quad \{\vec{p}_\uparrow, \quad -\vec{p}_\downarrow\}$$

These pairs are by no means independent, in the contrary strong
correlations exist: All Cooper pairs occupy the same quantum
state with the same energy. This state is often referred to as
a *condensed state*. The electrons are bound together to
form a state of lower energy. The same thing happens to the
atoms of a gas when they condense to form a liquid. This macroscopic occupation of one single quantum state is quite a new
phenomenon. Apparently it is responsible for all the peculiar
properties of a superconductor.

Now let us apply an electric field. Cooper pairs are accelerated.
They obtain an additional momentum which has to be exactly equal
for all pairs, otherwise one of these pairs would occupy another
quantum state which was excluded. This could only happen if the
energy available would exceed the binding energy of the pair
correlation. In this case the cooper pair will break open and be
destroyed. If this energy is not available Cooper pairs cannot
interact with the lattice. That, however, means zero resistance
charge transport. But this cannot last for ever. As we already
know if the momentum gain by the electrical field exceeds the
binding energy of the Cooper pairs they will break. The system
of Cooper pairs will be destroyed. The conductor returns to its
normal conducting state. This existence of a critical momentum is equivalent to a *critical current*.

1.1.3 The critical parameters and its correlations

The existence of a superconducting state is bound to the existence
of several critical parameters: temperature, current and field.

Neither of these must exceed a certain critical value to ensure
the maintenance of the superconducting state:

- the *critical temperature* T_c below which a superconductor becomes superconducting. It ranges from several mK to about
  23 K for Nb_3Ge.

- the *critical density* J_c which a superconductor is able to
  carry. That includes shielding as well as transport currents.
  Typical values are $J_c \approx$ several 10^6 A/cm^2.

- The *lower and upper critical fields* H_{c1} and H_{c2} above which the
  distribution of the superconducting state is changed or fully
  destroyed respectively.

The critical parameters are not independent of one another. A
qualitative relation between them is shown in fig. 3. The domain
of practical interest is between the origin and the surface. The
hope is to increase this domain. Domains for superconductors of
technical interest are shown in fig. 6. The observed temperature
dependence of the critical parameters H_{c2} and J_c can easily be
described by a quadratic law:

$$H_{c2} = H_{c2}(0) \left| 1 - (\frac{T}{T_c})^2 \right|$$

where $H_{c2}(0)$ is the critical field at zero K and T_c the temperature
at which the critical field becomes zero. An equivalent relation
holds for J_c. Each superconductor has different values of $H_{c2}(0)$
and T_c, and can be characterized by these values. In table 1 superconductors of technical interest are listed.

The current flowing in a superconductor produces a magnetic field
strength. If the current is sufficiently large the current density
at the surface will reach its critical value J_c (depending on the
temperature). The associated magnetic field strength at the surface is the critical field H_{c2}. The superconductor will return
to the normal conducting state when at any point on the surface
the total magnetic field owing to transport current or the

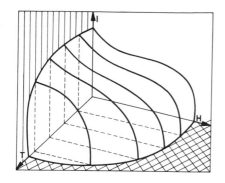

Fig. 3: Critical surface limiting
the stability area of superconductivity; J-H-T limiting
surface.

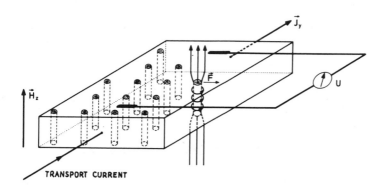

Fig. 4: Mixed state or Shubnikov phase of a superconductor
exposed to an external field and carrying a transport current.
A Lorentz force acts on each of the flux tubes initiating
flux motion and therewith inducing a voltage.

	T_c (K)	H_{c2} (T)		J_c (A/cm^2)
		$T = 0$	$T = 4.2$ K	
Pb	7.2	0.08	0.055	
Nb	9.3	0.19	0.15	
$Nb_{75}Zr_{25}$	10.9	13.7	10.0	$1 \cdot 10^5$ (at 5 T)
$Nb_{50}Ti_{50}$	9.5	14.5	12.0	$2\text{-}3 \cdot 10^5$ (at 5 T)
Nb_3Sn	18.1	24.5	23.0	$\begin{cases} 10 \cdot 10^5 \text{ (at 5 T)} \\ 4 \cdot 10^5 \text{ (at 10T)} \\ 0.5 \cdot 10^5 \text{ (at 8 T)} \end{cases}$
V_3Ga	14.1	20.6	19.8	$\begin{cases} 2 \cdot 10^5 \text{ (at 10T)} \\ 1.1 \cdot 10^5 \text{ (at 18T)} \end{cases}$
$Nb_3Ge^{2)}$	22.9		38	$1 \cdot 10^5$ (at 5 T)

Table 1: Some superconductors of technical interest.
Pb is of type I, all the others are of type II.

applied magnetic field exceeds the critical field. The stronger
the applied field and the higher the operation temperature the
smaller is the current carrying capacity of a superconductor.

1.2 Superconducting materials for magnet technology

1.2.1 Types of superconductors

None of the 40 metals known to become superconducting show critical
parameters which make them candidates for use in superconducting
magnets. Only a few have transition temperatures above the boiling
temperature of liquid helium (4.2 K): lanthanum (5.9 K), lead
(7.2 K), niobium (9.3 K), tantalum (4.5 K), technetium (8.2 K),
tin (5.3 K), and vanadium (5.3 K). Of these niobium has the
highest critical field but nevertheless does not exceed 0.15 T
at 4.2 K.

The discovery of superconductors which can withstand a considerable magnetic field is now 15 years old. Since that time superconducting magnet technology has made rapid progress. All of these superconductors are compounds or alloys, e.g. NbTi, Nb_3Sn, NbN, V_3Ga, V_3Si. They belong to a special type of superconductor, they are *type II superconductors*. Pure metals, except niobium and vanadium, which are becoming superconducting are of type I.

Type I superconductors completely exclude the magnetic flux until the critical field is reached. The superconductor is said to be in its *Meissner state*. It undergoes a magnetic transition from the Meissner state (B = 0) to the normal state ($B = \mu_0 H$) at a critical field $H_c(T)$.

1.2.2 Properties of type II superconductors

Type II superconductors are in the Meissner state only up to a lower critical field H_{c1}. Above this value their behaviour differs from that of a type I superconductor. The magnetic flux begins to rise gradually but remains less than the external flux. The normal state is not restored until a considerably higher field H_{c2} is reached. The entry of flux is in form of separate tubes of flux, each tube carrying a minimum amount of flux, the unit quantum of magnetic flux ($\emptyset_0 \simeq \frac{h}{2e} = 2 \cdot 10^{-15}$ Tm^2). This is a consequence of the surface energy becoming negative so that the interfaces become as extensive as possible.

Each flux tube is screened from the surrounding superconducting region by a circulating screening current. The tubes lie parallel to the externally applied magnetic field, their number being proportional to this field. They repel each other thus forming a regular pattern in an ideal crystal. The superconductor is said to be in the *mixed or Shubnikov state*. The pattern is known as the *Abrikosov lattice*.

The material remains superconducting as long as any continuous path of superconducting material exists. Whereas superconductivity

367

in type I is "quenched" on exposure to relatively low magnetic fields H_c. Type II superconductors possess two critical fields. Although the lower field H_{c1} is indeed low, the upper critical field H_{c2} can be very high thus fulfilling one of the main requirements of a material for use in superconducting magnets to withstand high fields.

1.2.3 Hard superconductors

The further requirement is that the material shall also sustain a large transport current. However in an ideal crystal the flux pattern starts to move when a transport current is applied because the flux lines experience a Lorentz force. A moving magnetic field pattern gives rise to an electric field and electric resistance which in turn is a source of energy dissipation (fig. 4) The superconductor has lost its unique property to carry a lossless current. It is in its *resistive state*. Ideal type II superconductors cannot carry technical currents.

Ideal refers to the specimen which has to be ideally homogeneous without any imperfections. Real materials have structural defects such as dislocations, precipitates of second phases, grain boundaries, martensitic transformations and voids. These interact with the magnetic flux. They present energetically favoured positions for the flux tubes and are able to *pin* them and prevent or restrict their movement. The forces acting on the flux tubes in this way are known as *pinning forces*.

Defects can be introduced and controlled by standard metallurgical processes such as cold-working, heat-treatment, ageing or sintering. The "ideal" material is hardened during this procedure. Thus technical superconductors are *hard superconductors*. The technical skill of the producer is directed to optimize kind, size, and separation of the defects to achieve a high critical current density besides maintaining other mechanical, electrical or thermal properties required.

368

In hard superconductors flux entry is delayed and flux escape is
partially prevented which results in hysteretic magnetization
curves. Whilst magnetization curves of type I or ideal type II
superconductors are reversible real type II superconductors
show irreversible magnetization curves. Charging or discharging
a magnet is thus unavoidably connected with hysteretic losses.

2. THE SUPERCONDUCTOR

2.1 Status of conductor development

2.1.1 Materials and conductors used presently

Notwithstanding the variety of superconductors known only very
few are used technically. Two types of superconductors are in use
for superconducting magnets: Alloys of niobium and titanium and
intermetallic compounds of niobium and vanadium with a β-tungsten
or A-15 crystal structure such as Nb_3Sn or V_3Ga.

NbTi-based superconductors are presently used almost exclusively
in magnet constructions. Titanium contents of 40 to 50 % by
weigth are employed. The superconductor is embedded in a matrix
made of normal conducting and ductile material such as copper or
copper alloys; in some cases aluminium is used as matrix material.
In most cases the superconducting wire consists of a considerable
number (up to many thousands) of thin superconducting filaments
(fig.5a). These multifilamentary conductors are fabricated by
cold-working of NbTi blocks to about centimeter size, embedding
these NbTi bars in copper jackets, and subsequent cold-working
(e.g. extrusion) together with suitable heat treatments until
the final dimensions of the wire or filaments have been obtained.
Cold-working introduces defect structures, heat treatment improves the bond between the copper matrix and NbTi filaments and
helps to optimize size, distribution, and kind of defects. These
will later act as *pinning centres*. In this way J_c-values of about
$2 \cdot 10^5$ A/cm^2 at 5 T are obtained. Critical fields at 4.2 K are

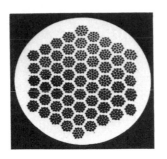 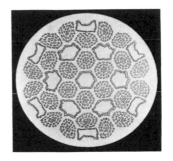

Fig. 5 a   Fig. 5 b

Cross section of multifilamentary conductors (micrographs):

a) NbTi multifilament wire made by Vacuumschmelze, Hanau.
Diameter 0.6 mm, 1159 filaments each 10 µm thick.
Current at 4.2 K and 5 T: 150 A[13].

b) Nb$_3$Sn multifilament wire made by Harwell, Didcot.
Diameter 0.43 mm, 888 (= 24x37) filaments, current at
4.2 K is 200 A at 5 T and 70 A at 10 T. Stabilizing
copper strands 14 %.

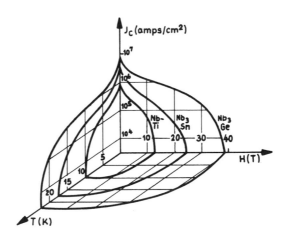

Fig. 6: J-H-T surface of NbTi, Nb$_3$Sn, and Nb$_3$Ge
respectively[2].

about 12 T. The conductor is ductile and can be produced in great lengths. These single strands of wire are then processed by cabling or braiding to the final conductor.

Among intermetallic compounds superconductors with highest transition temperatures were found. Among these the compounds with A-15 structure, a rather complex structure, seem to be especially favourable for high transition temperatures. This may be due to the long chains in which the transition metal is arranged, but is not yet fully understood. Nevertheless two superconductors technically used at present belong to this class: Nb_3Sn and V_3Ga. Both are extremely brittle. Strains of less than 0.2% may already cause current degradation.

Their critical temperatures and fields are above those of NbTi: 18.1 K and 14.1 K respectively 24.5 T and 20.6 T (see table 1). The critical current densities are high and exceed the NbTi-values especially at high fields. Below about 12 to 15 T Nb_3Sn and above V_3Ga has the higher critical current density.

There are two different procedures to process multifilamentary Nb_3Sn (V_3Ga) wires: the bronze route and surface diffusion route. The first uses niobium rods embedded in tin-bronze. The composite is worked down to its final size and then thermally treated (typicly 24 h at 750° C). Solid diffusion and final reaction to form a Nb_3Sn layer between the bronze and the rods takes place. In the second process, the composite is worked down to final size and then given a surface coating of tin, to be followed by a similar heat treatment. Most of the manufactures make use of the bronze route.

Nb_3Sn and V_3Ga multifilamentary wires are now commercially available (fig. 5 b). Because of the complexity of the processing Nb_3Sn is a factor of about 5 more expensive than NbTi conductors of equal length and currents that is per Am. V_3Ga is even more expensive because of the higher cost for the raw materials. Thus high field conductors for fusion applications will concentrate on Nb_3Sn.

2.1.2 Future prospects for advanced superconductors

The dream of the experts is to raise the transition temperature considerably to make the application of superconductors
economically more attractive. A considerable improvement which
seems within the range of technical feasibility of the near
future would be a superconductor operating at liquid hydrogen
temperature. It would enable savings in refrigeration cost and
offer the possibility to use hydrogen as a cooling fluid which
is cheap and of unlimited availability.

An even less ambitious increase in transition temperature would
be advantageous for technical application because of the wider
safety margin and freedom in choosing the operation temperature.

Three types of superconductors are in the focus of present
interest:

- A15-composites, especially Nb_3Ge or ternary composites of it.
  Thin films of Nb_3Ge have been prepared by sputtering techniques
  which remain superconducting up to approximately 23 K. These
  very high T_c's together with an extremely attractive combination
  of high field and high current density properties have made
  Nb_3Ge a very important potential candidate for a variety of
  applications. Of these, the most important one is large volume
  high field magnets for plasma containment in fusion reactors.
  A possible commercial process may be chemical vapor deposition
  known to be capable of preparing large quantities of material
  in a controlled manner.

  J-H-T surfaces for NbTi, Nb_3Sn, and Nb_3Ge are compared in
  fig. 6. The comparison shows the considerable improvement which
  may be achieved when Nb_3Ge will be prepared in sufficient
  quantities and at reasonable cost: The J_c-H-values of NbTi at
  4.2 K and of Nb_3Ge at 20 K are comparable.

- Ternary molybdenum chalcogenides. In some of these phases
  superconductivity has been found, e.g. Sn-, SnAl-, Pb-

molybdenum sulfides. These substances are high field
superconductors with critical fields (H_{c2}) in the range of
30 - 60 T. Highest T_c (14.4 K) was measured at Pb compounds
with H_{c2} (4.2 K) $\approx$ 50 T which extrapolates to $H_{c2}(0)$ = 60 T.

- Refractory metals, e.g. carbides and nitrides of
  Nb, Ta, Mo or Ti. They have a much simpler crystal structure
  (NaCl lattice) than the A15's and show very promising superconducting parameters: Thin films were measured with current
  densities of about 10^7 A/cm^2 and T_c-values up to 18 K ($NbC_{0.3}N_{0.7}$).

2.2 Stabilization techniques of superconductors

A hard superconductor carrying a transport current is intrinsically unstable, it is not in its thermodynamic equilibrium. By
the introduction of a defect structure acting as pinning centres
a flux density gradient is maintained. In case of local perturbations such as current or temperature fluctuations, mechanical movements of single wires, or an ac field component, flux
lines will move leading to so-called *flux jumps*. The changing
flux generates an electric field, energy is dissipated and the
temperature will rise locally. This leads to a reduction of
the critical current and, hence, to a further movement of flux
lines. New flux jumps occur raising the temperature again. In
this way a rapidly extending perturbation can occur driving
finally the superconductor normal, the magnet will *quench*.

Stabilization means to find possibilities to prevent flux jumps
or their consequences so that a perturbation will die out. The
crudest but very effective way of doing so is to provide a
second parallel electric circuit made of a low resistive normal
conducting material which can take over the full current in case
of a quench. The dimensions of this bypass conductor are chosen
such in spite of Joule's heating not to exceed the critical
temperature of the superconductor when carrying the full current.
Thus the overheated superconductor may return to its operation

temperature below T_c and in return take over the current again when the perturbation has disappeared. This kind of stabilization is called *crystatic* or *full stabilization*. The ratio of copper to superconductor is as large as 10 or 20. Overall current density is reduced drastically, because of the extra copper the conductor will get expensive.

In big magnet systems cryostatic stabilization is considered to be the only way to prevent any quench. On the other hand the conductor used in big magnets in any case has to be reinforced to withstand the magnetomechanic forces. So both will take place: the strengthening material will help to stabilize and the stabilizing material to strengthen the conductor.

The superconductor is divided into filaments with diameters ranging from a fraction of a millimeter to a few μm depending on the application.

Composite conductors may be *intrinsically stable* by using one of the following principles:

- *adiabatic stabilization*. It means that the succeeding temperature rise ΔT_2 is made smaller than the initiating one ΔT_1, thus the perturbation dies away. Taking into account the specific heat, the critical current density, and its temperature dependence a limiting penetration thickness can be calculated below which no flux jumps should occur. Thus the criteria of adiabatic stability can be complied with by providing filaments of small diameters (< 50 μm for NbTi).

- *dynamic stabilization*. It means that heat is removed more rapidly than produced by the motion of flux lines. Now the magnetic field penetrates into the superconductor 10^4 times faster than the heat is removed. Thus the superconductor considerably violates the stability criterion. The contrary is true for copper. It turns out that for small dimensions (< 50 μm for NbTi) the superconductor can be stabilized by embedding it into highly conductive copper.

The resulting conductor is a multifilamentary wire with many thin filaments (typically 5-25 μm) embedded into a low resistive copper matrix. Copper to superconductor ratio may be as small as 1 : 1 (figs. 5).

2.3 The behaviour of the superconductor when operated normally

2.3.1 AC losses

A superconductor exposed to varying field amplitudes especially to pulsed fields will no longer be lossless. A number of loss mechanism are known. Amongst them we are already familiar to *hysteretic losses*. Losses per cycle are proportional to the filament thickness, the critical current density, and a function of the field amplitude. In composites additional losses will occur: *eddy current* losses in the matrix and filament *coupling losses*. These occur in current loops forming between neighbouring filaments across the normal conducting matrix.

The understanding of loss mechanisms is well advanced. Calculation of losses is in agreement with experiments. Low loss conductors are available. Several thousands of very thin filaments (typically 5 μm) to reduce hysterestic losses; twisting of filaments (twist pitch several mm to several cm) to prevent filament coupling; resistive barrieres (cupronickel) between filaments or groups of filaments to supress transverse current flow and eddy currents; and finally full transposition of strands in the conducting cable to achieve uniform current distribution are the main characteristics of an ac superconductor.

2.3.2 Mechanical properties

Torus coils and their conductors will experience huge magneto-mechanical forces: radial compressing forces, bending forces due to the radial inhomogeneity of the torus field, axial forces due to intrinsic asymmetries or to fault conditions. Additional bending and shear forces will arise when poloidal

375

pulsed fields are taken into account. The radial force on the
inner straight part of one D-shaped coil of a power reactor will
be of the order of 10^9 N. The bending forces can be minimized
by suitably shaping the coils (D-shape). The tensile forces
have to be accomodated by reinforcing the coils appropriately.

The superconductor will degrade when subjected to mechanical
load. It was found that NbTi will quench when the strain exceeds
about 1.5 %. At smaller loads discontinuities or serrations,
in the stress-strain characteristic are observed. These serrations are one of the causes of *magnet training*, they lead to an
irreversible deformation and hardening of the material. Degradation of Nb_3Sn conductors will occur even earlier because of
their brittleness. The tolerable strain is ranging from 0.2 to
0.8 %. The design of the magnet has to be preceded by a thorough
stress analysis. In most of the designs the maximum strain does
not exceed 0.2 %. A few of the mechanical properties of superconductors and matrix materials are listed in table 2.

	Young's modulus 10^{11} Nm^{-2}	Yield point ($\sigma_{0.2}$ limit) 10^8 Nm^{-2}	Failure stress 10^8 Nm^{-2}
NbTi	0.6 - 0.8	-	13
NbTi (multifil. conductor)	0.8	3	10
Cu(RRR∿150)	1	1	4.5
Cu(RRR∿ 60)	1 - 1.2	4	-
CuNi	2	9.2	9.7

Table 2: Mechanical properties of superconducting and
matrix material[3]

2.3.3 Radiation damage

Superconducting magnets in a future fusion reactor will be exposed to fast neutron and gamma irradiation [4], which will produce lattice defects thus causing changes of the properties of
the superconductor, especially critical current density and
critical temperature will decrease. It is therefore necessary
to know the response of the possible superconducting material
to radiation exposure. This can play an important role in determining the magnet design characteristics, the amount of shielding,
and the economics of a reactor concept. The result of irradiation
measurements show that radiation damage of the superconductor
itself and of the stabilizing material will not be crucial.
This may be different for insulating materials which well could
limit the permissable dose, but few data exist on this. Another
limiting factor will be the thermal limitation due to heat
deposition of the absorbed neutrons or gammas within the coil:
conductor, cold structure, and cooling fluid [4].

A-15 superconductors start degrading about a factor of 10 earlier
than NbTi. This may be attributable to the complex crystal
structure of the A-15's where especially the Nb-chains will break.
Thus, future conductors with an NaCl-structure and transition
temperatures as high as the A-15's may prove to be superior.

NbTi multifilament conductors have been irradiated at 1He-tempe-
rature with fast neutrons and deuterons. Critical current density
did decrease[5,6,7,8] above fluences greater than a few times
10^{17} n/cm^2 by 10 to 20 %, with deuterons which are a factor of
about 20 more effective than neutrons the reduction at 10^{18} d/cm^2
was about 10 % (fig. 7), the reduction in T_c was only 3 %.

With Nb_3Sn and V_3Ga a different degradation behaviour was observed.
Reductions in either T_c, H_{c2} and I_c were observed after irradiation.
Little reduction in I_c is observed for neutron fluences below
about 10^{18} n/cm^2 and for deuteron fluences below about 10^{17} d/cm^2.
Above these values there is an apparant threshold where a rapid
reduction is initiated. At about 10^{19} n/cm^2 the I_c of Nb_3Sn was
reduced to 4 %[6]. At $7 \cdot 10^{17}$ d/cm^2 about 10 % of the unirradiated
value of a Nb_3Sn multifilamentary wire was found, a V_3Ga wire de-

graded to 25 % at 10^{18} d/cm^2 (fig. 7). Above the current threshold the T_c-reduction is as catastrofic as the I_c-degradation.

When copper (or aluminium) is subjected to neutron irradiation, the resistivity ratio decreases, which in turn spoils the stabilization properties, but fluences of 10^{18} to 10^{19} n/cm^2 do not seem to alter it appreciably. Another phenomenon is annealing which may partly cancel radiation damage effects. Presently it has not been studied very systematicly.

A reactor design has to take into account irradiation results to keep the degradation and warming up of the superconducting magne below tolerable values.

2.4 Superconductors for fusion magnets

Fusion magnets open quite a new domain of magnet technology. Stored energy of a single torus coil is a least a factor of 10 above that of biggest existing superconducting magnet systems, the big bubble chamber magnets. The conductor for such coils has to carry large currents and to withstand high mechanical and electrical tensions. Nevertheless the superconducting state has to be maintained in any normal and fault condition. No final solution has been found but a lot of proposals exist.

At normal operation the total current will be about 10 kA. The conductor has to be fully stabilized, average copper to superconductor ratio being about 10 : 1. The material to be used may be either NbTi or Nb$_3$Sn depending on the reactor design: The UMWAK[9] design is with NbTi and 8.3 T maximum field at the conductor, the Princeton Fusion Power Plant[10] is with Nb$_3$Sn and 16 T maximum field at the innermost conductor. The reinforcement has to withstand tensile stresses of 2 to $5 \cdot 10^8$ N/m^2 and to ensure a maximum strain in the superconductor which is less than 0.2 %. It will probably be a distributed steel reinforcement which may be integral part of the conductor or a steel ribbon wrapped together with the conductor.

The conductor is subjected to varying field amplitudes. The field ripple due to the $\dot{B}$ of the poloidal field windings will be in

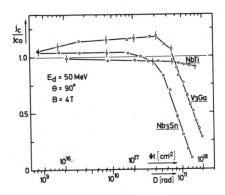

Fig. 7: Dependence of relative j_c-values after deuteron irradiation.

NbTi   (F 60 - 14/141, Vacuumschmelze)

Nb_3Sn (FSW - NS - 1 S, Furukawa)

V_3Ga (FSW - VG - 1 S, Furukawa)

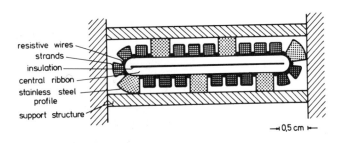

Fig. 8: Concept of a flat cable conductor for use in fusion magnets[11]

the order of about 0.5 T/sec without and one tenth of this value with shielding. Total ac losses have and can be kept small.

Conductor cooling may be provided by helium-bath-cooling or by forced circulation of supercritical or two-phase helium. Bath cooling with boiling helium under normal pressure has the advantage that a lot of experience exists. The practical maximum heat flux between conductor wall and helium which can be obtained is about 0.4 W/cm^2. But for long narrow cooling channels the permiss heat flux drops very fast to about a few percent of this value. Therefore only short cooling ducts are allowed. [3]

Forced cooling with "supercritical" helium (helium under pressure exceeding the critical pressure of $\simeq$ 2.2 bar), has the advantage that only the enthalpy of the cooling fluid is enhanced without any phase transition. Therefore no sudden jumps in the heat flux occurs in contrast to bath cooling. Because of the low enthalpy of supercritical helium high heat transfer rates demand high mass flow rates resulting in high pressure drops. A much higher cooling power at reduced mass flow rates and lower pressure drops can be achieved by cooling with forced circulation of two-phase helium, presuming instabilities due to vapor lock, choking flow, or density differences of the two phases, can be avoided. Additional advantage of forced cooling are that the cooling ducts are integral part of the magnet structure and the considerable reduction of the total amount of helium.

The principle assembly of a superconductor for fusion magnets may be as follows [11]. Several single conductors are fixed to a stainless steel carrier ribbon. Each of these will carry a current of several 100 A, is fully stabilized, directly cooled, and thermically decoupled. AC losses are small. Spacers from cooling ducts and a support structure gives additional mechanical stability and final shaping of the conductor (fig. 8). The inner part will be used for a small torus assembly TESPE, the conductor is being developed and tested.

Another concept has been known as "bundle" conductor. It follows an idea of Kafka and is being developed at MIT[12]. Several twisted strands are transposed and cabled, several of these cables are in turn twisted, transposed and bundled. This bundle is finally forced into a steel tube (fig. 9). Advantages are low ac losses, enlarged surface cooling, and simple winding procedure. Cooling is proposed to be forced supercritical helium cooling.

3. MAGNET TECHNOLOGY

3.1 Field,- force- and heat transport calculations

A lot of computer programs are available which allow to calculate magnetic fields, field distribution, field homogeneity, magnetic parameters like inductivity, forces on the conductor, stability of the structure subjected to magnetomechanical forces, heat transport, paths of particles in the fields etc.
A few examples may illustrate this[13]:

HEDO developed by IPP Garching can calculate air-core solenoids including D-shaped coils. MAGNET, NUTCRACKER, TRIM and others are twodimensional programs including iron and using a relaxation method. GFUN and its three dimensional version GFUN 3D, both developed by Rutherford Laboratory and both including iron, make use of an interactive display.

Thermodynamic data of helium can be calculated with HELTHERM, heat transport problems with HEHT 3 both are available from NBS. STRUDL (STRUctural Desing Language) is used for strength calculations.

3.2 Technology of components

Superconducting technology with NbTi conductors is well established for both d.c. and a.c. applications. It has been successfully used for building high field magnets (above 10 T), medium size, and large magnet units (up to about 1 GJ of stored energy).

Problemes encountered and solved for each special application are

- magnet impregnation and conductor insulation

- conductor joining

- winding technique

- structural reinforcement

- cooling

Techniques are now well established for magnets made with NbTi conductors and of medium size with fields of about 2 to 8 T. New
problems will arise for very high field magnets with Nb_3Sn or very
huge and complex magnet systems (stored energy above 1 GJ). The
biggest magnet units are bubble chamber solenoids which have a
very simple geometry compared to a torus magnet.

Coils are impregnated by epoxy resins thus fixing individual
windings but at the same time preventing the direct access of the
helium to the conducting wire. Appropriate design of cooling
channels or heat drains has to be ensured taking into account
thermal conductivity and heat transfer through the complex winding
structure. Thermal and mechanical properties of epoxy resins can
be matched to those of the conductor and the metallic structural
material. Quartz or alumina powder allows to match thermal conductivity. Structural fibres of glass, carbons or other suitable
material considerably enhances mechanical strength. Thermal contraction can be matched by playing with these fibres and their
arrangement. Low temperature properties of reinforced resins
promise to be superior to many other materials.

Conductor joints of superconductors have to fulfill the following
criteria: they must be at least as strong as the parent conductor, secondly the joint resistance should be as low as possible,
10^{-8} Ω being acceptable for high current joints. Several techniques
are in use: pinching under high mechanical pressure, soldering
or several welding technique like pulse welding, explosive
welding and recently ultrasonic welding.

Winding of the magnets strongly depends on the magnet concept, their geometry, and the conductor. It will be different whether an integrated or separate reinforcement will be used, whether cooling ducts are integral parts of the conductor or have to be provided by suitably winding the conductur (e.g. with separate spacers). It will depend on the concept of coil protection and the distribution of the coil package into single pancakes.

3.3 Status of superconducting magnet technology

Very high field magnet systems have been built and operated successfully. A 25 T hybrid system with a resistive core and a NbTi envelope producing a field of 8 T is in operation in the Kurchatov Institute, Moscow[14]. A 16.5 T magnet with a clear bore of 25.7 mm and made of a Nb_3Sn ribbon has been built by IGC[15].

DC and AC magnets for use in beam lines of accelerators and high energy physics experiments have been developed and are widely used[13]. Typical parameters are overall length of several meters, magnetic fields of about 5 T, field accuracies of 10^{-3} $\Delta B/B$ integrated field deviations, field configurations are dipole or quadrupole field geometries. AC magnets with rise times of a few seconds and overall losses of about 10 W/m have been developed and tested[16]. Large volume detector magnets bath and forced cooled are used at high energy laboratories like CERN, DESY or FERMI laboratory. Superconducting magnets for alternators or high speed levitated trains have been built and are used in several devices.

A number of superconducting magnets have been used in plasma research experiments as well. Confining to toroidal devices outstanding experiments have been made with levitated superconducting rings buried within the plasma, e.g.[17,18]. The superconducting toroidal system of the NASA Lewis 'bumby torus' consists of 12 coils with 22 cm inner diameter and capable of producing 3 T on their axis[19]. A stellerator prototype coil for the Wendel-

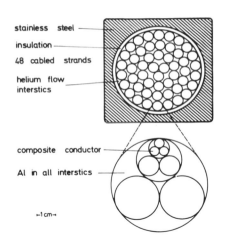

Fig. 9: Bundle conductor made of
Nb$_3$Sn wires. ORNL/MIT design[12].

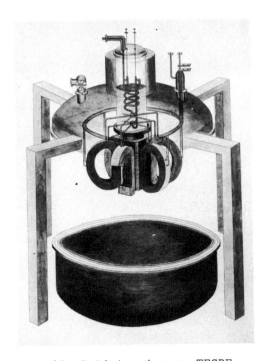

Fig. 10: Artists view on TESPE

stein VII experiment has been built by SIEMENS. Mean coil diameter is 1.04 m, designed field in the torus 6.2 T. It has
been successfully operated in a simulated torus experiment and
exposed to pulsed fields with field components up to 0.5 T with
a dB/dT of 100 T/s[20].

A superconducting tokamak T-7 using a superconducting torus is
investigated at Kurchatov Institute, Moscow[21]. Major diameter
of the plasma is 2.5 m, minor diameter 0.6 m, plasma current will
be 0.5 MA. The torus consists of 8 sections with 6 coils each.
Mean coil radius is 1 m. The conductor is a fully stabilized
NbTi hollow conductor cooled by supercritical helium. The toroidal
magnet system is designed to have a central field of 3 T and will
store 20 MJ of magnetic energy. It is being assembled and tested.

A superconducting toroidal magnet system TESPE is under construction at Karlsruhe. It will consist of 6 coils. Major torus
diameter is 1 m, inner coil bore 0.4 m. Central field is 5 T,
maximum field at the coils 8 T. The conductor is niobium-titanium.
It will be forced cooled by supercritical helium. An artists
view on TESPE gives fig. 10.

4. PROBLEMS OF LARGE SUPERCONDUCTING MAGNETS FOR FUSION

The superconducting magnet technology is one of the major technological breakthroughs of the last 10 years. Nevertheless a lot of
problems is left, especially for large superconducting magnet
systems. They may even not yet be realized up to now by magnet builders,
but will unavoidably arise when proceding to bigger and more complex systems underlying the constraints of another complex component, the plasma and its environment.

A major problem of fusion reactor magnets will be a mechanical
one: make magnets, cryostats, and support structure withstand the
magnetomagnetic forces which will exert steady state and dynamical
loads on the systems. There is first the selection and development

of suitable materials and their composites, and second the investigation and possibly necessary improvement of their long term behaviour, like fatigue problems, low temperature properties, radiation damage problems.

Another most important problem is coil protection. Huge amounts of stored energy have to be handled safely. The integrity of coils must be ensured under all possible circumstances. A lot of ideas have been brought forth. Passive devices such as external resistive shunts acting as energy dumps or active devices such as fast-acting switch-operated heaters embedded in the windings have been suggested[22]. External and internal energy dissipation may be foreseen.

Considering the cost of the magnet system 0.1 to 0.3 of the total cost of the reactor) economics seems to be equally important as technical problems. Both are strongly intercorrelated. Thermal losses and losses due to pulsed fields or field components have to be kept small to minimize the expense for the cooling system. Conductor development has to be directed to achieve a technically feasible but economic solution.

The problems of reliability of a large superconducting magnet system can only be suspected by existing experience but will hit magnet builders with its full weight. Looking to reactor safety programs for fission reactors the task cannot be underestimated.

All these problems are crucial in the sense that each of them may kill a technological approach to a future fusion reactor. Therefore we are in time to start with their solutions.

References

1) W. Buckel, Supraleitung, Physik Verlag 1972

2) J.R. Gavaler et.al., IEEE Transactions on Magnetics,
Vol. MAG-11, March 1975, p. 192

3) E. Seibt ed., KFK 2359, September 1976

4) G.M. McCracken and S. Blow, CLM-R 120 (1972)

5) M. Soell, S.L. Wipf, and G. Vogl, Proc. Conf. Applied
Superconductivity Annapolis 1972, p. 434

6) D.M. Parkin and A.R. Swedler, IEEE Transactions on Magnetics,
Vol. MAG-11, March 1975, p. 166

7) E. Seibt, IEEE Transactions on Magnetics, Vol. MAG-11,
March 1975, p. 174

8) P. Maier, H. Ruoss, E. Seibt, Verhandl. DPG (VI) 11, 712 (1976)

9) University of Wisconsion Fusion Feasibility Study Group,
UWFDM-112, October 1975

10) R.G. Mills ed., MATT-1050, August 1974

11) C.-H. Dustmann, H. Krauth, G. Ries, Proc. 9th Symp. Fusion
Technology, Garmisch-Partenkirchen 1976 (to be published)

12) M.O. Hoenig, Y. Iwasa, D.B. Montgomery, and A. Bejan, Proc.
ICEC 6, Grenoble 1976 (to be published)

13) W. Maurer ed., KFK 2290, May 1976

14) P.A. Tscheremnych et. al., Kurchatov Institute, IAE 2316,
Moscow 1973

15) IGC newsletter, Vol. 1, No 3 (1973)

16) GESSS-2, a report of the GESSS collaboration, May 1973

17) D.N. Cornish, 6th Symp. Fusion Technology (1970), p. 103

18) C.E. Taylor et. al., 4th Symp. Eng. Problems of Fusion
    Research, Washington 1971

19) J.R. Roth et. al., Proc. Appl. Superconductivity Conference,
    Annapolis 1972, p. 361

20) M. Pillsticker and P. Krüger, Proc. 6th Symp. Engineering
    Problems of Fusion Research, San Diego 1975

21) D.P. Iwanow et. al., Proc. Conf. Technical Applications of
    Superconductivity, Alushta, Sept. 1975 (to be published)

22) H.M. Long and M.S. Lubell editors, ORNL/TM-5401, revision
    of ORNL/TM-5109, revised January 1976

B. SUPERCONDUCTING MAGNET SYSTEMS IN EPR DESIGNS

A.F. Knobloch

Max-Planck-Institut für Plasmaphysik, Garching

Abstract

Tokamak experiments have reached a stage where large scale
application of superconductors can be envisaged for machines
becoming operational within the next decade. Existing designs
for future devices already indicate some of the tasks and problems associated with large superconducting magnet systems. Using
this information the coming magnet system requirements are
summarized, some design considerations given and in conclusion a
brief survey describes already existing Tokamak magnet development programs.

389

1. Introduction

The technology of superconducting magnets for fusion application
would have a slow start if it were dependant only on near term
experiments inevitably needing it. Its proper development
obviously is costly time-consuming and also risky since it must
rely on extrapolations from present plasma physics achievements.
It shares, however, these features with the Tokamak oriented
fusion technology as a whole. Therefore, in case there is a
strategy or only a confidence in progress towards a Tokamak reactor, there has to be an appropriate effort in this part of
technology, too. The appearance of experiments like TFTR or JET
gradually reveals the real engineering and sometimes political
tasks of designing and building a large integral Tokamak magnet
system and to make its several 100 MW pulsed power supply available. Beyond a power requirement of this order of magnitude it
may be hard in the future to convince anybody of the necessity
to increase further the electricity supply for fusion apparatus
while promising finally to produce power but not to waste it. In
addition to this argument time schedules for experiments like
TFTR or JET set the case for starting substantial work towards
fusion-related superconducting magnets for the next machine
generation.

Introducing the superconductor technology into the complicated
Tokamak fusion line in a sound manner requires due to almost unanimous opinion some previous technological feasibility demonstration by means of a large magnet test able to simulate all
relevant operational features of an integral Post-TFTR or Post-
JET magnet preferably in a conservative approach which means
today e.g. the choice of NbTi as the superconductor material and
a design for full cryogenic stability.[1] This demonstration should
be achieved in a separate technical test facility preceeding the
application in an experiment. Here one should mention that only
a superconducting toroidal magnet system not much below JET size
would yield extrapolable design information for a later TNS or
EPR machine[2] whereas much smaller superconducting magnets or test
facilities would not. While designing now and building the next
(and possibly last) generation of experiments with normal

390

conducting toroidal magnets[3] already much important know-how is
being gathered for a relevant specification and construction of
a large superconducting Tokamak magnet. This know-how includes
the specific field and stress calculations[4], structural design,
definition of the poloidal field configuration including the
plasma[5] maintenance considerations, aspects of radiation shielding,
remote handling and failure analysis. On top of this basic know-
how additional information is required for a safe design and construction of superconducting fusion magnets namely on conductors,
winding, cooling, monitoring and protection under the Tokamak
specific boundary conditions like rapidly varying external magnetic
fields, high currents and voltages, cyclic stresses and very large
dimensions and fields. It is this latter group of specific problems
which call for a basic development program for fusion superconducting magnets additionally to the costly task of establishing
confidence in these magnets by a relevant integral test set up[6].

The following lecture deals in the above mentioned sense with the
systems aspects of the integral magnet of future Tokamak machines
and will add some remarks on superconducting magnet development
programs and possible international collaboration.

2.   Requirements for next generation confinement experiments, TNS and EPR machines, comparison with reactor data.

2.1 Toroidal field coils

According to designs available at present the magnet requirements
of large future Tokamak machines are generally the following:
to establish a steady state toroidal field and pulsed poloidal
fields for the confinement of a plasma ring with a diameter of
$6 \div 26$ m, an aspect ratio of $3 \div 4$ and currents between 3 and
15 MA.

The toroidal magnet generally has to be designed for an axial
flux density between 3 and 6 T. Table I shows a data comparison
for a selected number of designs of next generation experiments
through EPRs up to full scale reactors. Due to the introduction
of a blanket and shield zone the usage of the toroidal field

TABLE I

	R (M)	a (M)	I_P (MA)	B_0 (T)	W_{MT} (GJ)	$\hat{P}_{MT}$ (GW)	N
JET	2,96	1,25/2,1o	4,8o	3,4o	1,45	0,33	32
GA ITR (TNS)	3,5o	0,8o		4,00		-	16
ORNL ITR (TNS)	5,00	1,25	3,3o	4,4o		-	
T 20	5,00	2,00	6,00	3,5o	6,00	1,2o	24
ANL EPR 1975	6,25	2,1o	4,8o	3,4o	15,6o	-	16
ANL EPR 1976	6,25	2,1o	7,58	4,32	30,00	-	16
ORNL EPR 76	6,75	2,25	7,2o	4,9o	29,00	-	20
JXER	6,75	1,5o	4,00	6,00	50,00	-	16
UWMAK II	13,00	5,00	14,9o	3,67	223,2o	-	24
UWMAK III	8,1o	2,7o	15,8o	4,o5	108,00	-	18
CULH. MK II	7,4o	2,1o/3,68	11,6o	4,1o	~57,00	-	20

decreases considerably for EPRs and reactors compared to experiments. All superconducting windings in EPRs and reactors have to be radiation shielded to the extent that the stabilizing material in the innermost windings during the accumulated operational time will suffer a resistance increase less than the unirradiated value at most. This condition is consistent with sufficient protection of the superconductor itself and of the coil and conductor insulation in the case of epoxy[18,19,20]. The number of toroidal coils is defined by the permitted toroidal field ripple at the outer plasma edge, in general about < 1 ÷ 2 % and on the required access between the outer coil parts for auxiliary heating devices and maintenance and repair concerning the plasma chamber and the blanket and shield region.

Table II gives some further toroidal field coil data, which show the size dependent increase of the forces, a major incentive to go for a noncircular coil shape adopted throughout. The constant tension, so-called modified D-shape, however, has been foreseen only for JET, one of the TNS designs, JXER and the ANL-EPRs. T20 and other designs deviate more or less from the modified D. This is an indication of both an adaptation to the inside space requirements and of economical considerations. As will be shown later, the constant tension D-shape couples the coil shape with the magn

TABLE II

	SHAPE	DIM. (M)	AT(MAT)	I_{COIL}(kA)	B_{MAX}(T)	$j_{AV}(\frac{kA}{cm^2})$	$F_T(10^6 N)$	$F_C(10^6 N)$	WEIGHT(T)
JET	MOD. D	3.12 x 4.90	1.42	66.4	6.9	1.6		18.4	12.0
GA ITR (TNS)	MOD. D	4.50 x 7.20	4.25	6.0	8.0	3.0			72.0
ORNL ITR (TNS)		4.50 x 6.50							
T 20	OVAL	5.40 x 8.30	3.60	120.0	7.8	1.6	15.0	61.2	66.7
ANL EPR 1975	MOD. D	7.70 x11.90	6.50	10.0	7.5	2.4	49.3	178.2	175.0
ANL EPR 1976	MOD. D	7.78 x12.60	8.37	60.0	10.0	3.7	82.7	358.0	208.0
ORNL EPR 76	OVAL	7.40 x 9.70	6.30	78.0	11.0	4.0	76-360	220.0	
JXER	MOD. D	7.00 x11.00			11.5				
UWMAK II	EXTENDED D	18.00 x27.50	9.99	11.5	8.3	6.0	43.8		710.0
UWMAK III	REDUCED D	12.50 x23.50	9.77	10.9	8.75	4.8	34.2		174.0
CULH. MK II		9.80 x17.50	7.50		8.0				

major diameter and will not generally be the most economical solution. In a torus of D-shaped coils the axial field tries to compress the array of straight inner coil sections radially towards the torus center. The corresponding force (see table II) is so large, that it almost predetermines the mechanical structure of a Tokamak machine, in that it calls for a strong central force bearing column or cylinder to support the toroidal field coils. For large superconducting toroidal field coils the magnitude of the centering forces almost rules out the possibility of completely separate dewars for each coil. The corresponding heat losses through the support to the warm environment would require excessive refrigerator power. This means that superconducting toroidal Tokamak magnets tend to require a so-called fitted joint dewar including the support cylinder at He-temperature and cannot easily be built sectionally (Fig. 1). Under emergency conditions when one or two adjacent toroidal coils fail and loose their current a large out of plane load occurs acting on the failed coils neighbours primarily. The corresponding deflection of the loaded coils has to be kept as low as possible and therefore following the principle of resisting forces where they occur, a strong outer shell structure connecting the coils in the circumferencial direction is called for. In the case of superconducting coils there is an additional consequence. Supporting the full

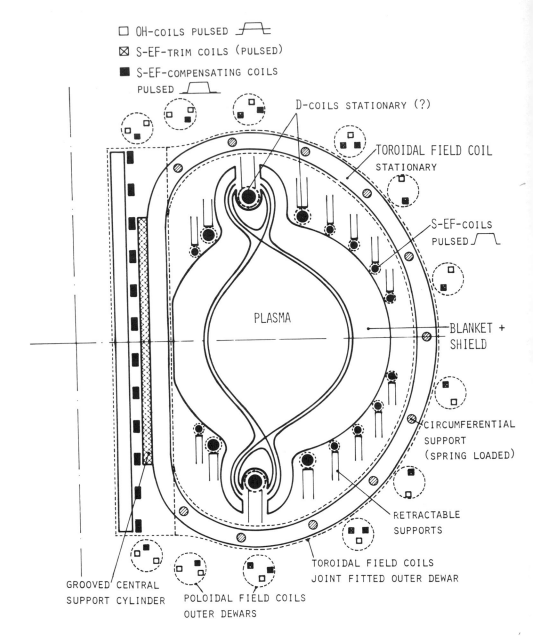

□ OH-COILS PULSED

⊠ S-EF-TRIM COILS (PULSED)

■ S-EF-COMPENSATING COILS
    PULSED

D-COILS STATIONARY (?)

TOROIDAL FIELD COIL
STATIONARY

S-EF-COILS
PULSED

PLASMA

BLANKET +
SHIELD

CIRCUMFERENTIAL
SUPPORT
(SPRING LOADED)

RETRACTABLE
SUPPORTS

TOROIDAL FIELD COILS
JOINT FITTED OUTER DEWAR

GROOVED CENTRAL
SUPPORT CYLINDER

POLOIDAL FIELD COILS
OUTER DEWARS

Fig. 1   E P R - MAGNET SYSTEM
WITH DOUBLE-O POLOIDAL DIVERTOR

out of plane load towards the warm environment in stationary
condition would again lead to intolerable refrigeration power.
Therefore in order to reduce the out of plane loading and also
the coil deflection during a failure of one coil an electrical
discharge scheme for the whole toroidal magnet is needed which
minimizes the transient differences in single coil currents
during discharge ideally to the extent that only the normal
operational deflection caused by the poloidal fields is reached,
and which assures that all but the failed coil causing the discharge can remain cold. Although the circumferential shell
structure has to be designed for the full out of plane load the
described failure of one or two coils has to be considered as a
singular catastrophic case which could lead to a major overall
revision. In this case a lateral movement of the coils inside
the dewars would be permitted to the extent that additional
supports come into action.

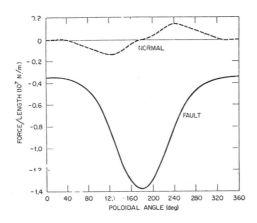

Fig. 2

Fig. 2 shows the full out of plane force distribution over the
circumference of a toroidal coil under fault conditions (one
coil without current in the ORNL EPR 1976) compared to the normal
pulsed out of plane load from the poloidal field system. The
minimum stiffness of the coil and shell structure is determined
by the limitation on the repetitive toroidal field magnet
deflection. Taking the maximum permitted relative vertical field
component of $3 \cdot 10^{-4}$ from the toroidal field magnet as defined
for T20 the permitted lateral coil deflection even in a very

large system is below 5 mm. In plane motions, however, in large
toroidal coils are larger. Cooling contraction or charging
expansion may lead to dimensional changes of the order of cm in
an EPR size Tokamak magnet. Since there are elastic dimensional
variations of such a size possible in a large toroidal magnet it
is almost obvious that an oscillatory or instable response of
such a system to mechanical excitation is quite possible. Recent
magnetoelastic analysis has shown that for large magnet dimensions, high fields, and in case of insufficient circumferencial
support there could be a circumferencial collapse or strong
deformation following a small out of plane deflection of a
single coil. The above mentioned outside shell structure can
prevent this. Of course the shell has still to provide access
openings for auxiliary plasma heating mainly, but these openings
have to be restricted to a width not allowing excessive deflec-
tions. Table III shows a number of calculated results for an
experiment and an EPR.[21,22,23,24]

TABLE III   CRITICAL BUCKLING CURRENTS AND FIELDS

MAGNET DESIGN	1st MODE	2nd MODE	3rd MODE	4th MODE	DESIGN DATA
TFTR I_c (MA) B_c (T)	2,77 3,96	5,68 8,12	11,82 16,89	19,9o 28,5o	3,22 4,6
ANL EPR 1975 I_c (MA) B_c (T)	1,55 0,816	3,17 1,67	6,46 3,41	10,83 5,71	6,5 3,4

SIDE VIEW OF COIL AND SUPPORTS

I_c --- CRITICAL BUCKLING CURRENT IN AMPERE-TURNS PER COIL

B_c --- CRITICAL MAGNETIC FIELD AT MAJOR RADIUS OF TORUS

A very important issue for toroidal magnet designs is the proper
choice of operational current and voltage with respect to the
safety discharge. Characteristic for many earlier superconducting
Tokamak magnet designs was the relatively low conductor current
compared to normal conducting magnets. This corresponded to the
foreseen difficulties of manufacturing stable high current conductors in long lengths. More recent designs propose parallel
operation of several conductors or single conductors for up to
60 kA since low currents at high stored energies could cause
insuperable insulation problems during an emergency discharge of
the torus magnet. As will be shown later, the safety discharge
conditions couple many of the magnet design data including the
permitted temperature rise in the failed winding, thus leaving
only a limited parameter choice.

Since very likely Tokamaks will remain pulsed devices as they are
now - even if confinement physics would become perfect, the
limitation on flux swing for inducing the plasma current indicates
this   and the existence of a self-sustained boot strap current
has not been demonstrated - under several aspects the toroidal
Tokamak magnet is not a DC device. With a superconducting winding
it has to be conceived of as operating in a difficult ac field
environment. In normal conducting Tokamak magnets the problems
from pulsed poloidal fields on the toroidal magnet like in the
superconducting versions arise from the bending and tilting
forces and additonal heat produced; but for a superconducting
magnet these forces and additional heat inputs may principally
limit the performance of the magnet or in other words, conductor,
winding and cooling principle of the superconducting toroidal
magnet has to be taylored according to the pulsed external field
pattern. There are reasons like state of the art and cost of ac
superconductors as well as space limitations to use normal conducting poloidal field windings in combination with the first
larger superconducting toroidal magnets. The spatial arrangement
of these poloidal windings, however, in any case will have to be
chosen such that a minimization of pulsed field effects in the
toroidal coils will be achieved.[5] There have been proposals to
use passive cryogenic or even superconducting shields[10] to protect
the toroidal field coils or ease their conductor design. It

appears that they either will cause additional overall heat production or in case of a superconducting shield shift the pulsed field interaction problem to a separate winding, which adds complication and can only compensate the pulsed fields for the toroidal coils for one single poloidal field pattern. This pattern is variable in time and must not be changed by toroidal field coil shields, which for efficient shielding would have to be designed for a minimum time constant like the burn cycle length. Insulating breaks in the shield would be required to allow the above mentioned safety discharge. Large local forces would be exerted on the shield especially in the vicinity of the central support cylinder. With or without a shield the stable operation of the toroidal magnet can only be assured if the pulsed field magnitude and rate of rise nowhere along the winding circumference locally can generate excessive losses and/or motion.

2.2 Poloidal field coils

The pulsed poloidal magnetic field system of a Tokamak (see Fig.1) besides of the plasma ring current itself can consist of the ohmic heating (OH) windings, the equilibrium field (EF) windings, a shield (S) winding, a field shaping (F) winding and a divertor (D) winding.[25] It is assumed here that only poloidal divertors will offer the chance of application in a reactor. It has been shown recently that the interesting radial bundle divertor concept for the technical reasons of necessary very local high fields and forces and radiation shielding problems is not reactor relevant

The OH-winding serves as the transformer primary to start the ring discharge and to induce the short turn plasma current, to keep constant its value over the burn period and to end the current flow.

Table IV is a list for existing designs showing the assumed volt-second requirements and their subdivision between OH and EF windings, the type of operating cycle together with the relevant times as well as the energies, voltages and powers involved. Also the flux density swing in the core is indicated. As can be seen from the list the start-up or recharge requirements for the OH

/5/ L.E. Feinendegen, Effects of Tritium on the Human Organism,
J. Belge de Radiologie 58 (1975), p. 147

/6/ Radioactive Effluents from Nuclear Power Stations in the
Community. Discharge Data 1969-74. Radiological Aspects.
Ed. by Commission of the European Communities, Directorate -
General for Social Affairs, Directorate of Health Protection,
Doc. No. V/356o/75 e, Dec. 1975

/7/ W. Friedrich, J. Knieper, H. Printz. P.F. Sauermann,
Untersuchungen über die Tritium-Kontamination der Atemluft
in der Umgebung von Experimenten mit großem Tritium-Inventar.
9. Jahrestagung des Fachverbandes f. Strahlenschutz, Alpbach
(Tirol, Austria), Oct. 6-8, 1975

/8/ H.G. Östlund, A.S. Mason, Atmospheric HT and HTO, Part I.
Experimental Procedure and Tropospheric Data 1968 - 1972,
Tellus XXVI (1974), p.91

/9/ Der Bundesminister für Bildung und Wissenschaft, ed., Emissions-
quellstärke von Kernkraftwerken. Schriftenreihe Kernforschung 6
(1972)

/1o/ C.E. Easterly, D.G. Jacobs, Tritium Release Strategy for
a Global System. Proc. Int. Conf. Radiation Effects and Tritium Technology for Fusion Reactors, Gatlinburg, Ten., Oct.
1-3, 1975. CONF-75o989 (1976), Vol. III, p. 58

/11/ H. Jahn, Tritium-Überwachung an einem Forschungsreaktor,
6. Jahrestagung des Fachverbandes für Strahlenschutz,
Karlsruhe, 17.-19. May, 1972

/12/ D.S. Myers, J.F. Tinney, P.H. Gudiksen, Health Physics
Aspects of a Large Accidental Tritium Release. In : Tritium,
ed. by A.A. Moghissi and M.W. Carter, Las Vegas, 1971, p.611

/13/ J.Darvas, Tritium Technology in Fusion Devices. Proc. Int.
Conf. Radiation Effects and Tritium Technology for Fusion
Reactors, Gatlinburg, Ten., Oct. 1-3, 1975. CONF-75o989
(1976), Vol. III, p. 1

/14/ C.L. Folkers, V.P. Gede, Transfer Operations with Tritium. UCRL-76729 (1975)

/15/ Summary of Proc. Workshop on Conceptual Design Studies of EPR-1. Oak Ridge, Sept. 9-1o, 1975. ERDA-89 (1975)

/16/ P. Columbo, R. Neilson jr., M. Steinberg, The Fixation of Aquous Tritiated Waste in Polymer Impregnated Concrete and in Polyacethylene. Proc. Int. Conf. Radiation Effects and Tritium Technology for Fusion Reactors, Gatlinburg, Ten., Oct. 1-3, 1975. CONF-75o989 (1976), Vol.III, p. 129

/17/ P.D. Gildea, Containment and Decontamination Systems Planned for the Tritium Research Laboratory Building at Sandia Laboratories, Livermore. Proc. Int. Conf. Radiation Effects and Tritium Technology for Fusion Reactors, Gatlinburg, Ten., Oct. 1-3, 1975. CONF-75o989 (1976), Vol. III, p.112

/18/ Amtsblatt der Europäischen Gemeinschaften, Nr. L 187/31, 12.7.1976

TABLE IV   O H / E F   WINDING DATA

	V (kV)	I (kA)	P_{MAX}(MVA)	E_{MAX}(MJ)	$\Delta\phi$(Vs)	ΔB(T)	$B(\frac{T}{s})$	ΔT_{RISE}(s)	COND.
T 20	28/26	120/120	3450/3100	220/460	71,5/46,5	± 4	4/	2	Cu
ANL EPR 1976	51/21	80/80	1910/420	1200/1500	85+4/50	± 5	6,7/1	2	NbTi
ORNL EPR 1976	69/	50/	3500/180			± 7	7/	2	NbTi/Cu
UWMAK II		65/	950	10400		+ 5,7 − 8,4	0,16/	10	NbTi
UWMAK III		10/5	900	8860		+ 6,9 − 8,6	0,77/	20	NbTi

coil in terms of $\dot B$ should become easier in large machines. This
may have an important influence on the experimental step at
which to introduce superconducting poloidal windings. Because of
their required large flux density variation the OH coils have to
be placed outside the toroidal field coils in such a spatial
distribution, that the field of the OH windings in the plasma
region is of the order of 10s of Gauss (Fig. 1). This means that
most of the OH coils will form a straight long cylindrical coil
inside the torus magnet with the remaining ones distributed outside along the torus magnet surface in such a distance as not to
cause large local pulsed fields. The straight cylindrical OH
magnet section is one of the most critical parts of the overall
magnet system since it influences directly the machine design and
the Tokamak scaling. In order to achieve economically interesting
regimes all Tokamak designs are being squeezed towards a low
plasma aspect ratio, which mainly involves the inside blanket and
shield dimension and the flux density swing in the machine core
to produce the required flux swing which in many designs contains
a safety or uncertainty margin of 2. This then leads to as high
as possible fields in the inner cylindrical OH coil. Because of
its moderate radial dimensions, even considering fatigue during
the machine life, tensile stress in this coil is not a major

problem. Consequently one finds the highest $\dot{B}$ values in Tokamak
designs in the self field of the central OH coil. These
characteristics have led so far to almost an exclusion of iron
cores in the very large Tokamak concepts and e.g. in JET to a
highly saturated core. There are estimates for iron core versions
of OH coils which seem to indicate that a complete iron core with
return yokes will yield no major advantage, but heavily burden
the design and construction with thousands of tons of magnetic
plate.[27] One concern, however, remains about the air core concept,
and that is the range of the poloidal magnet dipole field, which
e.g. for the UWMAK II design is still about 30 G at a distance
of 100 m from the reactor center line. This may call for some
kind of flux shield close to the reactor. Another effect of the
pulsed poloidal fields of course is the necessity of insulating
breaks for about 1000 V in all metal structures like mechanical
supports, blanket and shield structure and cryostats arranged
around the torus axis. Despite these insulating breaks which
only avoid direct OH transformer short turns detailled eddy
current loss assessment especially in the cold support structures
is important in order to minimize excess heating there. Fig. 3
shows some eddy current decay time constants at different
temperatures.[28] Recent more detailed studies of EPR magnets and

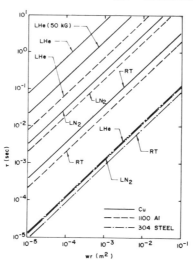

Fig. 3   Decay time constant for currents in a cylindrical
shell of radius r and thickness w.

their support structure have revealed the eddy current problem in large Tokamak magnet structures as a major design constraint.[29]

The equilibrium field winding which may be combined with a shield winding to be explained later has to provide an essentially plasma current proportional but adjustable vertical field with a spatial distribution according to the plasma current distribution which controls the plasma position according to the plasma current and radius, plasma current distribution and the poloidal plasma pressure. Table V lists the vertical field values required for some Tokamak designs together with the quantity $\frac{I_p}{R}$ mainly determining the vertical field. Since the vertical field should

TABLE V	$B_V(T)$	$I_p/R(10^4\frac{A}{CM})$
JET	~0,47	1,62
T 20	0,42	1,2o
ANL EPR 1976	0,46	1,21
ORNL EPR 76	~0,34	1,o7
UWMAK II	~0,44	1,15
UWMAK III	~0,85	1,95

increase in large Tokamaks especially for noncircular plasma cross sections the requirements of $\dot{B}$ in the EF coils may not much easier to be met in future apparatus unlike the OH windings. Another difference between the OH windings and the EF windings concerns their maximum field impact on the toroidal field coils. Whereas with an outside OH winding the maximum local pulsed field in the region of the toroidal field winding can be restricted to about 0.1 T[9], about 3 - 7 times higher fields have to be expected from a simple EF winding almost regardless whether this winding is placed outside or inside the toroidal field coils. For symmetry the equilibrium field windings have to be distributed above and below the plasma ring. Placing them outside the toroidal field coils would be desirable for

easier fabrication and assembly. Great care has to be taken concerning their spatial distribution under the constraints of
access for pumping ports, cooling ducts and auxiliary heating
systems but also avoiding excess local field peaking in the
toroidal field winding due to large EF ampere turn concentrations
close to it. Fig. 4 shows this for one of the EPR designs.[9]
Complete penetration of the vertical field generated by the EF
coils across the lower and upper parts of the toroidal field
coils (see Fig. 5)[13] will cause there undesirable heating and

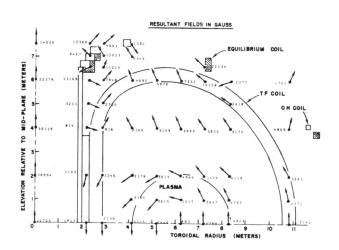

Fig. 4

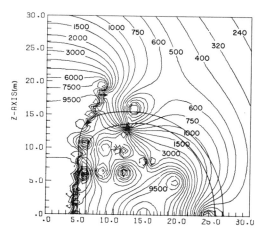

Fig. 5

local twisting and overall bending moments. The latter can be rather large (see also Fig. 2). Table VI gives some corresponding figures for Tokamak designs. In UWMAK II the maximum repetitive

TABLE VI
MAXIMUM OUT OF PLANE FORCE / LENGTH
ON TOROIDAL COIL FROM POLOIDAL FIELDS

ANL EPR 1976	1.0×10^7 N/M
ORNL EPR '76	1.4×10^6 N/M
UWMAK II	1.2×10^7 N/M

lateral toroidal coil movement due to the combined forces from all the poloidal coils has been calculated to be about 1 cm which could mean about $7 \cdot 10^{-4}$ relative vertical field component from the toroidal coils. This is twice as much as has been permitted in the T20 design. In future Tokamaks of the flux conserved type[30] or with a non-circular cross section, the required vertical fields will go to the upper limit just mentioned. This would mean that the corresponding ac losses in the toroidal field magnet and the pulsed forces would become then rather large. The ORNL EPR indicates therefore the possibility to place the EF winding inside the toroidal magnet which of course has its great draw-backs like winding them in place and the necessity of vertically movable supports to enable maintenance and repair in the plasma chamber, blanket and shield region (Fig. 1). The special feature of the ORNL EF winding[5] is that it serves at the same time as a shielding winding for the plasma currents poloidal field, thus reducing the maximum vertical field component in the toroidal field windings. The combined so-called shield-equilibrium winding has a current distribution consisting of a first shielding part corresponding to the shielding currents in an infinitely conducting shell around the plasma and a second part of the EF winding with a zero poloidal net current. In order to decouple it from the OH winding

the S-EF winding is series connected with a decoupling winding
placed close to and with a similar current distribution as the
OH winding. This additional winding with a sum current equal to
the plasma current causes some space problems in the region of the
straight inner OH coil section. The S-EF winding thus defined will
only correctly shield the toroidal field coils for one definite
plasma current distribution and one single value of the poloidal
plasma pressure. This is the reason for the reduction of the
vertical field by only some factor (about 3 - 5) whose exact
value depends on the range of β_{pol} and plasma current profile to
be corrected for. The correcting part comes from an additional
EF trim winding outside the toroidal coils.Fig. 6 shows a
comparison between the poloidal flux line pattern of an outside
EF and an inside S+EF winding scheme.[11] Although proposed for the

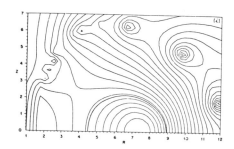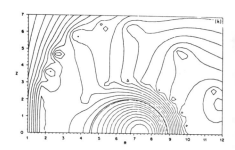

Fig. 6   Poloidal flux line pattern

          with EF winding   with S + EF winding

ORNL-EPR with normal conducting inner S+EF windings, this scheme
applying inner poloidal superconducting windings seems to be
inescapable for a pulsed full scale reactor, especially should a
non-circular plasma cross section turn out feasible. In this
case of a preferably vertically elongated plasma cross section
additionally a field shaping winding, roughly a quadrupole
winding with a zero poloidal net current is required. Since the
quadrupole field components have a rather short range and the
required fields are relatively high, this F winding can only be
situated inside the toroidal magnet. The F winding is needed to
control the plasma cross section independent of the plasma current

distribution. For a certain current distribution it can be combined with the EF or S+EF winding. It is assumed that this does not considerably increase the pulsed fields at the toroidal magnets, but the spatial distribution of the pulsed fields will be different and may lead to additional local twisting moments on the toroidal field coils.

In addition to this machine concepts of the flux conserved Tokamak type will require the possibility that single conductors or single coils of the EF or F+EF winding can be controlled independently such as to conserve the poloidal flux like in an infinitely conducting shell surrounding the plasma.

Since the blanket and shield dimension will separate the plasma surface from the location of the EF or F+EF winding, there will be in the case of the flux conserved Tokamak the need for a vertical field trim winding to center the plasma ring and a quadrupole trim winding to enforce the required plasma cross section despite of the distance between plasma surface and EF or F+EF winding.

As already mentioned a poloidal D winding seems to be compatible with an EPR or reactor situation. For topological reasons, in order to save magnetic energy and to limit the external field loading on the toroidal magnet the D winding will have also to be placed inside like the S+EF winding. The topological aspects more specifically concern the geometry of the particle collector plates, the shape of the diverted particle streams and the location of the stagnation line. A localized poloidal divertor as shown in Fig. 1 has a winding of zero net current. It creates, however, locally rather high fields. Should it turn out necessary to substantially vary the divertor winding current in time, then the outer two D windings could be distributed to shield the toroidal magnets against the divertor fields. Concerning its radial position such a distributed divertor winding could be integrated in the S+EF winding.

For the divertor windings as well as for the other poloidal coils the local magnetic fields at the windings and the field variations are rather large, typically about half the values occurring at the inner OH-winding.

2.3 Mechanical structure, safety

All poloidal windings arranged inside the toroidal magnet are
subject to an additional constraint when compared to the outside
ones: they have to operate in a high parallel field approaching
the maximum toroidal flux density at the inner circumference of
the torus magnet. Besides of the need to design these windings,
if superconducting, especially for these high parallel fields,
there is a mechanical influence on them from the toroidal magnet.
With a necessarily limited number of toroidal field coils there
is a large spatial modulation of the toroidal field. Since the
large radius inner poloidal windings are rather close to the
toroidal coils they will see considerable radial field components
which produce in plane and out of plane forces on the poloidal
field coils (Fig. 7).[11] This calls for a rather stiff mechanical

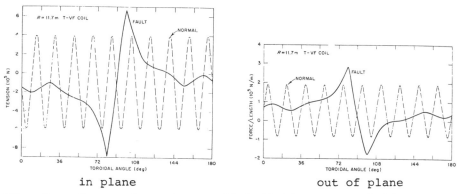

in plane   out of plane

Fig. 7   Forces from the toroidal field on one poloidal
            field coil

structure especially for the support of inside poloidal coils
which otherwise can experience a helical instability as recent
magnetoelastic calculations have shown (Fig. 8).[24] In the case of
of a failure in the toroidal magnet system like loss of current
in a coil the interaction forces on the poloidal coil system would
become even larger. Since, however, it is very unlikely that under
such circumstances the plasma ring could remain stable any further
but would rapidly disintegrate and possibly damage the first wall
it is inferred here that in any case of a major failure in the

406

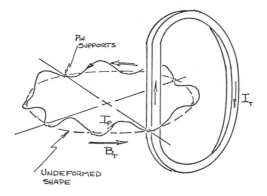

Fig. 8   Helical "instability" of poloidal coil

integral magnet system calling for a discharge of the toroidal
magnet, the poloidal magnet system would have to be discharged
first. This implies further consequences on the ability of the
toroidal coils to stand faulty condition for some time, that is
to be built with a high degree of electrical stability and to
have an early fault detection scheme.

The possible failure modes in the poloidal field coil system
are manifold in principle since they involve the plasma
behaviour itself. Up to now the experience tells that a disruption may cause a plasma current decay several times faster
than the initial current rise. In this situation the S+EF winding as described above may help in preventing a disruption or
to slow it down. Emergency shielding to slow down the local
field variations in order to protect the superconducting inner
poloidal windings has not been considered so far. As has been
calculated for JET, the sudden loss of plasma current with the
other poloidal currents unchanged will lead to approximately
doubling of the tilting moments on the toroidal coils. Such an
effect can be smoothed out by an inner S+EF winding.

To cope with failures occuring in the poloidal coil system means
very roughly the following: First the plasma current must be
driven to zero. This can be done either by the OH or the inner

407

S+EF winding, provided the latter has been decoupled from its compensation winding. Both can provide roughly the volt seconds for a plasma current decay. After that a safety discharge to protect the failing superconducting poloidal winding from being overheated can follow. The relatively high voltages necessary in the poloidal field system (see Table IV), however, increase the likelyhood of a dielectric failure there. Such failures of course can lead very rapidly to asymmetries in the poloidal field pattern implying again the danger of the plasma hitting the wall. Therefore a series connection of poloidal field windings symmetric about the machine mid-plane and redundant parallel coils to preserve the S+EF field pattern may be required. The definition of the whole poloidal magnet system needs a lot more experience with windings closely coupled to the plasma especially for non-circular plasma cross sections.It may be appropriate to build the first TNS or even EPR size machines relying on inside normal con-ducting poloidal coils like proposed in the ORNL and GA designs.[31,32,33,34]

Designing the inner poloidal windings and laying out their manufacturing process has as a prerequisite the detailled logistics of the whole machine assembly-procedure. Maintenance and repair operations for the plasma chamber, blanket and shield have to be specified before inner poloidal coil design and they cannot cover inner superconducting poloidal windings in an easy way, since so far no obvious solution for a sectional superconducting poloidal field coil[35] design is known. Reflecting the problem of dielectric failures in the toroidal or poloidal magnets and remembering again the possible catastrophic effect of a rapid field distortion one must state that the electrical insulation in the toroidal and poloidal magnets must not involve the slightest likely risk. This may have the consequence that only high current hollow conductors or more generally such winding concepts can be permitted which decouple superconductor cooling from electrical insulation.

The toroidal and poloidal magnets both determine together the basic mechanical structure of a Tokamak. The croygenic envelope of the toroidal magnet will partially include also the OH-coil following the favourable tendency of low aspect ratio. The toroidal coils will be essentially hanging from and pressing against the grooved vertical cold support cylinder and need an additional support from below against the dewar wall. Depending on the stiffness of the coil together with its helium dewar another support from above may be required. These additional supports have to take into account in their design the dimensional changes of the toroidal coils depending on temperature and current which should remain symmetric about the reactor mid plane. The main lateral support of the toroidal coils themselves will be most likely against their dewar walls via sliding thermally insulating struts designed for carrying the transient out of plane forces during safety discharge or the poloidal field interaction forces which ever are larger.[13,14] Retractable vertical support pillars from above and below for the inner poloidal field windings seem to be the most compatible solution for horizontal access requirements of maintenance and repair operations in the vacuum chamber, blanket and shield region. In order to stiffen the poloidal coil system it will be necessary to mechanically connect the retractable supports in the radial direction. The necessary supports to resist gravitational forces will be separate for the toroidal and poloidal coils. The central column besides of carrying essentially the weight of the toroidal coils can serve as a mechanical restraint for the outer poloidal coil system in separate or common annular cryostats.

The overall picture would not be complete without mentioning that a high reliability of the toroidal magnet system is called for because it can hardly be dismantled and repaired during the lifetime of the machine without causing excessive downtime. To dismantle any of the poloidal coils except for the upper few seems almost equally impractical. The safe operation of a complete Tokamak magnet will have to be ensured also by redundancy in the cooling capacity and by an efficient sensing system to monitor the regularity of the electrical and mechanical behaviour. The repetitively pulsed operation can produce cumulative effects such as fatigue or excessive motion. An automatic fast failure evaluation and emergency discharge system seems inescabable.

So far a rough description of the requirements for future Tokamak magnet systems. This description combines a large number of different features to be found in existing designs. By putting them into a suitable combination, one may gradually evolve a viable Tokamak EPR or reactor magnet configuration. The above is an early attempt to crystallize it.

Fig. 9 TOKAMAK MAGNET SYSTEMS AND DESIGNS FROM JET TO EPR

JET

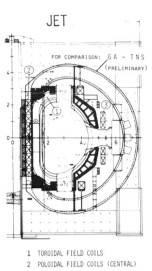

1 TOROIDAL FIELD COILS
2 POLOIDAL FIELD COILS (CENTRAL)
3 POLOIDAL FIELD COILS (EXTERNAL)

T 20

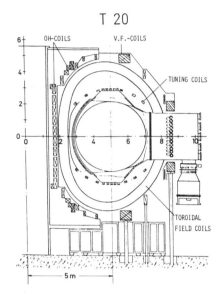

ORNL-EPR 1975

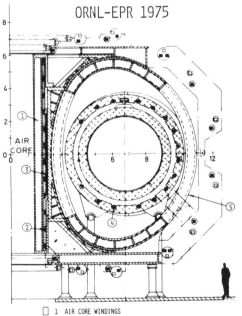

☐ 1 AIR CORE WINDINGS
■ 2 AIR CORE DECOUPLING WINDINGS
☒ 3 V.F. TRIM WINDINGS
4 SHIELDING WDGS FOR T.F. COILS / V.F. WDGS
5 T.F. COIL (CONDUCTOR, BOBBIN, DEWAR)

ANL-EPR 1975

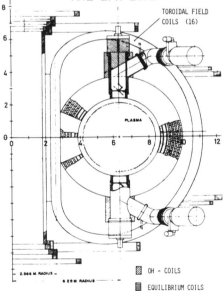

▨ OH - COILS
▤ EQUILIBRIUM COILS

SCALE

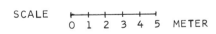

0 1 2 3 4 5 METER

410

3.   Some design considerations

3.1 General

Today no realistic detailed EPR design is possible yet, but the
EPR design studies, many of them from the United States but also
from Europe and Japan, have brought forward interesting proposals for future technical solutions. These proposals together
with some existing detailed work on fusion relevant superconductors and background experience from building much smaller
than EPR-magnets still form the narrow basis for the design of
large superconducting Tokamak magnets. Reference has been made
and will be further made to the existing designs. Fig. 9 - 11
show JET, TNS, T20 for comparison at equal scale as the ANL,GA,JXER
and ORNL EPRs[11,36] and at a scale further reduced by a factor of 2
FINTOR EPR, UWMAK II and III together with with Culham MKII
reactor design to illustrate their size and magnet configuration.
Quite characteristic for the present situation there is still a
large variety in proposed reactor data versus physical machine
size to the extent that the geometrically largest EPR design
approaches full scale reactor design dimensions and the most
compact full scale reactor design has dimensions only slightly
larger than most of the EPR designs. One notices that the
toroidal magnets look all rather similar whereas the poloidal
field systems still show considerable differences. Before having
a look to some selected aspects of magnet design one should
again bring to mind a few basic conditions:

- the superconducting toroidal Tokamak magnets we are looking
  forward to have a very large stored energy in the range of 10
  to maybe several 100 GJ. Especially in the case of geometrically linked poloidal windings they can hardly be
  repaired without almost completely devaluating the corresponding experiment or reactor because of excessive downtime.

- structural stresses of the order of 200 $\frac{MN}{m^2}$ or more will occur
  in these large magnets. The stresses are largely determined
  by the permitted strain in the stabilization material for NbTi
  superconductors and by the permitted strain in the superconductor itself for Nb_3Sn.

411

Fig. 10   TOKAMAK MAGNET DESIGNS

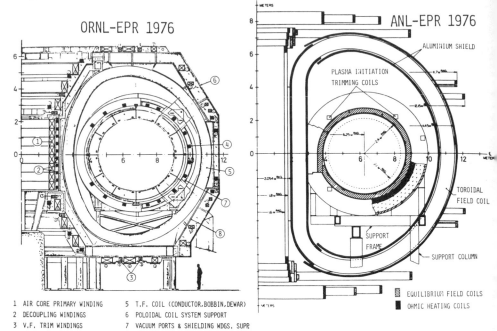

ORNL-EPR 1976

ANL-EPR 1976

ALUMINIUM SHIELD

PLASMA INITIATION
TRIMMING COILS

TOROIDAL
FIELD COIL

SUPPORT
FRAME

SUPPORT COLUMN

1 AIR CORE PRIMARY WINDING   5 T.F. COIL (CONDUCTOR,BOBBIN,DEWAR)
2 DECOUPLING WINDINGS   6 POLOIDAL COIL SYSTEM SUPPORT
3 V.F. TRIM WINDINGS   7 VACUUM PORTS & SHIELDING WDGS. SUPP.
4 SHIELDING / V.F. WINDINGS   8 SHIELD & BLANKET SUPPORT

▨ EQUILIBRIUM FIELD COILS
■ OHMIC HEATING COILS

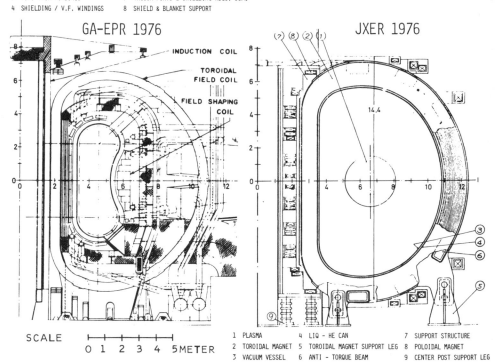

GA-EPR 1976

INDUCTION COIL

TOROIDAL
FIELD COIL

FIELD SHAPING
COIL

JXER 1976

(7)(8)(2)(1)

14.4

SCALE   ├─┼─┼─┼─┼─┤
        0 1 2 3 4 5 METER

1 PLASMA   4 LIQ - HE CAN   7 SUPPORT STRUCTURE
2 TOROIDAL MAGNET   5 TOROIDAL MAGNET SUPPORT LEG 8 POLOIDAL MAGNET
3 VACUUM VESSEL   6 ANTI - TORQUE BEAM   9 CENTER POST SUPPORT LEG

Fig.11 TOKAMAK MAGNET DESIGNS FROM EPR TO FULL SCALE REACTOR

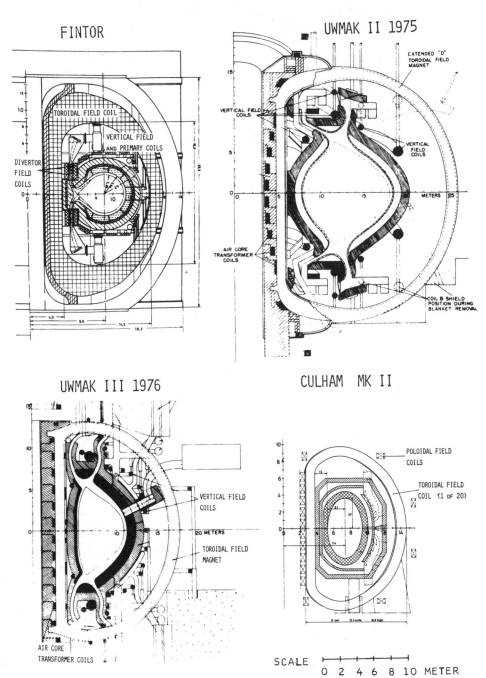

FINTOR

UWMAK II 1975

UWMAK III 1976

CULHAM MK II

SCALE 0 2 4 6 8 10 METER

- Superconducting magnet electrical insulation schemes are subject to the restriction that He gas will be present in any emergency situation and is a very poor insulator.
- With a low voltage restriction a large magnet's energy can only be removed from the system at very high current which is the operational current.
- No rapid and large temperature excursions in large magnets will be tolerable, otherwise plastic deformation due to differential strain will occur. Excessive temperature excursion must be prevented by supplying enough stabilizer, sufficient cooling and proper energy removal in an emergency case.
- Superconductors fulfilling the above requirements will have substantial dimensions like heavy electrical machinery normal conductors.
- The maximum out of plane load by one toroidal coil failed has to be taken by the mechanical structure.
- The elastically stored mechanical energy in large magnets is considerable. Conductor movement or slippage in the winding therefore must not occur.
- The magnet's stored energy per unit winding volume increases with magnet size.
- The pulsed poloidal field pattern will be variable in time and space. Efficient control of high power density plasmas will only be achieved by poloidal windings as close as possible to the plasma in a distance controlled for super-conducting poloidal windings by the blanket and shield thickness. This means that a considerable part of the poloidal field magnets is placed inside the toroidal field magnet.
- In a working large Tokamak the plasma current must be turned off before an emergency discharge of the toroidal field magnet.
- The refrigerator electric power should remain in the range of percents of the rated reactor power.

These points indicate that the Tokamak machine size plays an important role in the sense that below a certain size relevant

414

models for large magnets cannot be built. This can e.g. be simply seen from the fact that in a Tokamak for a given a, A and toroidal field the plasma current density scales like $\frac{1}{R}$. Therefore increasing the dimensions eases the current density requirements in other windings, too. The transition point where fully stable superconducting Tokamak toroidal magnets with a maximum magnetic field of 8 T become possible for shear spatial reasons, lies around a torus radius and coil diameter of 3 m. Since the forces increase with magnet size, the size dependant relationship between current density and stress is important. Smaller magnets have to be built partially stabilized; they will quench rapidly at any surplus heat input, protection will scarcely be possible without quenching all the rest of the coils, too, and without excessive refrigeration such an event causes long interruption times.

One should just state that JET size is about the smallest stable superconducting Tokamak magnet coil one should consider, which in turn would leave a further range for scaling up in linear dimensions by a factor of 2 for the TNS or EPR and after that maybe by another factor between 1 and 2 for the full-scale reactor magnet.

3.2   Toroidal coil shape, strain

Regardless of the size, the appropriate shape for large toroidal coils is at first glance the constant tension D-shape (modified D) which can be described according to Fig. 12 in an approximate analytical expression taking into account a finite coil number:[38,39]

$$\frac{\varrho}{R} = \left[\frac{\varrho_2}{R_2}\left(1+\frac{1}{N}\right) - \frac{1}{N}\ln\frac{R}{R_2} \right] \Big/ \left(1+\frac{1}{N}\cos\phi\right)$$

As has been shown this analytical expression agrees quite well with an iterative solution developed elsewhere[40] (at least for N > 8). There are problems, however, with this shape; mainly because the width of the coil determines its height depending on the inside torus radius and therefore may yield a magnet too high and storing too much energy for the purpose. This shows up

415

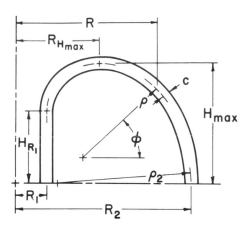

Fig. 12

in the T20 coil design or the ORNL-EPR coil (see Figs. 9, 10, 11)
where the inner straight section has been replaced by a curved
one. The price to be paid for this is an increase of tension in
the bent inner section and a consequent shear zone at the
transition of the bent inner section to the outside D-shape. In
the UWMAK II and III coil designs the straight inner section is
retained but it is connected to a D-coil which is too broad in
UWAMK II or too narrow in UWMAK III when compared to the
practical D-shape. This leads to additional or less tension in
the straight section and bending moments in the curved part of
the coil. Quite naturally such deviations from the constant
tension case, where of course even in the modified D-shape the
tensile stress is never really constant but shows deviations of
$\pm$ 10 to $\pm$ 20 % across the coil cross section, requires a special
fixation for almost any turn of the winding in order not to sum
up the shear forces but to transfer them directly into the force
restraining structure. This can e.g. be solid grooved coil disks
like in the UWMAK reactor designs or a honey comb structure as
proposed in the ORNL-EPR.[11] In any case the wall thickness and con-
ductivityof this force restraining structure has to be adjusted
to the magnetic field variation of the poloidal fields for
limitation of eddy currents. This may rule out high strength
aluminium structures for EPR size machines with their assumed
rapid field variations.

416

An interesting coil shape resulting from using the modified D-shape for every winding layer is presented in the FINTOR I design.[27] Apart from the above mentioned coupling of the shape to the torus radius with such a coil shape additional magnetic energy is stored in the extended upper and lower coil cross sections.

Since a variation of the conductor and load bearing winding cross section along the circumference of the coil is not feasible because there is no conductor joining method producing a perfect super-conductor transition, because such a variation would be exceedingly expensive and also since shear stress in a winding should be small in order to exclude the possiblity of conductor slippage, a coil shape close to the D-shape in principle makes the best use of the material spent. The amount of material in the conductor is given first by the stabilizer requirements. Depending on field and coil size additional restraining material such as stainless steel or high strength aluminium has to be added. Within the given strain limits there is a certain limited choice of coil shapes deviating from the practical D-shape as mentioned above. The locally per-mitted strain limit, which sets essentially the mechanical design, is given for the very ductile NbTi superconductor only by the properties of the stabilizer material (for copper 0.1 - 0.15 %), for Nb_3Sn the superconductor itself very likely will determine the upper strain limit, which is a more risky situation. Since it appears that the permitted strain on Nb_3Sn is very low, it may turn out that it will be hard to achieve more than 12 T maximum field in large magnets.

Copper as a stabilizer material shows a strong magnetoresistance effect. Therefore in order to minimize the material in the coil stabilizer grading in the conductor has been proposed,[42] e.g. by using conductors with 3 different fractions of copper stabilizer for the higher, medium and low field regions at the inner part of the coil. Because of the strong $\frac{1}{R}$ dependance of the toroidal flux density this means, however, that with copper stabilizer grading but necessarily constant conductor cross section along the coil circumference still too much stabilizer material is being spent. It also means that in terms of magnet stability the inner coil section is the most critical one. In the case of an emergency discharge this section would be heated preferably, thus causing differential strain. An equal fraction of stabilizer in the whole winding could be envisaged with aluminium whose magneto-resistive effect saturates at low field.[43] Also its lower resistance at 4 K compared to copper may reduce the amount of material needed Since high resistance ratio aluminium is very soft, its application depends on more information about its repetitive resistance-strain behaviour. In any case it needs e.g. stainless steel restraint, possibly directly bonded to it in order to enforce an apparent elastic behaviour possibly at strains which permit full usage of the stainless steel like proposed in the UWMAK III design.[44]

3.3 Safety discharge

Protection of fully stabilized magnets by a safety discharge is governed by two simple equations:[45,6] The sum voltage over all the coils connected in n parallel branches is per branch

$$V_{branch} \simeq \frac{\sum V}{n} = j^2 \, \frac{W_{mt}}{n \, J \, f(\theta)}$$

W_{mt} — total stored energy

$f(\theta)$ — protection function for given temperature rise

and the relationship between stabilizer current density j and conductor current J is

$$J = \left(\frac{K q}{\varsigma}\right)^2 \cdot \frac{1}{j^3}$$

K — geometry factor

ς — stabilizer resistivity

$\dot{q}$ — heat flow to the coolant

The corresponding time constant of the magnet discharge is

$$\tau \simeq \frac{2 W_{mt}}{J \sum V}$$

Connecting the coils and/or conductors in parallel branches can help to reduce the maximum voltage level at safety discharge at the expense of multiplying the number of separate power supplies.

A safety discharge will be started after a voltage sensing system monitoring the sum of resistive and conductor internal inductive voltage has detected a persistent or increasing normal conducting

418

zone, which in a fully stabilized magnet can occur e.g. by local loss of coolant due to channel blocking, excessive energy input to a conductor by movement or a rapid field change causing excessive losses beyond the design limit. Fig. 13[9,10,11,14] gives a couple of discharge schemes suggested for the toroidal magnets of above

Fig. 13

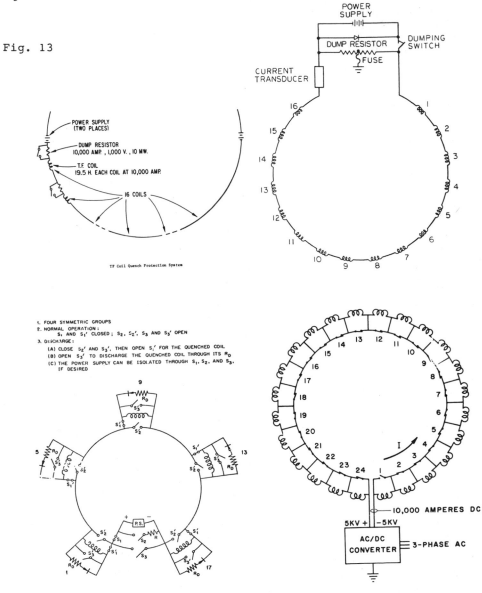

419

mentioned designs. Tab. VII lists the relevant data for the
safety discharge according to these schemes. None of the proposals presents a transient analysis so far. Fig. 14 taken from
the JET design[46] shows, how the maximum out of plane load in discharging the faulted and the other coils together can be
diminished using the mutual coupling of the toroidal coils.
J_2 is the current in the faulted single coil suffering a low
resistance short at the terminals, J_1 is the current in the
other series connected coils shorted by a switch with a time
lag ΔT.

The safety discharge of the poloidal coil system has yet to be
studied in detail.

TABLE VII	I_{COIL}(KA)	$V_{MAX\,COIL}$ (V)	τ(s)	PARALLEL BRANCHES	LEAD PAIRS	COILS DISCHARGED
ANL EPR 1975	10	1000	195	1	16	ALL
ANL EPR 1976	60	2000	506	1	1	ALL
ORNL EPR 76	78	1300	28	4	4	ALL
	78	300	10	4	4	5 OF 20
UWMAK III	10	10000	32	1	24	1 OF 24
	10	417	2280	1	1	ALL

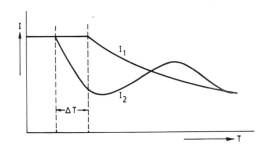

Fig. 14

3.4 Cooling

The field of cooling for a fully stable magnet is still under
broad discussion and even the definition of full cryostability
still has to be settled. In detail it requires to take into
account any possible superposition of steady state and transient
heat loads, the transient temperature distribution and the
transient cooling properties in the different spatial directions
relative to the conductor. A thorough general treatment of these
items does not yet exist. There appear 2 main possibilities,
each of which having two alternatives: firstly the immersion of
the magnet into a bath of liquid or superfluid helium is possible.
Liquid helium at 4.2 K may provide a heat flux of 0.3 W/cm^2 of
cooled conductor surface for vertical surfaces and a temperature
difference between conductor surface and helium of about 0.5 K,
but this in a large magnet would appear to be a rather uncertain
figure, since for a large surface to be exposed any surface
orientation can occur in a coil, and other than vertical orientation may decrease the possible heat flux by at least an order
of magnitude. The consequence is to use only the vertical surfaces for cooling and this preferably leads to pancake toroidal
coils like in the UWMAK designs. Immersion in superfluid helium
at 1.8 K would drastically improve superconductor performance but
at the expense of a high relative room temperature refrigerator
power and at the risk of large sealing problems. All solutions
where the winding is simply immersed in the coolant may cause
voltage problems since the cooling surface cannot be insulated
and thus the conductor may in fact be insulated against the
restraining structure just by a sliding distance of the order
of the thickness of the conductor insulation. This in an
emergency case with He gas production with a corona inception
gradient of 200 V/mm may lead to internal arcing, about the
worst fault which can happen to a superconducting magnet.[43]

The second possibility of forced cooling in hollow conductors
or ducts between conductor bundles allows to increase the
cooling heat flow by increasing the cooled surface and a consequent reduction of the stabilizer material. There is, however,
only a short maximum length of conductor which may go normal

after a sudden heat input and still will recover after that.
Also there is of course a restriction in the pressure drop from
channel inlet to outlet, which determines the pumping power
needed. Since pumping must be done at He-temperature, the
pumping power adds to the refrigeration power. There are two
possibilities of forced flow cooling. One is two-phase helium
at a pressure drop of about 0.3 bar with 4.5 K inlet temperature
and 4.2 K outlet temperature. One can reach 0.4 W/cm^2 at 0.5 K
temperature difference between conductor surface and helium in
cooling channels $\sim$ 150 m long.[47] This is about the same cooling
rate per unit surface cooled as for 4 K helium bath cooling
but now the orientation of the cooled surface is rather
irrelevant and the conductor can be fully insulated. Another
advantage of the scheme is that in case of a local sudden energy
input almost, like in the case of bath cooling - but somewhat
reduced according to the two phase situation - latent heat of
liquid helium will be available. In the case of two phase forced
He cooling a typical length of conductor which can recover from
a normal transition is about 5 m. Such a figure is favourable in
that it sets the lower limit of the normal zone detection
voltage to about 0.3 V. There are, however, questions about the
applicability of this scheme because of the two phase situation.
The second possibility of forced cooling applies high pressure
supercritical helium with a minimum outlet pressure of about
3 bar.[11] Supercritical helium applies naturally in the case where
an increase of the heat transfer is possible via an increase of
the helium wetted perimeter per unit conductor length, e.g. when
a finally subdivided conductor is needed in order to lower the
eddy current losses. In the case of such a bundle[48,49] conductor the
pressure drop along a conductor length of about 400 m may be
5 bar, thus calling for a large pumping power.[11] The maximum length
which can recover from a sudden heat pulse is at least of the order of several
meters. Of course rather high average current density can be achieved in a
force cooled bundle conductor but due to this high current density the discharg
in case of emergency must be rather fast.

In order to give a rough impression about the energetics involved
in superconductors and their cooling,[50] table VIII gives a comparisc

TABLE VIII		ANL EPR 1975			ANL EPR 1976		ORNL EPR 1976			FINTOR	UWMAK III	
COND.		TF	PF	PF	TF	PF	TF	TF	PF	TF	TF	PF
COOLING		BATH	2PHASE	BATH	BATH	BATH	SUPCRIT	SUPCRIT	SUPCRIT	2PHASE	BATH	BATH
S C		NbTi	NbTi	NbTi	NbTi	NbTi	Nb$_3$Sn	NbTi	NbTi	NbTi	NbTi	NbTi
I	kA	10	40	40	60	40	19.5	19.5	25	10	11	14.5
N$_{FIL}$		425262	1904877	1889360		1756500	677040	19530	5621114	2760	182	182
D$_{FIL}$	μ	10	5	5	10	5	1/10ø	40	4	260	380	380
SUDDEN HEAT LIMIT							100	300				
HEAT LIMIT		250	~2250	250			300	700		1184	~250	~250
AC LOSS PER CYCLE	mJ/cm^3	12.2	18.3							7.56		20.1
AC LOSS PER TRANSIT							48	144				
PUMPING LOSS		—	—	—	—	—	82	82		—	—	—
COND 0.1MM 50 MN/M		100	100	100	100	100	100	100	100	100	100	100
STORED EN/COND. VOL.	J/cm^3	97.5			226		232	232		137	394	56.8

for the ANL-EPR, ORNL-EPR, FINTOR and UWMAK III. Looking into
the heat inputs per unit volume of conductor it is interesting
to note that the capability of all 3 cooling systems represented
in the list of Table VIII are rather similar. Owing to a 10 %
helium fraction the ANL and UWMAK III toroidal coil conductors
can take a maximum heat load of 250 mJ/cm^3 and the ORNL EPR
shows values of the same order. Its forced cooling scheme has
been optimized versus pumping power. Despite this fact, the heat
load from pumping is of the same order of magnitude as the ac-
losses. For all 4 designs the heat input from ac-losses per cycle
is in the range between 3 and 20 $\frac{mJ}{cm^3}$. These are averaged data.

Locally the difference between heat input and cooling limit may be much smaller. It should also in comparison be born in mind – which throws a light on possible difficulties with conductors and winding schemes which cannot be absolutely fixed – that at a transverse pressure of e.g. 30 MN/m^2 a conductor movement of 0.1 mm will cause a heat input of about 100 $mJ/cm^{3 50}$. Conductor and winding designs such as the two phase helium cooled poloidal field cable conductor for ANL-EPR or the FINTOR toroidal field hollow conductor, however, can offer 1000 to 2000 $\frac{mJ}{cm^3}$ at the expense of a large helium inventory. It may be further of interest to note that the stored energy per unit conductor volume, which of course increases with the machine size, varies between 90 and 400 J/cm^3 thus again pointing to the crucial safety discharge problem.

As can be seen from table IX showing the single loss contributions, one important conductor design criterium for Tokamak magnets is to minimize the ac-losses and possibly about equalize the single loss contributions. As can be seen, going from EPRs

TABLE IX   HEAT LOADS IN SUPERCONDUCTING MAGNETS (W AT 4 K)

	ANL 1975	ANL 1976	ORNL 1976	UWMAK III
TF COILS	WITHOUT FIELD SHIELD	WITH AL-SHIELD		
CONDUCT. SURFACE RAD. HYST.+ EDDY RADIAT.	} 208 55832	160	1400 500 1300	1430 375 0 (?)
PUMPING SHIELD	256 — —	1500 4800 + 136000 (18 K)	16000 6380 —	600 —
OH + EF COILS				
CONDUCT. SURF. RAD. HYST.+EDDY RADIAT. PUMPING	} 2090	} 331 3594 80	1500 1600 1360 200	1600 231 1740
TF COILS LEADS PF COILS LEADS CONDUCTOR JOINTS TRANSFER LINES		120	8240 (EF COILS NORM.COND.) 780 L/H (4 PAIRS) 120 L/H (2 PAIRS)	1718 6730 300 800

424

to the full scale reactor eases the conductor design, since less and larger superconductor filaments will suffice. The proposed poloidal conductors for currents between 15 and 40 kA - one ORNL EPR[36] again foresees paralleling them up to 175 kA - in the EPR-designs require millions of filaments compared to about 200 for UWMAK III. The very high hysteresis losses of the earlier ANL-EPR toroidal field conductor shows the problems connected with outside vertical field windings and the high pumping power in the coil system of the ORNL-EPR seems to indicate that super-critical helium cooling of bundle like conductors if at all can only work with more space for the coolant flow, that is at a lower overall current density. This might call e.g. for a con-figuration like the Karlsruhe conductor[51] as shown in the previous lecture. The large radiation heating in the ORNL EPR magnet certainly would be reduced by 10 cm more shielding thickness to about an order of magnitude less. Also it would appear that the helium ducting losses in large installations can easily come into the kW-range and should be considered.

Figs. 15-19 give a collection of conductor and winding cross sections for a few selected designs.[9,10,11,14,27] Looking at the various con-ductor and winding proposals one should mention that a thorough comprehensive study of the different cooling schemes also in-cluding transient cooling and their application to different conductor and winding configurations[11,44] is a task which has just been identified. For the detailed design of a fully stabilized magnet there are still many alternatives and we do not yet know the optimum solutions.

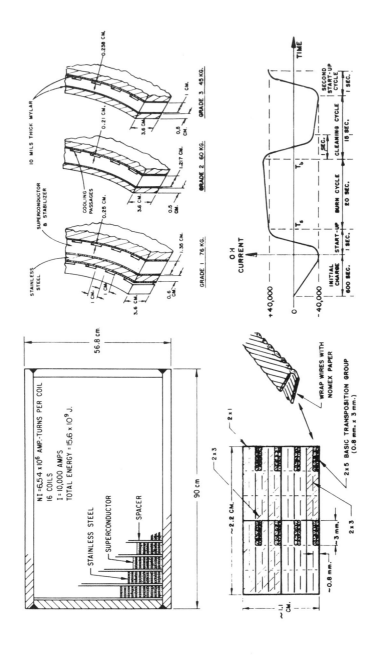

Fig. 15   ANL-EPR 1975: TOROIDAL COILS WINDING AND CONDUCTOR CROSS SECTIONS
OH-COIL CONDUCTOR CROSS SECTION AND OH-CURRENT VS. TIME

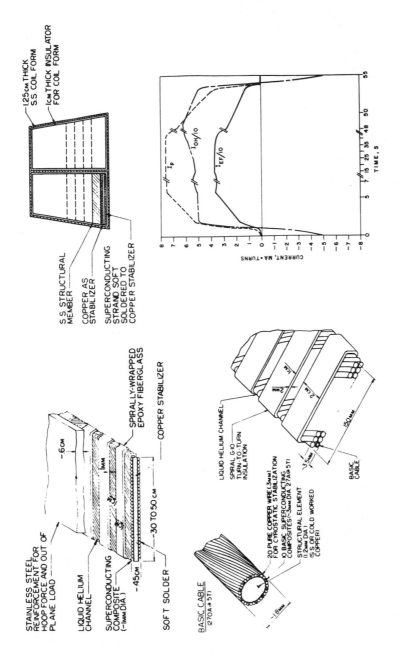

Fig. 16   ANL-EPR 1976 : TOROIDAL COILS WINDING AND CONDUCTOR CROSS SECTIONS
OH-COIL CONDUCTOR CROSS SECTION AND OH-CURRENT VS. TIME

427

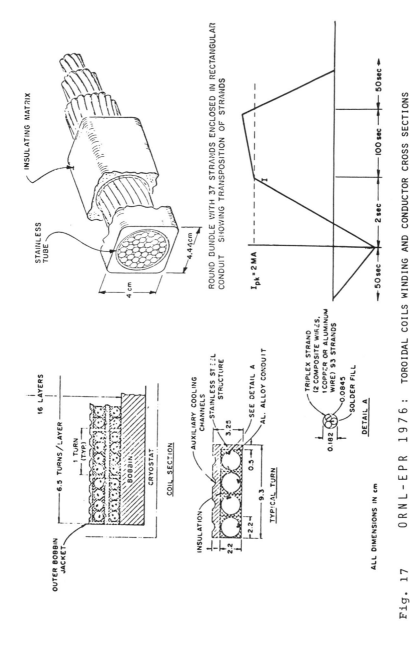

INSULATING MATRIX

STAINLESS TUBE

4.44 cm

4 cm

ROUND BUNDLE WITH 37 STRANDS ENCLOSED IN RECTANGULAR CONDUIT SHOWING TRANSPOSITION OF STRANDS

$I_{pk} = 2 MA$

50 sec — 2 sec — 100 sec — 50 sec

16 LAYERS

6.5 TURNS / LAYER

1 TURN (TYP)

OUTER BOBBIN JACKET

BOBBIN

CRYOSTAT

COIL SECTION

AUXILIARY COOLING CHANNELS

STAINLESS STEEL STRUCTURE

SEE DETAIL A

AL. ALLOY CONDUIT

INSULATION

3.25

0.5

9.3

2.2

2.2

TYPICAL TURN

TRIPLEX STRAND (2 COMPOSITE WIRES, 1 COPPER OR ALUMINUM WIRE) 93 STRANDS

SOLDER FILL

0.0845

0.182

DETAIL A

ALL DIMENSIONS IN cm

Fig. 17   ORNL-EPR 1976 :   TOROIDAL COILS WINDING AND CONDUCTOR CROSS SECTIONS

OH-COIL CONDUCTOR CROSS SECTION AND OH-CURRENT VS. TIME

428

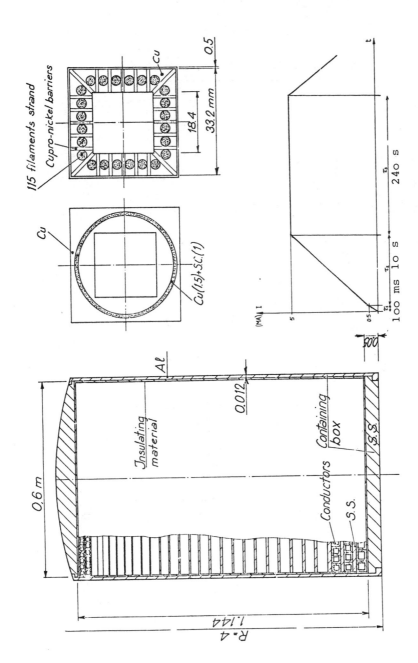

115 filaments strand
Cupro-nickel barriers
Cu
0.5
18.4
33.2 mm

Cu
Cu(1.5)+S.C.(1)

I (MA)
5
0.5
100 ms
τ_1
10 s
τ_2
240 s
τ_3
t
300

Al
Insulating material
0.012
Conductors
S.S.
Containing box
S.S.
0.6 m
R = 4
1.144

Fig. 18   F I N T O R 1 9 7 6 : TOROIDAL COILS WINDING AND CONDUCTOR CROSS SECTIONS

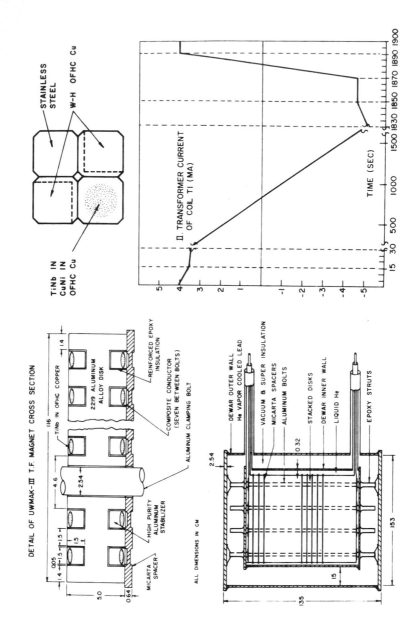

Fig. 19 UWMAK III 1976 : TOROIDAL COILS WINDING AND CONDUCTOR CROSS SECTIONS
OH-COIL CONDUCTOR CROSS SECTION AND OH-CURRENT VS. TIME

3.5 Reliability

The availability of an integral Tokamak magnet system necessarily depends on the reliability of all its components. An easy repair scheme in case of a major magnet failure can hardly be thought of. It has been said that the toroidal and poloidal magnets must operate without any major failure during the plant lifetime. Warming up of the magnets to room temperature may occur once per 1 or 2 years for annealing radiation induced damage in the conductor stabilizer and for routine inspection. The refrigeration plant should have enough spare capacity to bridge the failure of one major refrigerator section.

For safety reasons one of the most important components in a large magnet system is a redundant monitor system to detect any irregularity as soon as possible. Especially important is the detection of effects like fatigue or gradual movement which may accumulate over many cycles. For coincidence of 2 signals required this might require up to 5 sensing devices for one quantity monitored in order to ensure cancelling of erroneous signals. Thousands of sensing devices will thus be necessary.[43]

3.6 Power supplies

Power supplies for the pulsed poloidal coils of a Tokamak magnet naturally very much depend on the current program required. With superconducting coils, where minimization of instantaneous energy and forces is essential for the OH circuit there are two principal schemes: Inserts in Figs. 15 and 19 show the typical shape for an EPR,[9] where most of the required volt seconds are spent for the current rise and in contrast give the shape for a full scale reactor with a long burn pulse where the volt second requirement of the burn pulse is much larger than for the current rise.[14] In the first case the large power demand occurs during the plasma current rise and fall and a high power fast storage device is required for these periods whereas in the case of a long burn pulse the large power demand occurs in the recharge phase of the OH coil. In these phases GVA power requirements may

431

occur for short periods of time which have to be supplied by some
special storage device which for cost reasons may be a homopolar
generator. The remaining pulse power requirements could in
principle be taken from the grid via invertor rectifiers. Recent
experience, however, indicates that the use of solid state controlled rectifiers directly fed from the grid may not be a
straightforward solution even if the pulsed power demand would
not be a problem for the grid.[52] The possible influence from the
grid on the rectifier in emergency cases and vice versa calls for
a link which electrically decouples the grid from the pulsed
magnet system such as a motor-generator set preferably storing
energy in the mass of the generator rotor itself or a superconducting storage coil.[10,11] The reliability of the overall Tokamak
magnet system strongly depends on the reliability of the pulsed
power supplies also. These have to be almost absolutely reliable
since the safe controlled shut down of the plasma current is the
first and most important step in the sequence of any safety discharge of the overall magnet system.

3.7 Scaling

Having touched a number of technical design problems associated
with Tokamak magnets in conclusion it may be interesting to look
briefly into the connection between Tokamak magnet and overall
Tokamak scaling. One can set up a simplified geometrical model
of a Tokamak according to Fig. 20.[53,54] Some basic assumptions about
the plasma quality (essentially $\beta_{pol} = \sqrt{A}$; $q = 2.5$) together with
technical constraints such as mechanical stress limits, full
stability of the toroidal magnet, flux density swing in the
transformer core and space requirement for the blanket and shield
lead to a set of consistent parameters for any size of reactor
subject to the constraints imposed. Comparing toroidal magnet
stored energies vs. reactor power for circular and modest-
elliptical ($\frac{a'}{a} = 2$) plasma cross sections (Fig. 21) shows for
the machine size between EPR and full-scale reactors the strong
dependance of Tokamak magnet requirements on the further
achievements in plasma fusion experiments and materials development. With an elongated plasma cross section NbTi magnets at

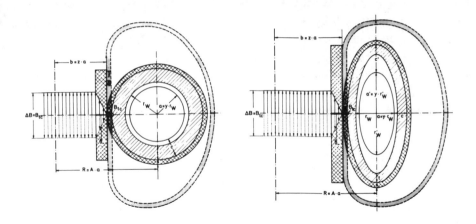

Fig. 20　　　　Tokamak reactor - scaling model

Cylindrical plasma cross section　　Elliptical plasma cross section

(CCS)　　　　　　　　　　　　　　　(ECS)

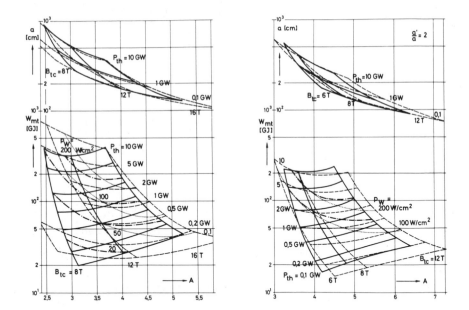

Fig. 21　　Plasma radius and toroidal magnet stored energy
　　　　　vs. plasma aspect ratio

(CCS)　　　　　　　　　　　　　　　　　　(ECS)

433

least for full-scale reactor size would fully suffice under the restrictions imposed by the first wall load limit to be envisaged at present. Higher magnetic fields calling for Nb$_3$Sn would seem to be required for low power reactors rather than for large ones in order to increase their low power wall loading. Fig. 22 shows the plasma current and the plasma poloidal field energy vs. reactor power according to the same scaling. When taking into

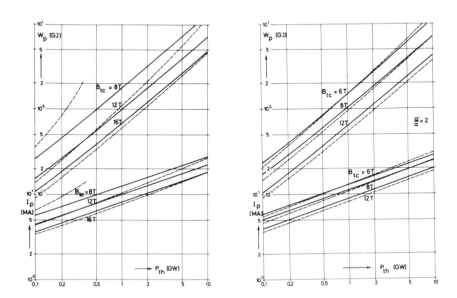

Fig. 22   Plasma current and poloidal magnetic energy
            vs. reactor power
            (CCS)   (ECS)

account that the required rate of plasma current rise in large reactors may remain the same as in smaller ones, it is again seen that the current rise time in large reactors will become longer thus rendering the ratio of pulsed power to reactor power more favourable for large units.

From the scaling the conclusion would follow that the interesting range of maximum fields in the toroidal magnet is up to 12 T for the non-circular plasma cross section which is to be favoured for its higher power density. Normal conducting OH and EF-coils seem appropriate in TNS or EPR machines mainly because of the rapid field variations called for. They should be arranged such as to minimize by their spatial distribution the poloidal field components imposed on the toroidal coils. The application of normal conducting OH coils may imply a larger plasma aspect ratio than shown in the scaling because of the otherwise very high field in the transformer core associated with an asymmetric flux swing for loss minimization.[35] Also the normal conducting OH-winding solution depends on the availability of a reliable circuit breaker for high repetition rate (every 30 - 100 s), since it might be still feasible even in a TNS to dump the transformer energy like in JET.[55]

Questions of efficiency and magnet costs are beyond the scope of this lecture. It may suffice to say that with reasonable assumptions about the energy recovery percentage in the poloidal field system the energy balance of a full-scale Tokamak reactor is not specially burdened by the losses in the magnets.[56] An assessment of the costs is even more difficult at this time. So far cost comparisons for magnets of different designs lack in compatibility and can only give a very rough indication. For an order of magnitude orientation Tab. X gives some cost figures from published designs.[10,13,14,16]

TABLE X   MAGNET COST INCL. REFRIG. AND POWER SUPPLIES IN M $

	TF	OH	EF	REFRIG.	POW.SUPPL.	TF, OH+EF	Σ
GA ITR (TNS)	48						
ANL EPR 19/6		100		20	~16		136
UWMAK II	162	21		19	4	33	229
UWMAK III	77	6	21	19	2	33	157

The obvious manifold tasks in large Tokamak magnet development have caused some corresponding program activities in several countries which will be briefly reported.

4. Programs for Tokamak magnet development

The European associated laboratories in the frame of the third pluriannual program 1976/80 which up to the final agreement on JET is only partially approved have established a development program for superconducting magnets the first two years (1976/77) phase of which has been agreed upon. This first definition phase of the program will be devoted to extending the basic understanding of superconducting materials and their use in large magnet systems and to the detailed assessment of the alternative ways of demonstrating the feasibility and reliability of large magnets. Subsequent phases would involve the design construction and operation of a large demonstration magnet at an estimated (1976) capital cost of 12 MUA corresponding to 17.5 M $ So far NbTi conductors for 10 kA at 8 T have been designed and prototypes of them constructed. Preliminary studies of the form for a demonstration magnet have been made, the choice of which will be the main result of the first program phase.[57] It is quite possible that opportunities for international collaboration in the testing of large coils will strongly influence both the form of a demonstration magnet and the time scale of its development. The ongoing discussion also include the possible construction of a complete Tokamak experiment with superconducting toroidal field coils. An organizational structure for the first 2 years phase is being formed. Table XI[58] shows a list of tasks to be covered and the associated manpower foreseen. The work during the definition phase is intended to be carried out mainly in the so-called GESSS laboratories (Karlsruhe, Rutherford, Saclay) under agreement with the corresponding fusion laboratories (Garching, Culham, Fontenay-aux-Roses) and by groups in Frascati and Petten. The program foresees an investment expenditure for test equipment of 1.5 MUA ≜2.2 M $ during the first 2 years.

Presently negotiations are under way concerning possible international cooperation through IEA at the proposed large coil test facility in the United States. This facility is part of the American Tokamak superconducting magnet program[59] to be carried out in or under guidance of the Oak Ridge National Laboratory supervised by ERDA. The program provides for the early fabrication

TABLE XI

EUROPEAN SUPERCONDUCTING MAGNET PROGRAM 1976 / 77

Summary of manpower requirements during first two years

TASK	Professional Manpower (pmy)	Sub-total
1. Outline Studies of post-JET Experiment	3	
2. Alternative Designs		
2.1 Full torus	4	
2.2 Single coil or cluster	4	
3. Definition and Preliminary Design	4	15
4. Specific Tests		
4.1 Conductors	7	
4.2 Computer programmes	6	
4.3 Cryogenics	4	
4.4 Electrical insulation	3	
4.5 Protection	5	
4.6 Test facilities	5	30
5. Supporting Activities		
5.1 Filamentary A-15 conductors	4	
5.2 Alternative stabilisation and cooling	2	
5.3 Energy transfer	3	
5.4 Radiation damage	2	11
TOTAL		56

of large EPR relevant coils by industrial subcontractors (Large Coil Project (LCP)). The aim is to utilize the experience and creativity existing in industry at every stage of the project, from conceptual design through fabrication of the coils. When completed, the industrially fabricated coils will be assembled in a test facility to provide a demonstration of the viability of superconducting magnets for EPRs. This test facility is referred to as the Large Coil Test, but the specific nature of the LCT has not been decided and still is the subject of studies. Concurrently with LCP, ORNL will undertake basic development of advanced fabrication techniques and novel coil designs. In the superconducting magnet development program (SCMDP), details are given for the design, fabrication and testing of Large Coil Segments which are similar in bore to the coils produced in the LCP but reduced in cross section so that thin coil would be an appropriate description. The large size fabrication experience will enable ORNL to provide guidance, technical input and evaluate the industrial participation in the LCP. The interrelation between the large segment tests and the LCP is shown in Table XII for the next two fiscal years.

TABLE XII

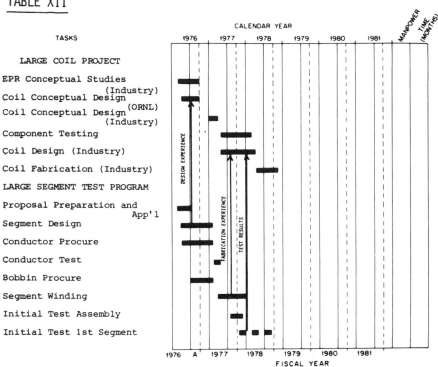

Interrelation of SCMDP large segment tests and LCP in industry

The small scale experiments in the frame of SCMDP are in direct support of the above activities as well as providing a base for the poloidal field requirements. The initial pulse field experiment will be on a small scale and are aimed at selecting the conductor and winding configuration which can meet the rate of rise constraints with acceptable losses. The second phase will require increase in size to achieve the field magnitude as well as the rate of rise of field simultaneously. The third phase will

438

be construction of a model coil which will be large enough to require structural constraint. The model coil can be scaled to the prototype poloidal field coil when a reference design is accepted for application in a particular superconducting Tokamak.

TABLE XIII

ORNL-Superconducting magnet program. Summary of ORNL-manpower requirements during first two years 1976/77

	Professional manpower (pmy)
Subprogram A: System Design	2,0
Subprogram B: Coil Design	5,0
Subprogram C: Conductor Selection and Test	12,5
Subprogram D: Radiation Effects on Supercon- ducting Coils	2,3
Subprogram E: Coil Protection, Eddy Current Shielding, and Power Supply	3,7
Subprogram F: Structural Analysis and Materials Investigation	9,0
Subprogram G: Cryogenics and Refrigeration	2,0
Subprogram H: Subsize Coil Fabrication	6,6
Subprogram I: Large Coil Project	5,0
Subprogram J: Coil Testing and Evaluation	8,9
	57,0

Table XIII gives a summary of the tasks and manpower foreseen at ORNL throughout 1976 and 1977 for SCMDP and LCP. The cost of the LCP has been evaluated in some detail for the compact torus version leading to a total investment cost of 28 M $. The program parts shown here represent only the Tokamak oriented part of the US superconducting magnet program. There is a strong inter-connection between this part and the mirror- and energy storage oriented program parts.

Japan having a very strong national Tokamak program is considering the future activities for superconducting magnets in fusion. Japan together with the European Communities and Switzerland takes part in the IEA-discussions on international cooperation. In Japan the main effort will be directed to the development of superconducting magnets for a Tokamak device coming after JT60.[60] Preliminary figures for such a machine which is called fusion core mock-up test facility have been quoted as follows. Preliminary dimensions are 6 m major radius, 1.2 m plasma minor radius and the axial magnetic field is 5 T. The first priority in Tokamak application is given to the development of superconducting toroidal coils, followed by Ohmic coils and EF coils. In future the superconducting energy storage and transfer device for Tokamak and other applications will also be studied. The development work will be carried out mainly at JAERI in collaboration with national laboratories having experience in superconducting magnet development. Although a long term program is not yet formulated, these organizations are now making plans for the next year (1977); development of a large current stabilized superconducting conductor and construction of a simulation facility of Tokamak toroidal coils at JAERI the design of which has already started. Poloidal effects and high field materials are studied at ETL and NRIM. The industri will support these activities on contract basis with these laboratories.

In the USSR already smaller Tokamaks like T-7 and possibly T10 in a second phase will be built with superconducting toroidal magnets. The current Soviet reference design (EPR) proposes D-shaped toroidal magnet coils built with NbTi conductors (B_{max} = 7.8 T), the plasma relevant data being I_p = 6.3 MA, R = 7.2 m, B_{t_o} = 4 T.[61]

Of course in other countries fusion oriented technology programs including superconducting magnets are being considered, too.

In the frame of the US superconducting fusion magnet program, whic seems to be the most elaborate and strategy based of the so far known programs the already mentioned large coil test facility plays a major role in early involvement of industry and conclusive feasibility demonstration for fusion application. The concept whic

was preferred among 5 alternate arrangements the so-called
compact torus foresees 6 approximately D-shaped coils arranged in
a symmetric toroidal geometry with a slightly tougher magnet
aspect ratio compared to a Tokamak experiment. Fig. 23 shows the
compact torus coil of the LCP and for comparison the JET toroidal
magnet cross section. The compact torus definition has been
arrived at by means of the following procedure: [62]

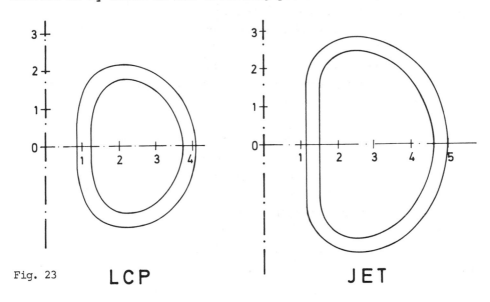

Fig. 23   **LCP**   **JET**

1. The basis is the reference EPR design.
2. By decreasing the coil size, but keeping the current density
    and maximum field at the inner coil circumference the stresses
    will be somewhat higher than in the EPR and no extrapolation
    in current density coil be necessary later.
3. The result is that with 6 coils and a reduced magnet aspect
    ratio the circumferential force distribution can be held
    within 10s of % of that in the EPR size coil.
4. Starting from an EPR design with 8 T maximum field and
    2500 A/cm^2 current density in the winding, inner coil
    dimensions of about 2.5 m horizontal diameter and about
    3.5 m vertical diameter - which are within the capability of
    existing machining equipment - can simulate EPR coil con-
    ditions relatively closely. Further conditions are
    - The conductor and coolant characteristics are the same as
      for EPR. Local heat inputs by built in heaters shall test

441

the coils response and also simulate radiation heating.
- The EPR field force and strain pattern in the coils shall
  be realized.
- The pulsed poloidal field effects have to be simulated,
  either locally or integrally. This point is under discussion since it depends on the choice of the poloidal
  configuration assumed for the EPR and strongly influences
  the conductor and winding design.
- An appropriate safety discharge scheme has to be tested
  in LCP.
- Another point under consideration is the dewar configuration for the compact torus. The most realistic test conditions would be provided by a joint fitted design with
  a common inner part housing the cold central support ring,
  however, it appears that in an overall large dewar housing
  the compact torus a convenient flexibility including the
  test of coils with fitted dewars is provided.
- The LCP has to demonstrate reliable operation.

LCP has been offered for contributions by one or more coils from
other countries. Fig. 24 shows a preliminary sketch of the large
coil test facility as it looked in August/September 1976.[62]

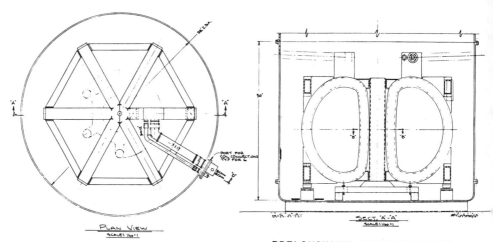

Fig. 24

PLAN VIEW

PRELIMINARY CONFIGURATION
OBSOLETE 9-2-76

442

5. Conclusion

The technology of superconducting Tokamak magnets provides challenging tasks both for specific developments in laboratories and heavy conductor and hardware construction in industry. The need to increase the efforts in this field is widely acknowledged at least to the extent that the generation of toroidal experiments after JET, TFTR and JT-60[63] provided this generation of experiments is going to be successful will require a superconducting toroidal magnet.

Relevant experience can be gained from a demonstration setup with torus relevant coils in a Tokamak relevant environment and at a size comparable to e.g. JET coils, but not much smaller. While the specific nature of such a demonstration set-up is still being considered, Table XIV[64] including preliminary LCP data briefly

TABLE XIV

COIL SHAPE	DIMENSION	DESIGNED	OPERATION	$\hat{P}_{MT}$(GW)	B_{MAX}(T)
LCP	2,5 x 3,5	1976	1980 ?	-	8,0
JET	3,12 x 4,9	1975	1981 ?	0,33	6,9
GA ITR (TNS)	4,5 x 7,2	1976	1985 ?	-	8,0
T 20	5,4 x 8,3	1975	1985 ?	1,20	7,8
ANL EPR	7,7 x 11,9	1975	1992 ?	-	7,5
ANL EPR	7,78 x 12,6	1976	1992 ?	-	10,0

summarizes the present situation. The justification for substantial efforts in superconducting magnet technology for fusion lies in the belief in the success of the coming toroidal fusion experiments.

Acknowledgements

The author gratefully acknowledges the help of all those who
provided up to date information on design studies and programs,
in particular, C.C.Baker, E.Bertolini, P.Casini, R.W.Boom,
G.Bronca, R.Hancox, P.N.Haubenreich, M.Lubell, S.Mori, J.Powell,
S.T.Wang, M.N.Wilson. He thanks also Miss Künzner und
Miss Schubert for bringing the puzzling figures and tables into
a handy shape and preparing the typed manuscript.

References

1   D.N.CORNISH and K.KHALAFALLAH: UKAEA Culham Lab., Proc.
     8th SOFT, Nordwijkerhout 1974

2   P.L.WALSTROM, T.C.DOMM: Proc. 6th Symp. on Engin.Problems
     of Fusion Research, San Diego, 1975

3   R.PÖHLCHEN and M.HUGUET: Proc. 5th International Conf. on
     Magnet Technology (MT-5), Rome, 1975

4   H.PREIS: IPP III/24, April 1976

5   F.B.MARCUS, Y-K.M.PENG, R.A.DORY and J.R.MOORE: Proc.
     6th Symp. on Engin. Problems of Fusion Research, San Diego,
     1975

6   European program on superconducting magnet technology for
     plasma and fusion applications, EURATOM Subcommittee on
     superconducting magnet technology, 1975

7   JET-Team, private communication

8   Tokamak 20 Moscow 1975 (translation) UWFDM-129

9   W.M.STACEY et al: ANL/CTR-75-2, June 1975

10   W.M.STACEY et al: ANL/CTR 76-3

11   M.ROBERTS: ORNL/TM-5574, 1976 (to be published)

12   K.SAKO, T.TONE et al: IAEA-CN-35/I3-1, 1976
      to be published

13   UWFDM-112: UWMAK II

14   UWFDM-150: UWMAK III

15   J.T.D.MITCHELL and A.HOLLIS: UKAEA, Proc. 9th SOFT,
      Garmisch-Partenkirchen, 1976

16   J.PURCELL: private communication

17   Mc ALEES: private communication

18   J.F.GUESS et al: ORNL TM5187

19   C.A.M. van der KLEIN: RCN-240

20   G.M.Mc CRACKEN, S.BLOW: CLM-R 120

21   F.C.MOON: Journal of Applied Physics, Vol.47, No.3,
      March 1976

22   F.C.MOON and C.SWANSON: Journal of Appl.Physics, Vol.47,
      No. 3, March 1976

23   C.SWANSON and F.C.MOON: "Magneto-Elastic Stability of
      Toroidal Magnet Designs for Proposed Fusion Reactors"
      (AEC No. AT (11-1)-2493

24   F.C.MOON: Progress Report for Period Sept.1975 - March 1976

25   A.KNOBLOCH, K.LACKNER: IPP-Report (to be published)

26   H.J.CRAWLEY: CTRD Collective Tokamak Reactor Design,
      Collection of Contributed Papers, October 1975

27   FINTOR-I, Ispra 1976 (to be published)

28   Z.J.J.STEKLEY and R.J.THOME: 4th International Conference on Magnet Technology, Brookhaven, 1972

29   C.K.JONES, C.H.ROSNER,Z.J.J.STEKLY: private communication

30   J.F.CLARKE: ORNL/TM 5429

31   P.H.SAGER,Jr. and E.R.HAGER: GA-A13826, March 1976
      J.R.PURCELL, W.CHEN, R.K.THOMAS: GA-A13910, April 1976

32   C.C.BAKER et al: GA-A13694, Oct.1975

33   J.A.DALESSANDRO: GA-A13922, April 1976

34   GAC Fusion Engineering Staff: GA-A13534, July 1975

35   F.ARENDT et al: Proc. 8th Symp. on Fusion Technology, Noordwijkerhout, 1974

36   M.ROBERTS, E.S.BETTIS: ORNL-TM 5042

37   J.Parain: Proc. Applied Superconducting Conf, Stanford 1976

38   R.W.MOSES, Jr. and W.C.YOUNG: Proc. 6th Symp. on Engin. Problems of Fusion Research, San Diego 1975

39   R.W.BOOM et al: UWFDM 158

40   S.T.WANG, J.R.PURCELL, D.W.DEMICHELE and L.R.TURNER: Proc. 6th Symp. on Engin.Problems of Fusion Research, San Diego 1975

41   P.M.RACKOV, C.D.HENNING: Final Report, Oct. 1975, ORNL-IGC-Contract 22X-69944V

446

42   K.H.SCHMITTER: International School of "Fusion Reactor Technology", Erice, 1972

43   P.BEZLER et al: First annual report Magnet Safety Studies Group, BNL, 1976 (to be published)

44   M.A.HILAL, R.W.BOOM: IEEE Transactions on Magnetics, Vol.MAG-11, No.2, March 1975, p.544-547

      M.A.HILAL, R.W.BOOM: Proc. 6th Symp. on Engineering Problems of Fusion Research, San Diego, 1975, p.111

      M.A.HILAL and R.W.BOOM: Proc. 9th Symp. on Fusion Technology, Garmisch-Partenkirchen, June 1976,UWFDM 166

45   B.J.MADDOCK et al: Proc. IEE Vol.115, 1968

46   H.PREIS: IPP 4/109, March 1973

47   G.PASOTTI and M.SPADONI: "A New Approach to the Cooling of Superconducting Magnets for Fusion Reactors" ICEC 6, Grenoble, 1976

48   W.KAFKA: IPP 4/70, March 1970

49   M.HOENIG: ICEC 6, Grenoble 1976

50   M.N.WILSON and C.R.WALTERS: RL-76-038, April 1976

51   C.-H.DUSTMANN, H.KRAUTH, G.RIES: 9th Symp. on Fusion Technology, Garmisch-Partenkirchen, 1976

52   K.H.SCHMITTER: private communication

53   A.KNOBLOCH: Proc. 6th Symp. on Engineering Problems of Fusion Research, San Diego, 1975, p.58

54   A.F.KNOBLOCH: Proc. 9th Symp. on Fusion Technology,
      Garmisch-Partenkirchen, June 1976

55   P.H.REBUT: Proc. 8th Symp. on Fusion Technology,
      Noordwijkerhout (The Netherlands), June 1974
      EUR/JET R 7

56   J.DARVAS et al: Report Jülich 1304, 1976

57   R.HANCOX: The Euratom Fusion Technology Program;
      ANS-Conference Richland, Sept. 1976, to be publ.

58   EUR FU/LG26/AHSC1

59   H.M.LONG, M.S.LUBELL (Ed.) : Program for Development of
      Toroidal Superconducting Magnets for Fusion Research.
      ORNL/TM-54o1, 1976

60   S.MORI: private communication

61   B.B.KADOMTSEV, T.K.FOWLER: Physics Today, November 1975
      W. HEINZ: private communication

62   P.N. HAUBENREICH: private communication

63   A.H. SPANO: Nuclear Fusion 15, 1975

64   E. KINTNER: 9th Symposium on Fusion Technology, Garmisch-
      Partenkirchen, 1976

Materials Problems and Possible Solutions
for Near Term Tokamak Fusion Reactors

by

G. L. Kulcinski

Nuclear Engineering Department
University of Wisconsin
Madison, Wisconsin 53706

September 17, 1976

1. Introduction

As the world-wide fusion community progresses beyond the TFTR/JET/T-20 phase of Tokamak devices,[1] it will begin to encounter the first serious DT neutron damage problems in reactor structures. These problems will first show up in the experimental power reactors (EPR's) which are presently designed to produce some electricity (hence high temperatures) by converting the kinetic energy of the high fluxes of the 14 MeV neutrons and 3.5 MeV helium ions to heat and eventually useable power.[2] It has been a common fault to dismiss the neutron damage problems in EPR's as insignificant compared to those in the next generation of Demonstration Power Reactors (DPRs) which are required to produce electricity on a steady-state basis. However, closer examination of the anticipated operating conditions of EPR's reveals that there could be some very serious problems with present day materials.

It is the purpose of this paper to clarify the magnitude of the problems that might arise from neutron damage and to put into perspective the methods and facilities that might be used to solve these problems.[3] First, a brief review of some of the fundamental aspects of radiation damage from neutrons will be given for the non-materials scientist* followed by a current listing of the anticipated radiation environment of the various near term (TFTR, JET, T-20), EPR, and DPR designs. The reader should note that such designs are highly fluid and may change considerably in the future (in fact due to the very problem we will be discussing). Next, the present and future facilities that could be used to test CTR materials will be reviewed and their utility in providing pertinent fundamental and engineering data will be discussed. Finally, some conclusions and recommendations on the near term reactor materials problems will be presented.

2. Background Information on DT Neutron Damage in Potential EPR and DPR Materials

When an energetic neutron strikes any solid material, it produces damage in a variety of ways. It is convenient to think of the damage process as broken up into primary and secondary responses of the material (see Figure 1). The primary responses are the displacement of atoms from their equilibrium sites via elastic, inelastic and nonelastic events, and the transmutation of some elements into different elements, or

---

* This chapter may be skipped by those familiar with the radiation effects field.

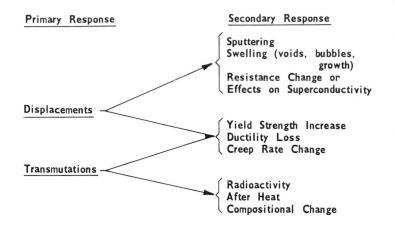

Figure I

Schematic Of The Response Of Materials To Neutron Damage

Table I

Typical Displacement and Gas Production Rates in Metals

Material	Fusion Reactor First Wall 1 MW/m^2			First Fission Test Reactor-EBR-II (max)		
	dpa/yr	appm He/yr	appm H/y	dpa/yr	appm He/yr	appm H/y
SAP[a]	17	410	790	76	7.9	50
316SS[b]	10	200	540	44	4.7	270
Nb	7	24	79	28	1	6.6
Mo	8	47	95	30	1.8	3.5
V	12	57	100	54	0.5	14
C	10	2700	Neg.	5	130	Neg.
Be[c]	(d)	2800	130[e]	(d)	3300	Neg.

(a) SAP = Sintered Aluminum Product, 5-10% Al$_2$O$_3$ in Al

(b) SS - Stainless Steel

(c) ~ Typical of 5 cm from first wall

(d) Displacement cross section not available

(e) Tritium

TYPICAL DAMAGE ENERGY FUNCTIONS
FOR POTENTIAL FUSION REACTOR MATERIALS

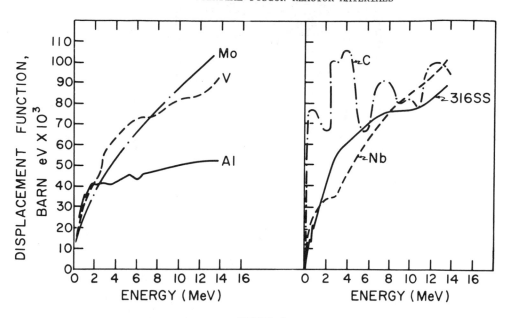

FIGURE 2

PRIMARY KNOCK—ON SPECTRA FOR
COPPER IN VARIOUS NUCLEAR FACILITIES

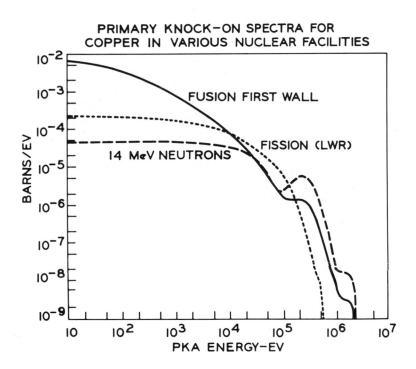

FIGURE 3

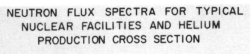

NEUTRON FLUX SPECTRA FOR TYPICAL NUCLEAR FACILITIES AND HELIUM PRODUCTION CROSS SECTION

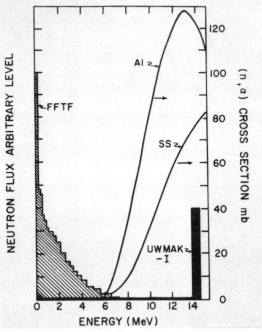

FIGURE 4

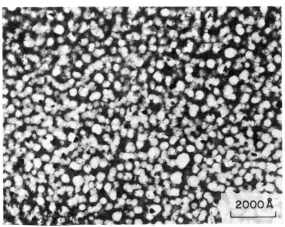

FIGURE 5

Solution treated 316 SS (fuel pin cladding) irradiated at ~525°C to 1.3×10^{23} n/cm^2, swelling ~25%. Courtesy H. R. Brager, HEDL

452

isotopes of the same element. These primary interactions can then produce secondary responses in the solid via the migration of the defects to internal sinks, or the formation of microdefects which change the physical properties, dimensions, and mechanical response of the material. Let us turn our attention to the primary interactions first.

2.1 Primary Responses

The cross section for displacing atoms is a rather complex function of neutron energy due to the multitude of reactions that can take place. In general, the displacement cross section increases rapidly with neutron energy up to ~1 MeV and rises somewhat slower after that (see Figure 2). It can be seen from that figure that the displacement cross section at 14 MeV is only 3-4 times the displacement cross section at 1 MeV for high Z elements. For low Z elements the displacement cross section is relatively constant with energy above 1 MeV. This latter effect is due to the anisotropic scattering at high neutron energies and is particularly important in that the absolute displacement rates in CTR materials are usually less than in fission reactors (the geometry of the source and first structural wall also plays a big role here).[4] Table 1 lists some typical values normalized to 1 MW/m^2.

The next important feature of DT neutron damage is the primary knock-on atom (PKA) energy. Figure 3 illustrates that the number of PKA's with energy greater than some energy E_n is almost always greater for a fusion neutron spectrum than for a fission spectrum. The difference is particularly pronounced above 500 keV. There is evidence that certain processes in metals are influenced by the energy of the PKA's. For example, resolution of precipitates in Ni-Al alloys is quite pronounced after 2.8 MeV nickel irradiation[5] whereas the particles actually grow after electron irradiation[6] (the PKA energy is $\leq$ 50 eV in the latter case). The production of vacancy loops inside of displacement cascades is also increased as the PKA energy is increased.[7] Therefore, it is important to know not only the rate at which atoms will be displaced in a fusion neutron environment, but also how they are displaced with respect to the actual region of the displacement spike. This will be important later when we discuss simulation techniques .

The next most important primary reaction is that producing the insoluble gas, helium. Figure 4 shows how the helium cross section varies with neutron energy and selected values for CTR materials are quoted in Table 1. We have also plotted in Figure 4 the typical fusion and fission neutron energy spectra. Note that while the fusion spectra extends well into the (n,α) reaction energies, the fission spectra has only a very few neutrons in that energy regime. Consequently, the helium production rates in fusion reactors can be extremely large and helium is known to be particularly devastating to mechanical properties in metals at very high temperatures.[8]

Transmutations which lead to solid elements are very system and spectrum dependent and we will discuss that later in our discussion of secondary responses.

2.2 Secondary Responses

The direct transfer of energy from the incident particles to the surface atoms or the indirect transfer of energy via focussing collisions is known to cause atoms to be sputtered off the surface. The rate of wall erosion due to neutron sputtering can be expressed as follows:

$$\frac{\Delta d}{t} = \frac{S_n \phi}{N_A}$$

where Δd is the wall thickness reduction

t = time
S_n = neutron sputtering coefficient
ϕ = flux of neutrons
N_A = atomic density

For $N_A \sim 8 \times 10^{22}$ atom/cm^3
$\phi = 2 \times 10^{14}$ n/cm^2/sec per MW/m^2.

$$\frac{\Delta d}{t} = 0.16 \ S_n \ \text{cm per MW-yr/m}^2 \quad *$$

Recent measurements of S_n vary from 10^{-3} to 10^{-5} for 14 MeV neutrons[9] and even at the highest rate, the rate of wall thinning is 2 microns per MW-yr/m^2. We shall see later that the maximum wall loading anticipated for a 10 year EPR operation time is ~10MW-yr/m^2 and hence less than 0.02 mm will be removed during the component's lifetime.

The production of voids in metals as a result of irradiation has been studied for almost 10 years. There are several good references which summarize the progress in the field and the reader is urged to consult them. [10-11] At first glance, the production of equal amounts of vacancies and interstitials should not result in any volume change. However the high mobility and extreme insolubility of interstitials combine to make the formation of interstitial loops energetically favorable. These loops in turn attract more interstitials than vacancies so that in a very short time there is an excess of vacancies in the matrix. Given sufficient nucleation sites and high enough temperature for the vacancies to move, the vacancies precipitate into three dimensional aggregates (voids) which act as further sinks for more vacancies. See Figure 5 for an example of voids in stainless steel[12] Swelling values of over 100% have been observed thus far and almost every known structural alloy shows some measure of swelling.

The important parameters in swelling are listed below.

'Temperature - It appears that the irradiation temperature must be above 20% and less than 50% of the absolute melting point (T_{mp}). Below 40% of T_{mp} the growth of voids is limiting and above 40% the nucleation is limiting.

'Stress - There appears to be little stress effect at less than $40\%T_{mp}$, but above that val tensile stress lowers the critical radius required for nucleation and enhances swellin

'Helium Gas - Some evidence suggests that it has little effect on the total swelling at low temperature but it can significantly alter the number density of voids. Above 40% T_{mp}, He gas stabilizes void nucleii and greatly enhances swelling

* using both sides of the first wall

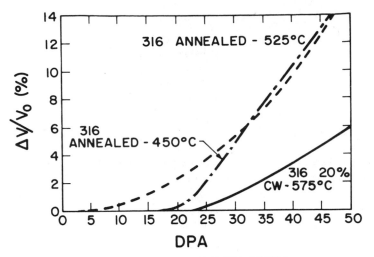

TYPICAL SWELLING IN AUSTENITIC STEELS

FIGURE 6

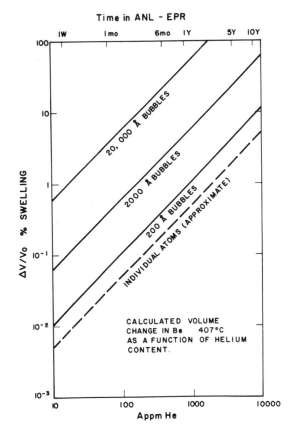

FIGURE 7

. Alloy (Transmutations) - In general, the more pure the metal, the higher the swelling under given temperature and damage conditions; therefore, generation of impurity atoms can reduce swelling.

. Total Damage - The swelling of most metals is proportional to the number of displaced atoms to the nth power. In some metals there is an incubation period where critical void nucleii size must be attained and this may be as high as 1-10 dpa in alloys or as low as 0.01 dpa in pure metals. After the incubation dose is exceeded most metals swell with n = 0.8 to 1.2 (See Figure 6 for typical behavior in austenitic steels).[13]

The significance of all of this to near term Tokamak fusion reactors is that as we progress into the electrical production mode,the operating temperature of most structural alloys will have to be raised to above 0.25 T_{mp} and well into the void formation regime. The use of high pressure coolants and rather large thermal stresses will tend to accelerate swelling above 0.4T_{mp}. As the wall loadings are raised, significant helium gas generation will occur and enhance high temperature swelling. As the solid transmutation products build up it is possible that swelling will be reduced. Since the useful component lifetime will be inversely proportional to the swelling rate it is imperative that a more fundamental understanding of this phenomena be developed to allow long term safe operation.

Dimensional instabilities can also be caused by the formation of helium filled gas bubbles which grow in the presence of an excess of vacancies by balancing the internal gas pressure, P, with the surface tension as follows:

$$P = \frac{2\gamma}{r}$$

where γ = surface energy

r = radius of the bubble.

The swelling induced in a solid under this equilibrium situation is simply:

$$\frac{\Delta V}{V_0} = C_{He}[\frac{rkT}{2\gamma} + b]$$

where C_{He} is the number of helium atoms per unit volume

k is the Boltzmann's constant

T is the temperature

b is the Van der Waal's constant

For typical values of the materials constants (γ = 1000 ergs cm^{-2}, $b = 4 \times 10^{-23}$ cm^3 atom^{-1}) the above expression for Be transforms to

$$\frac{\Delta V}{V_0} (\%) = C_{He}[6950 \, rT + 4] (1.236 \times 10^{-4})$$

where r is expressed in $\overset{\circ}{A}$, T in $°K$, and C_{He} is in appm.

Figure 7 shows how the theoretical swelling depends on various levels of helium in Be. At small bubble radius the swelling is low because of the high surface tension. As

456

the bubble size increases the number of vacancies per gas atom increases, and the swelling is enhanced. Temperature has little intrinsic effect on the swelling but the bubble mobility is enhanced at high temperature promoting coalescence and hence large bubble radius. The significance of this particular example is that Be has been proposed [14] as a coating on the first wall to protect the plasma from contamination and if it undergoes large dimensional changes, it may not adhere to the metallic substrate.

Some non metals like graphite undergo another type of dimensional instability directly connected to the anisotropic crystal structure [15-17] Vacancies and interstitials tend to precipitate on different crystallographic planes which makes the single crystals grow anisotropically and even randomly oriented, fine grain material will swell. The general behavior is for the polycrystalline material to shrink initially while all of the as fabricated porosity is being filled. Eventually this process saturates and the specimens begin to grow very rapidly in the so called runaway mode (Figure 8). The damage level to achieve this is 1-10 dpa and the cross over point ($\Delta \ell = 0$) appears to decrease with temperatures below 1000 °C and increase with temperatures above 1000 °C.

The significance of this phenomena for tokamak reactors is that there have been several proposals to reduce the impurity problem by lining the surface facing the plasma with carbon either in a sheet or coating form. If the carbon grows during irradiation it could buckle, tear, or crack the layers, or cause the coatings to peal off thus eliminating the effectiveness of the protection. One way to alleviate this problem is to use the carbon in a loose 2 dimensional weave which could retain considerable flexibility under large dimensional changes. [18]

Resistance changes in metals irradiated at low temperature has been studied for more than 20 years. It is a well established fact that, at low fluences, the amount of resistivity increase is directly proportional to the number of defects produced, inversely proportional to the temperature of irradiation, and moderately dependent on the type of particle used to damage the specimens. The numbers range from 2-15 micro-ohm-cm per % Frankel pairs at low temperature until relatively high damage levels ($\sim 10^{-4}$ - 10^{-3} dpa). [19] Saturation at damage levels higher than this is due to overlapping damage zones and annealing of existing damage by the displacement spikes (See Figure 9).

The practical significance of this effect in tokamak fusion reactors is that neutrons leaking out of the blanket and shield zone could increase the resistance of normal magnets or increase the resistance of the stabilizer in superconducting magnets. The effect is not so important at room temperature and above, but can be quite serious at 50° K and lower. The higher resistance increases the I^2R losses to the cryogenic cooling system and could eventually require that the magnets be annealed (at RT to 300 ° C) to remove the damage.

Irradiation of superconducting filaments is known to reduce T_c and J_c in some alloys. Nb_3Sn is particularly sensitive (Figure 10) in the 10^{-3} to 10^{-2} dpa range whereas the NbTi system is relatively immune up to about $\sim 10^{-2}$ dpa. The practical significance here is that the continued use of S/C coils in a radiation field could cause the coil to

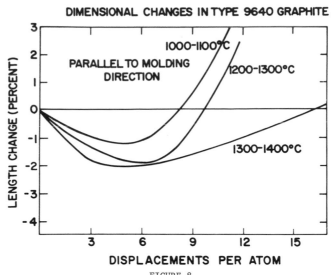

DIMENSIONAL CHANGES IN TYPE 9640 GRAPHITE

FIGURE 8

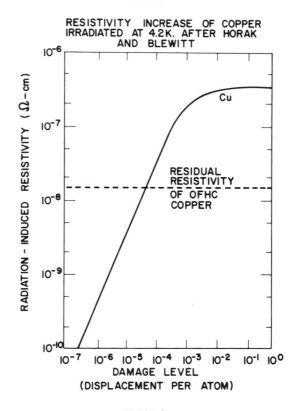

FIGURE 9

458

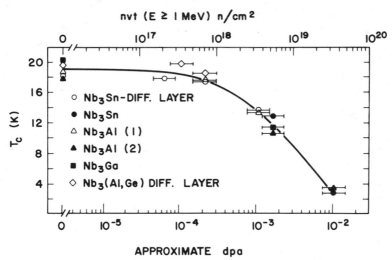

EFFECT OF NEUTRON IRRADIATION ON CRITICAL TEMPERATURE OF A-15 SUPERCONDUCTORS

(AFTER SWEEDLER et al)

FIGURE 10a

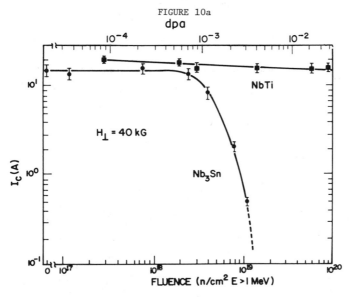

EFFECT OF NEUTRON IRRADIATION ON THE CRITICAL
CURRENT IN NbTi AND Nb$_3$Sn.

FIGURE 10b

459

EFFECT OF NEUTRON IRRADIATION ON THE YIELD STRENGTH
AND DUCTILITY OF 304 STAINLESS STEEL - (AFTER FISH ET AL. - HEDL)

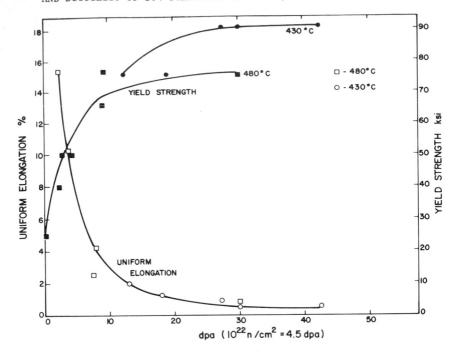

FIGURE 11

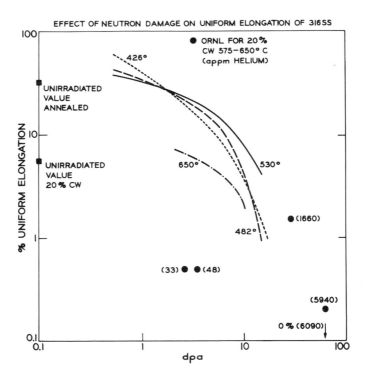

FIGURE 12
460

go normal due to drops in T_c or J_c, thus jeopardizing the integrity of the magnet and the continued operation of the reactor.

It is well known that irradiation produced defects in metals can act as significant barriers to dislocation motion, thus causing the metal to become stronger. These clusters can range all the way from small loops at low temperatures to voids at high temperatures. The general effect is to raise the yield strength to as much as 3-5 times the unirradiated value. While such strengthening appears beneficial on the surface, it is usually accompanied by a drop in ductility which is detrimental. (See Figure 11)

The hardening of the matrix in the grains of metals causes the deformation under stress to take place along the grain boundaries. When the deformation is localized in this manner, premature failure can occur. The generation of large amounts of helium, which has a tendency to collect on the grain boundaries, tends to aggravate an already serious situation. The bubbles interfere with grain boundary sliding causing micro cracks to form on the grain boundaries, which eventually link up to cause intergranular fracture. This fracture can occur at uniform elongations of <1% (see Figure 11, 12) and a reactor which is made of metal in this state is very subject to crack formation. Any rapid loading or unloading of the stresses could cause brittle fractures to occur, destroying the vacuum conditions and jeopardizing the integrity of the blanket structure.

The significance for fusion reactors is that because of the high helium generation rates, (Figure 4 and Table 1) the probability of brittle fracture at high temperature is very large. This puts an added penalty on high temperature ($> 0.4\ T_{mp}$) operation for DT fusion devices and also has a great bearing on the lifetime of a first wall component.

The plastic deformation of materials at high temperatures, under high stresses below the elastic limit, is well known and called thermally induced creep. This phenomena can also occur during neutron irradiation because of the large concentration of vacancies and interstitials available for dislocation motion. The thermal creep rate is extremely temperature dependent but the irradiation induced creep is relatively independent of temperature (see Figure 13). The consequence of this is that above 0.5 T_{mp} thermal creep is dominant and below that temperature irradiation creep is dominant. The level or irradiation creep is proportional to the displacement rate and can be beneficial or detrimental for fusion reactors. On the one hand, it will tend to relieve stress concentrations built up because of differential swelling, thermal expansion or fabrication difficulties. On the other hand, the overall dimensions of the systems will change making insertion or extraction of some components very difficult. Warping of certain coolant channels could also adversely effect the temperature profiles in the blankets. Fig. 13 also shows how the irradiation creep rate depends on total damage.

The constant cycling of temperature or stress levels can cause the premature failure of certain metallic components by a process called "fatigue." The generation and propogation of cracks at alternating stress and stain levels in the elastic regime is well known and the number of cycles to cause failure in metals usually decreases as the magnitude of stress or strain is increased. Figure 14 shows how the fatigue failure in 316 SS at 593 °C is affected by the strain range per cycle. [14] The effect of irradiation

461

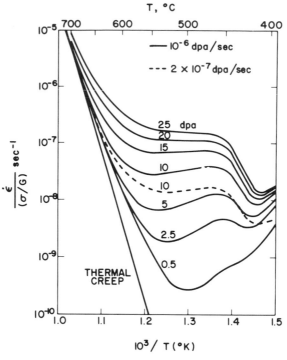

IRRADIATION CREEP RATE FOR SA 316 AS A
FUNCTION OF FLUENCE (W.G. Wolfer)

FIGURE 13

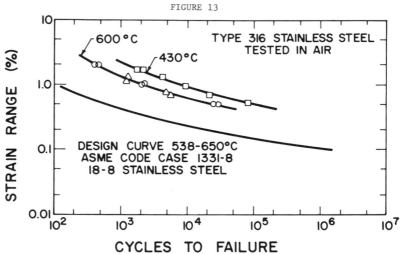

FATIGUE LIFE OF ANNEALED 316 STAINLESS STEEL AT 593°C UNDER
VARIOUS TEST CONDITIONS ($\dot{\epsilon}$ = 4 × 10^{-3}s^{-1}). [14]

FIGURE 14

on fatigue life is rather unclear at this time but in some cases, the generation
of dislocation loops tends to reduce crack propagation and extend useful lifetime.

The significance of this phenomena is readily apparent when we remember that tokamak
devices normally operate in a pulsed mode,with as many as 100,000 or more cycles per
year anticipated for some EPR designs.

The transmutation reactions which occur in practically all neutron irradiated elements
will convert some stable isotopes into unstable isotopes. The decay of these isotopes can produce large radiation fields which prohibit unprotected personnel from doing
maintenance on the reactor components. The total induced activity levels range from
1-5 curies per thermal watt of energy generated and can take years to decay (Figure 15).
Thus, extremely conservative design policies need to be followed to insure long term
safe operation and these policies may tend to penalize one material much more than
another (i.e. steel versus vanadium).

The decay energy of the radioactive species will generate heat, which, if not
properly accounted for in fission reactors,can actually cause some metallic components
to melt. However, the afterheat density in fusion reactors of ~1 watt per cm^3, is
usually at least an order of magnitude lower than for fission reactors and normally is
not a serious problem.

Finally, the production of some isotopes in certain metals can cause significant
changes in their composition and formation of secondary phases. For example, the solubility of Zr in Nb is ~10% at normal operating temperature and in a typical CTR neutron
spectrum the production rate is 0.1 to 0.2 atomic percent per $MW-yr/m^2$. The composi-
tional changes may be particularly important if the recently observed irradiation induced
precipitation phenomena are a general rule of CTR alloys [20-21]

3. Summary of Selected Operating Parameters and Damage Conditions for Near Term Fusion Reactors

As noted in the introduction, we are assessing the next three generations of reactors
while they are in a very fluid state of design and any parameters which are quoted must
be carefully referenced. Future readers of this paper should be sure that key parameters
have not changed because of a reassessment of the very problems we are addressing. It
is not the purpose of this paper to completly review the future tokamak reactors, but
rather to highlight the potential materials problems associated with neutron damage.
Therefore, we have included in Table 2 an abbreviated list of pertinent parameters for
the analysis to be performed in Section 4. We will only consider devices that will
burn a substantial amount of D&T (e. g., not JT - 60).

The dates of first DT operation in tokamaks vary from 1983-6 for the first generation
of devices to 1991 for the EPR's and 1998 for the DPR (logic III, USERDA [22]). Thermal
power levels during the burn will be quite low in TFTR (20 x 10 MW) and increase to 400-600
MW for the EPRs and probably ~1500-1700 MW for the DPRs. Taking into account the fraction
of the energy produced by neutrons and wall area,we find that the first wall neutron loadings vary from a maximum of 0.25 MW/m^2 for the first generation devices to 0.6-0.8 MW/m^2 for
EPRs and probably 1-3 MW/m^2 in the DPRs. When the burn cycle and plant factors are
included we find the integrated first wall loadings range from a low of $10^{-5} MW-y/m^2$

463

Table 2

Summary of Selected Parameters of Near Term DT Tokamak Systems

Parameter	First Generation (23)			Experimental Power Reactors			Demonstration Power Reactors	
	TFTR	JET	T-20	ANL (24)	GA (25)	ORNL (26)	GA (27)	ORNL (28)
Reference								
Est. Date DT OP	1983	1983	1986	1991	1991	1991	1999	1999
Inst. Power—MW_t	10	15	55–140	638	410	410	1660	1700
First Wall Area—m^2	110	220	400	592	~390	600	350	350
Neutron MW/m^2	0.1	0.05	0.25	0.56	0.83	0.55(e)	1.2	3.0
Burn Time—sec	1	20	5–20	55	105	100	517	1200
Downtime—sec	300	600	240	15	20	15	60	50
Max Plant Factor	1-04	6-04	0.03	0.5	0.2	0.7	0.7	~0.7
Cycles per Year	1000	5000	5+04	2.3+05	5+04	1.9+05	3.8+04	1.8+04
14 Mev n $cm^{-2}y^{-1}$	2+16	4+16	1+19	4+20	2+20	1+21	1×10^{21}	3×10^{21}
Total n $cm^{-2}y^{-1}$	NS	NS	4+20	6+21	4+21	1+22	1.2+22	NS
Est. DT Reactor Life—y	1	1	2	10	10	10	25	25
First Wall Exp. $MW-y/m^2$	1-05	3-05	0.015	2.5	1.7	3.9	21	53
Low Z Liner	None	None	None	Be	SiC/C	C	SiC	NS
Liner T_{max} °C	--	---	---	380	1450/1650	1780	1570	NS
Met. Struc.	305SS	Inconel	SS	316SS	316SS	316SS	Inconel	Adv. SS
Coolant				H_2O	He	H_2O/He	He	Metal Salt
Coolant, Inlet/Outlet °C				40/310	325/575	40/121(H_2O) 200/370(He)	350/650	NS
T_{max} First Wall—°C	80	50	100/500	500	600	125/540	700	400
Stress 1st Wall—MPa	----	----	----	69	200	NS	200–400	NS
Blanket Materials				SS/H_2O	SS/C	SS/C/K	(a)	SS/K/Na/Li

Table 2 (con't.)

	First Generation			Experimental Power Reactors			Demonstration Power Reactors	
	TFTR	JET	T-20	ANL	GA	ORNL	GA	ORNL
Blanket Thickness-cm				28	25	52	70	NS
Shield Materials				SS/B_4C/Pb	SS/B_4C/LiH/Pb	SS/Pb/H_2O	(b)	NS
Shield Thickness Inside/Outside, cm				58/97	40/100	48/48	70/70	NS
Magnet Materials-TF	Cu	Cu	Cu	NbTi/Cu	NbTi/Cu	NbTi/Nb_3Sn/Cu	NS	NbTi or Cu
Magnet Materials-VF	Cu	Cu	Cu	"	Cu	Cu/NbTi(d)	Cu(c)	NbTi or Cu
Magnet Materials-OH	Cu	Cu	Cu	"	NbTi/Cu	NbTi/Cu	NS	NS
T_{max} Magnets-°K-TF	RT	RT	RT	3.0	4.2	4.0	NS	4.2 or RT
T_{max} Magnets-°K-VF	RT	RT	RT	4.2	313(c)	RT/4.2(d)	NS	4.2 or RT
T_{max} Magnets-°K-OH	RT	RT	RT	4.2	4.2	4.2	NS	4.2 or RT
DPA rate-Liner-y^{-1}	--	--	--	NS	---	NS	--	--
DPA rate 1st Wall-y^{-1}	1-04	3-04	0.07	2.8	2.2	4	9	23
DPA rate TF coil-y^{-1}	--	--	--	8-06	3-05	2-04	5-05	NS
appm-He-y^{-1}-liner	--	--	--	780	210	1200	1600	--
appm-He-y^{-1}-1st Wall	0.002	0.006	1.4	54	37	129	110	420
MRad-y^{-1} -Insul.	--	--	--	40	100	100	---	---

$1 + .04 = 1 \times 10^{+4}$

NS - not stated

NA - not applicable

a) Inconel/C/Li_7Pb_2/Li_4SiO_4

b) ss/B_4C/C/Pb/Al

c) F coil

d) Cu shield VF coil, NbTi Trim VF coil.

e) Based on personnal communication, J. Clarke

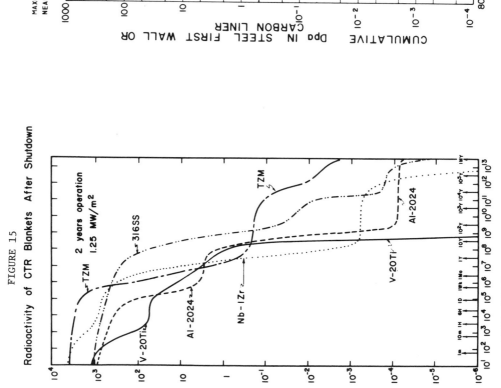

FIGURE 16

MAXIMUM PROJECTED DISPLACEMENT DAMAGE IN METAL STRUCTURES OF
NEAR TERM TOKAMAK FUSION REACTORS.

FIGURE 15

Radioactivity of CTR Blankets After Shutdown

per year for first generation DT devices to 0.2 to 0.4 $MW\text{-}yr/m^2$ per year for EPRs and ~1-2 $MW\text{-}yr/m^2$ per year for DPRs.

The estimated lifetime of the reactors (using DT) are somewhat arbitrary, but we have assumed a 1-2 year lifetime for the first generation devices with tritium, 10 years for EPRs, and 25 years for DPRs. These assumptions yield total first wall exposures of 10^{-5} $MW\text{-}yr/m^2$ for first generation devices, 2-4 $MW\text{-}yr/m^2$ for EPRS and 20-50 $MW\text{-}yr/m^2$ for DPRs.

Most EPR designs now call for low Z liners to reduce plasma contamination. These range from Be in the ANL reactor to SiC and C in the GA design and carbon in the ORNL design. Specific proposals for low Z liners have not been made for the DPRS, but they will probably be even _more_ necessary there. These liner temperatures are envisioned to range from 400°C for Be to ~1500°C for SiC and 1700-1800°C for carbon.

The metallic structure for the first generation reactors includes various austenitic alloys such as 305 SS and some specific Inconel compositions. The EPRs uniformly employ 316 SS and the DPRs are likely to utilize some advanced Fe-Cr-Ni alloy (i.e., Modified 316 SS, Inconel 718). A common coolant in all the first generation and EPR reactor designs is water and helium coolant is also used in the GA and ORNL EPR design. The DPR is likely to use liquid metals or He. The maximum first wall structural temperature is <100°C in the first generation designs (except for some parts of T-20) and ranges from 380 to 650°C in the EPR design. The ORNL design cools the first solid wall to 125°C and parts of the blanket immediately behind that operate at high temperatures (up to 540°C). The DPR reactor designs run the first wall at 400-700°C. Stresses in first walls were not stated in all designs, but in the higher temperature EPRs they are 70-200 MPa.

There are other materials besides the structure and coolant in the blanket. The EPR designs use C, K, and borated water to reduce the neutron energy. The blanket thickness ranges from 20 to 52 cm in all the designs.

The shields of the near term reactors contain Pb, B_4C, steel, C, and some borated water. The shield thicknesses in these designs range from 50-100 cm thick.

All the magnets for the first generation designs are copper and it is only when one gets to the EPR and DPR designs that superconducting TF, VF, and OH coils are used. The GA EPR design still uses a normal F coil. The superconductors are NbTi with varying amounts of copper stabilizer and the ORNL design even calls for some Nb_3Sn. Operating temperatures of the normal coils are ~ 40°C while the S/C coils run at liquid helium temperatures.

The maximum damage rates in the liners of the EPRs are on the order of ~2-4 dpa/yr for C and SiC (no values for Be have been calculated). The maximum helium production rates are ~1200 appm per year for carbon and ~200 for SiC and ~800 for Be. The values would be somewhat higher for the DPRs.

The first wall (assuming steel) of the first generation reactors will suffer only modest damage levels up to a maximum of 0.07 dpa per year while those in the EPRs experience from 2-4 dpa per year with an associated helium production of 40-80 appm/yr.

467

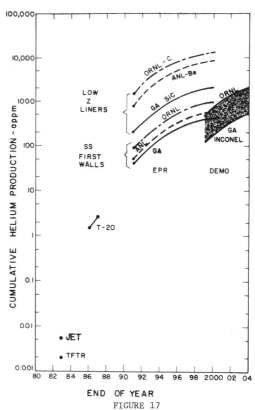

MAXIMUM PROJECTED HELIUM CONTENT IN SELECTED MATERIALS FOR NEAR TERM TOKAMAK REACTORS

FIGURE 17

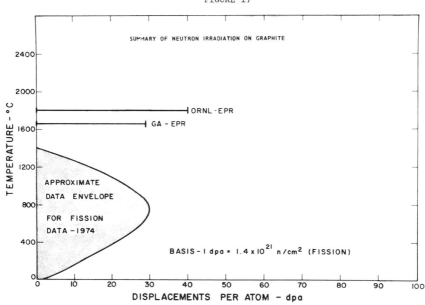

FIGURE 18

468

These values are increased to 10-20 dpa per year in the DPR in addition to 100-400 appm He per year in the first wall.

Damage to the TF coils is very low in the first generation designs ($<<10^{-7}$dpa/year) and the TF coils of the EPR and DPR designs may experience 10^{-5} to 10^{-4} dpa/year. Calculation of damage to the organic electrical insulators has been performed and it is found to range from 40 to 100 MRad/year.

The time dependence of the displacement and gas production damage in the low Z liners and steel first walls is shown in Figures 16 and 17 where cumulative values are displayed. It can be seen that significant displacement damage in the metallic first walls does not occur until 1991. The damage in the ORNL EPR is about twice as high as the ANL and GA design. For reference, note that it takes ~1 year to generate ~20 dpa in the current fast neutron test reactors which is essentially end of life for the ANL and GA designs and equal to <5 years in the ORNL design. The damage level in the ORNL DPR exceeds that in the ORNL EPR in approximately 2 years after startup and it will reach the end of life level for the GA-EPR design in approximately 1 year.

The major point about low Z liners is that the Be coating in the ANL design could contain ~800 appm He after one year and ~8000 appm after 10 years. The helium in the C liner of the ORNL design could approach 12,000 appm after 10 years of operation and the SiC of the GA design may contain 2000 appm after 10 years.

The displacement damage in the TF coils will reach ~8 x 10^{-5} dpa in the ANL design and 2 x 10^{-3} dpa for the ORNL design after 10 years of operation. The damage rate in the normal F coils of the GA design is approximately 0.01 dpa/year and would amount to 0.1 dpa at the end of life.

To put this data into perspective, we have plotted in Figures 18, 19 and 20 the current status of data for these materials. It is noted in Figure 18 that both of the EPR designs that use carbon as a liner material are using it in a damage-temperature regime that has never before been explored. The GA carbon liners, if left in for the life of the reactor, would achieve an exposure about as high as the highest reported in the literature and at temperatures of ~200-300°C above the highest reported data. The carbon in the ORNL design would be subjected to about the same as the highest reported exposure to carbon up to now at temperatures of ~300°C above the highest tests to date.

The situation for the displacement damage in 316 SS is much better (Figure 19). Steel has been tested at high temperture to damage levels as high as required for all EPR designs. Low temperature damage data is not available.
More serious is the lack of a large body of data on high helium contents. At low temperatures, the thermal neutron production of He from Ni-59 has generated ~500 appm in some LWR cladding. As one goes to higher temperatures where fast reactors operate, the amount of helium generated drops considerably. If this was the only data available, the EPR designs would exceed it by factors of 10-100. Fortunately, limited high temperature data is now available in thermal reactors (shown in Figure 19). This information should help to determine the effect of helium on dimensional stability and mechanical integrity. Unfortunately, the helium tends to enhance swelling (typical

469

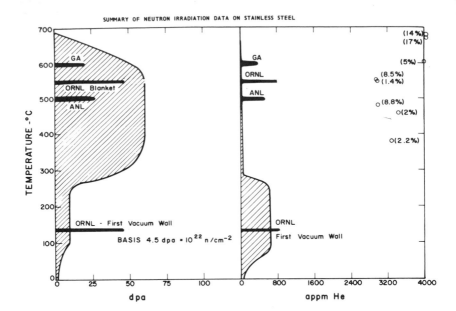

FIGURE 19

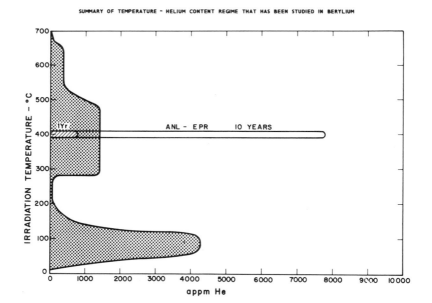

FIGURE 20

470

values are shown next to high helium data points in Figure 19) and reduce the ductility as we have seen.

Finally, Be metal has been irradiated to ~4000 appm (~5 years in ANL EPR) at temperatures ranging from 40-100°C. There is very little data from ~100 to 300°C and then information is available from 300-500°C for up to 1500 appm He. There is little information available above 700°C and limited data at 500-650°C (~400 appm He). The Be coating in the ANL-EPR design is envisioned to operate at ~400°C and will accumulate helium at the rate of 780 appm per year. Hence, the end of life will exceed 5 times the present data base.

4. Potential Problems of Neutron Irradiated Materials in Near Term Tokamak Reactors

4.1 First Generation Reactors

From the analysis in Section III it is apparent that very little, if any, materials degradation will take place in near term tokamak reactors. The low temperature, low wall loading, low plant factor, and relatively short device life combine to minimize any neutron damage effects. The only area that could present problems might be that of electrical insulators in or near the first wall. So little is known about the fundamental effects of 14 MeV neutrons (or even fission neutrons) on electrical conductivity that we hesitate to say that there will be "no" problems in the first generation tokamak reactors.

4.2 Experimental Power Reactors

The first major problem that may arise is associated with the low Z liners. In the case of Be it is the dimensional instabilities caused by the generation of large amounts of helium and with carbon it is anisotropic growth of the crystallites.

Figure 7 shows how the swelling in Be varies with the amount of helium generated and the size of the bubbles. The approximate time in the ANL-EPR is given at the top of the graph. Obviously it is important to keep the helium from agglomerating into bubbles because they then must capture many vacancies to retain equilibrium with the surface tension, $2\gamma/r$. The operating temperature of ~400°C is low enough to keep the bubble size to probably ~200 Å or lower although high stresses are known to promote bubble agglomeration. The questions to be addressed for Be are; will there be any transients, thermal gradients, or hot spots in temperature thay may cause the Be to exceed a design temperature of 400°C (and thereby promote bubble formation, movement and coalescence); what is the maximum expansion that a coating can experience and still avoid cracking, peeling or flaking; and what effect will thermal (stress) cycling have on the precipitation of Be into bubbles and promoting larger values of swelling?

The carbon "shingles" in the ORNL design have somewhat the same problems as the Be except they may go through a shrinkage stage first before progressing into runaway swelling and possible fracture. The dpa level at which this runaway swelling will occur is very much dependent on the material texture and temperature.[15-17] At 1400°C this may occur at damage levels of as low as 10-20 dpa for some graphites. Intuition would lead us to believe that this runaway swelling will be reached at a higher dpa level at 1800°C but no firm data exists and it would be safe to say there is some doubt whether

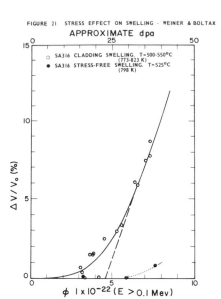

FIGURE 21   STRESS EFFECT ON SWELLING · WEINER & BOLTAX

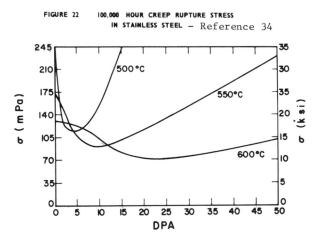

FIGURE 22   100,000 HOUR CREEP RUPTURE STRESS
IN STAINLESS STEEL – Reference 34

472

the shingles will last the full lifetime of the reactor (~ 40 dpa). Further research is desperately need nere.

The pessimism for the carbon shingles in the ORNL design can be somewhat tempered in the GA design because of the lower wall loading (1.7 MW-yr/m^2 vs 4 MW-yr/m^2 in the ORNL design). This damage level of ~ 17 dpa can almost be tolerated at 1300 - 1400° C and it is anticipated that the runaway swelling at 1650° C will occur at dpa levels of > 20.

Even if the runaway swelling in carbon were no problem in the GA and ORNL designs, one might legitimately worry about the high helium generation rate (up to ~1200 appm per yr in the ORNL design) and what effect several thousand ppm Helium may have on the physical and mechanical properties of carbon. Fortunately, it appears that helium can readily diffuse out of graphite above 1200°C. Thomas, et al,[30] have shown that there is ~100% remission of implanted helium at 1200°C, Holt,et al,[31] have measured helium diffusion distances in graphite and find that randomly migrating He can travel 100 to 1000 microns in 100 sec at 1700 °C. These high diffusivities, coupled with small crystallite size (10 microns) should allow most of the neutronically generated helium to escape. Therefore, as long as the carbon is operated at high temperature ($\geq$ 1200 °C) we would tentatively conclude that there should be no severe problem with helium build up as there is in the case of metals.

The SiC plates in the GA-EPR will suffer some linear expansion during irradiation and this may amount to 1 to 1.5% at 1350 °C.[32] Such expansions in a small plate are probably tolerable if it is unconstrained. A more serious problem may arise from the ~300 appm helium that is generated per year. If this helium does not diffuse out of the SiC (like it does in graphite at those temperatures), then there may be a considerable volume change associated with bubble formation. Such volume changes could amount to 1-10% depending on the bubble radius.

Summarizing the situation with respect to low Z coatings and liners one can say that, aside from surface effects and chemical reactions, the neutron induced dimensional changes will be the most critical problems. The effect of stress on bubble agglomeration in Be, the runaway swelling threshold for C at 1600 °C and 1800 °C, and the release of helium in SiC are all areas that must be assessed.

The next class of problems that could arise have to do with the high operating temperatures of the first metallic walls in the reactor designs. These temperatures range from 500 °C in the ANL design to 600 °C in the GA design.* The phenomena of concern here are swelling, creep, fatigue, and ductility. Fortunately, the dpa rates are relatively low in the ANL and GA EPR design (2-3 per year) such that even after 10 years the swelling values predicted from fission reactors and simulation studies amount to a few percent of less (Figure 6). However, when high helium gas generation rates are taken into account (Figure 19) the levels could be as high as 9%. There is one word of caution that needs to be inserted here. All of the previous data comes from essentially stress free material and the recent discovery of stress enhanced swelling [33] may cause even higher swelling to be produced (see Figure 21). If swelling values exceed 5%, there will probably be regions of high stress generated at joints, welds, or in regions of high swelling gradients. The situation may be even more critical in the ORNL EPR where the fluence levels are even higher.

---

* The very first 316SS wall to intercept particles from the plasma in the ORNL design is cooled to ~125°C. However, immediately behind that wall is a first wall running at much higher temperature (~540°C) and the neutron loading is not very much lower.

Irradiation creep is certain to take place in the highly stressed portions of the first walls. As can be seen from Table 2, the stress levels can be 70-200 MPa. Figure 13 predicts that at 500 °C and 200 MPa, creep strains of 1-5% could occur at the end of life of some of the components (if they could stand that much strain without failing!) This will be even more aggrevated at the higher stresses levels that could occur at welds joints or corners. These strains can act to relieve stresses built up due to differentia swelling or fabrication defects, but they might also impose difficult disassembly problem on the maintenance crews. Therefore, the importance of tight dimensional stability must be carefully reviewed before the reactor begins to operate.

Another measure of the "low" stress, long term behavior of irradiated steel is its creep rupture strength for its useful life. In this case it should be ~100,000 hrs. f EPRs if the first walls are to last for the lifetime of the plant. The effect of dis-placement damage with only a small amount of helium generation is given in Figure 22. [34 Note that the minimum of the creep rupture stress is 110 MPa for 500 °C at ~5 dpa (1/2-1 years in an EPR), 85 MPa at 550 °C and ~10 dpa (~ 1-2 years in an EPR), or 70 MPa at 20 dpa (~ 2-4 years at 600 °C). Clearly the present EPR designs do not meet these criter because their design stresses are 70-200 MPa. One might anticipate that the simultaneous generation of helium would aggravate this already serious situation and data is urgently needed in this area.

The fatigue problems in the EPRs(and possibly even the T-20) must be carefully in-vestigated because of the large number of cycles envisioned per year. These range from 1000 in TFTR to 230,000 in the ANL-EPR. As shown in Figure 14, if the ASME design code were to be followed for the 538-650 °C case,such a variation in cycles implies limits on strains of 0.5 to 0.15%. The actual failure levels in irradiated steel at 600 °C appear to be 1.5 to 0.5%. Given all the complexities of vacuum joints,coolant channels, and penetrations in the blanket couples with temperature and damage gradients, it will be ver difficult to insure that the strain per cycle in all the blanket structure is <1%.

Finally, with respect to the ductility of the structure, it is an open question as to whether the stainless steel first walls will be able to retain a reasonable amount of ductility (measured by uniform elongation) to withstand abnormal strains of as much as 1% during shutdown, start up, or other unforeseen transients. Figures 11 and 12 showed that above 20 dpa and temperatures of 428 to 650 °C, uniform elongation values of austenitic steels drop to below 1%. As the helium content and irradiation temperature is increased, this 1% level is reached earlier. The 600 °C operating temperature of the GA design may be particularly vulnerable to early failure due to this mechanism. While this appears to be one of the most limiting features of the EPR (and perhaps DPR) designs, more quantitat information on the ductility of metal irradiated under typical stress-temperature cycle condtions is required. The tolerable level of ductility needs to be clearly established by designers if one is going to be able to assess useful lifetimes for metallic component

474

The final area of concern for neutron damage in the tokamak EPRs is the effect on the magnets. We have seen from Table 2 that the maximum displacement rates in the TF coils range from a low of 8×10^{-6} in the ANL design to 2×10^{-4} in the ORNL work. From Figure 9, we see that even after 10 year, the resistivity of the Cu in the ANL coil will roughly double due to radiation damage (assuming no intermediate anneals). On the other hand, the resisitivity in the GA copper toroidal field coil will double in a few years and that in the ORNL TF coil it will double in a few months. It appears that the ANL and GA designs can reasonably recover the damage by appropriate annealing every few years but the situation in the ORNL design is more serious. If the TF coil in the ORNL design were never annealed over the 10 year operation period, the resistivity of the Cu would increase by a factor of ~10! This would clearly be unacceptable in the event of an accident.

It appears that none of the EPR designs will produce enough damage in the NbTi superconductor to significantly alter its T_c and J_c values (see Figures 10a and 10b). However, the Nb_3Sn windings in the high field region of the ORNL magnet could suffer serious degradation in the T_c and J_c values. If this damage were never annealed out (the exact temperature required for this is unknown but it is probably 300-400°C for 95% recovery).The T_c value might be reduced by a factor of more than 10. Such degradations are probably not tolerable while still maintaining high reliability and therefore high temperature annealing or increased shielding is probably required.

The exposure level of 40 to 100 MRad per year to the organic thermal and electrical insulators does not seem to be a major problem.* The threshold for damage to mylar appears to be 8000 MRads, a factor of 4 above the anticipated 10 year exposure level in the ORNL design.

In summary, swelling in the EPR materials appears to be on the borderline between 'manageable' and a serious problem. The effect of stress on swelling will be extremely important here. The problem's introduced by irradiation creep are definitely quite serious, fatigue failures are definitely possible, and the high temperature in the GA blanket structure are **quite probable** and the high temperature in the GA blanket The Nb_3Sn and Cu stabilizer in the ORNL design appear to be vulnerable to damage.

4.3 Demonstration Power Reactors

Because of the extreme fluidity of the DPR designs at this time, it is not meaningful to analyze the operating parameters in great detail. However, there are a few qualitative remarks that can be made. On the positive side, the longer burn time will decrease the number of thermal cycles and therefore reduce the possibility of fatigue failure. We should also be able to better quantify the allowable stresses, strains and creep rates so that potential solutions can be tested before actual reactor operation.

On the negative side, the combination of higher wall loadings (1-3 MW/m^2) higher plant factors (approaching 70%) and longer plant lifetimes (up to 25y) will undoubtedly require frequent blanket replacement. It does not appear that integrated wall lifetimes of 35-50 $MW/y/m^2$ (250-500 DPA and 5000-10,000 appm He in steel) will be achieved in any presently known material operated at temperatures high enough to produce net electricity. Even if metals were operated at temperatures well below the onset of helium embrittlement, creep rupture lives of 200,000 hr at stresses approaching 100-200 MPa at 300-500°C will be very

---

* This conclusion only applies to the TF coils. The F coil in the GA design represents a much more severe problem. No specific exposure levels are known at this time but they are most likely to be in the 10^{10} Rad/year range for the present design.

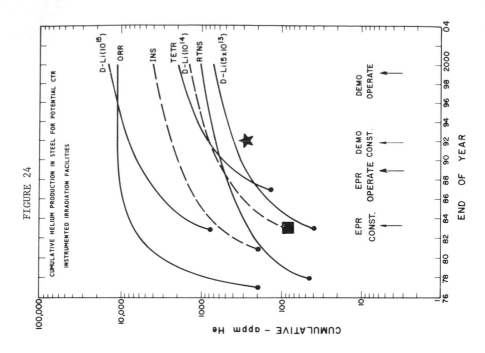

FIGURE 24

CUMULATIVE HELIUM PRODUCTION IN STEEL FOR POTENTIAL CTR
INSTRUMENTED IRRADIATION FACILITIES

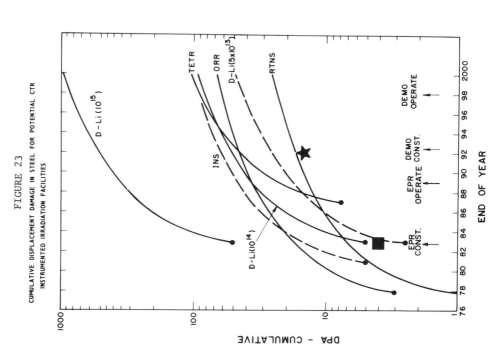

FIGURE 23

CUMULATIVE DISPLACEMENT DAMAGE IN STEEL FOR POTENTIAL CTR
INSTRUMENTED IRRADIATION FACILITIES

476

difficult, if not impossible to achieve. In addition even if we could operate below the void formation temperature (~ 0.25 T_m) swelling due to high helium contents would limit the lifetimes to less than required.

The low Z liners also will face a higher frequency of replacement for the same reasons outlined above. However, such replacements, while difficult do not necessarily appear to be prohibitive both in terms of cost or time involved.

In summary, the materials problems in the DPR's will definitely force fusion reactor engineers into a design which places greater emphasis on accessibility than presently required by the EPR engineers.

5. Potential Facilities That Could be Used to Test Materials for Future Tokamak Reactors

It is not the purpose of this paper to review the irradiation testing field as that has been the topic of a recent conference.[40] There are 5 types of potential sources that could apply to the in-situ testing of Fe-Ni-Cr alloys for the experimental demonstration power reactors. These are (see Table 3);

A) The Rotating Target Neutron Source (RTNS) at the Lawrence Livermore Laboratory. An up graded version producing 2×10^{13} n cm^{-2}s^{-1} (14 MeV) in a small (~ 1 cm^3) volume is currently under construction and scheduled for operation in 1978.[36]

B) The Intense Neutron Source (INS) at the Los Alamos Scientific Laboratory.[37] This source will produce up to 10^{14} n cm^{-2} s^{-1} (14 MeV) in a volume of few cubic centimeters and it should be operational in 1981.

C) A neutron stripping source such as the D-Li source proposed by workers at Brookhaven National Laboratory.[38] The neutron spectrum is rather broad and the testing volume can be stated as a function of the flux of total neutrons. Table 3 shows that ~ 10cm^3 can be subjected to 10^{15} total neutrons cm^{-2}s^{-1} whereas a volume of ~ 300 cm^3 will experience fluxes above 10^{14} n cm^{-2} s^{-1} and ~ 1000 cm^3 will see a flux of 5×10^{13} n cm^{-2} s^{-1}. This source is presently being designed and if funded in the U.S., may operate in 1983.

D) Because of the production of helium by thermal neutrons in Ni containing alloys one can consider the use of thermal test reactors. Our requirement that such facilities have sufficient testing space for in situ studies narrows the number of reactors down to those like the Advanced Test Reactor (ATR) [47] in Idaho Falls or the Oak Ridge Research REactor (ORR) [39] at the Oak Ridge National Laboratory. The flux levels here are not too meaningful because of different neutron spectra but on a displacement basis,such thermal reactors can produce damage at a rate three times higher than the RTNS and produce helium at ~ 6 times the rate in RTNS. These values combined with a few hundred cm^3 of testing volume show the utility of such facilities. It would be relatively easy to modify such reactors for in situ testing in a year and such facilities could be operating in 1978.

E) A typical DT fusion reactor spectra could be provided by a driven (power and tritium consuming tokamak reactor similar to the Tokamak Engineering Test Reactor (TETR) described elsewhere.[34] The displacement rates are only twice that in thermal facilities and the helium production rates are comparable or slightly

477

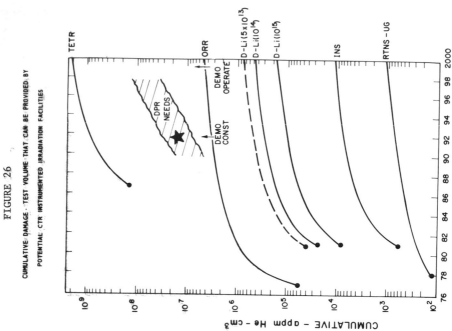

FIGURE 26

CUMULATIVE DAMAGE - TEST VOLUME THAT CAN BE PROVIDED BY
POTENTIAL CTR INSTRUMENTED IRRADIATION FACILITIES

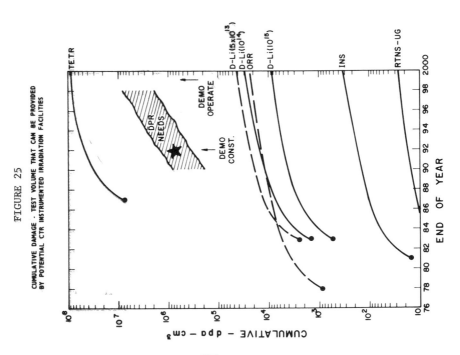

FIGURE 25

CUMULATIVE DAMAGE - TEST VOLUME THAT CAN BE PROVIDED
BY POTENTIAL CTR INSTRUMENTED IRRADIATION FACILITIES

478

less. However such a facility could provide a very large (a few million cm^3) volume for in situ testing. The drawback of this device is that even if the plasma physics and technology systems behave as anticipated it would be ~1987 before such a reactor could start to provide data.

When one combines the instantaneous damage rates with the anticipated plant factors and projected starting dates,he can calculate maximum cumulative damage levels that can be achieved. Such information is plotted in figure 23 and 24. As a point of reference we might compare this information with what might be required by designers of DPRs. It is reasonable that these designers would want information about at least one year of operation in a DPR and this information should be available by the start of construction of the reactor. Applying such criteria to the DPR we see that an average value between the GA and ORNL design would indicate that in situ tests of an austenitic steel up to 15 dpa and containing 300 appm He would be required by 1992. The interesting point about figures 23 and 24 is that all of the source considered for the DPRs and all except the TETR for EPRs, can produce the damage levels required by the proposed date of reactor construction. The appropriate displacement damage and helium production levels can be produced by the times listed in Table 4 for the two types of reactors.

However, just achieving the appropriate damage level is not sufficient for a successful materials test program. Many temperatures, environment's stresses, alloy variations and back up samples must be tested in order to obtain a clear picture of the material's response to irradiation and to develop theoretical models. It is impossible at this time to establish a specific value for the number of samples that need to be tested and the total cumulative volume required for those tests. However, a recent estimate [34] for the PPR reactor arrived at approximately 5000 different samples that would need to be tested to provide a proper base to construct a DPR. At an average of 10 cm^3 per test (some post irradiation tests would require only a few cm^3 and some complex insitu tests might require 20-30 cm^3), this translates into a requirement for a damage-test volume product of about 10^6 dpa - cm^3 and ~ 2x10^7 appm He-cm^3 by 1991. Such a number could be off by a factor of 2 in either direction but it is doubtful whether it is off by as much as a factor of 10. The anticipated DPR requirements are plotted in figure 25 and 26 along with the cumulative damage-test volume product of the facilities listed in Table 3. The first obvious point in these figures is that none of the currently proposed facilities can provide the necessary data by 1991. The RTNS-UG is woefully inadequate in this respect and although such facilities cost only ~ 1-5% of the other facilities, even 20-100 such devices would not solve the problem. The next point is that a single INS is not adequate and more than 1000 would be required to produce the desired number of test specimens. The various D-Li combination are much better suited to provide this damage-volume ratio but again even under the best of circumstances, the product is a factor of 100 too low. The thermal reactors are not much better with regard to displacement damage but can provide data for four years before stripping sources are cosntructed. The situation with respect to helium is much better and the thermal reactors can provide information within a factor of 5 of that required (that is, if the samples

479

Table 3

Possible Test Facilities to Provide Instrumented Data on Fe-Cr-Ni alloys
for Fusion

Facility (Flux)	dpa/yr*	appm He/yr*	Instrumented Test Volume cm^3	PF%	Op Date	Re
RTNS-UG (2 x 10^{13})	1.3	50	1	90	Jan '78	3
INS (10^{14})	6.5	250	3	80	Jan '81	3
D-Li (10^{15})	65	1000	10	80	Jan '83	3
D-Li (10^{14})	6.5	100	300	80	Jan '83	3
D-Li (5 x 10^{13})	3.3	50	1000	80	Jan '83	3
ORR	4.3	300**	300	70	Jan '78	3
TETR	11	200	2,000,000	70	Jan '87	3

*100% PF

** First Year

Table 4

Date When 1 Year Equivalent Insitu Data Could Be Produced (End of Year) By Projected
Neutron Facilities For at Least One Specimen

	EPR		DPR	
	Max dpa	Max appm He	Max dpa	Max appm He
RTNS-UG	1980	1979	1990	1983
INS	1981	1981	1983	1982
D-Li (10^{15})	1983	1983	1983	1983
D-Li (10^{14})	1983	1983	1985	1986
D-Li (5x10^{13})	1984	1984	1988	1990
ORR	1978	1978	1982	1979
TETR	1987[a]	1987[a]	1988	1988

(a) system not applicable

480

are removed every 4 years after the appropriate dpa damage is accumulated).

The only facility that could provide more than enough displacement and helium accumulation information is a DT plasma device like the TETR which operates at an average $1MW/m^2$ over a year with a large testing volume. Such a device could be built with mid 1980 state of the art plasma physics and fusion technology if present research plans are met. The key to the advantage of such a device is the large ($\sim$ several million cm^3) volume of test area available. The cost of such a test facility may be in the 500 million dollar range so any judgement on this device versus multiple thermal fission, neutron stripping or solid target test facilities would have to balance the damage-test volume levels with cost and probabilities that such a device could be successfull built.

In summary, while current and proposed test facilities could produce a few specimens tested under realistic environments by the time the data is needed for a DPR, there is no facility except a large scale D-T reactor which can provide the appropriate number of samples. Further investigation will better define what is required in terms of a test matrix and what is ultimately possible from proposed test devices. Continued up dating of this information may provide the fusion community with more options to provide the necessary information.

6. Conclusion

The irradiation damage problems for the near term fusion devices are not expected to significantly effect their operation or the safety of such experiments. By the time the currently proposed experimental power reactors are built there will be several problems that could arise if these reactors are required to operate at high temperature. A great deal of testing and theoretical model development is required if such reactors are to operate in an efficient and safe manner. The situation is definately more critical for the DPRs and even if appropriate materials testing information can be provided, it is unlikely that any CTR first wall material will last the lifetime of the reactor. Therefore early fusion devices should be designed in a manner such that they can test remote maintenance techniques that will be required for later DPR and commercial reactors.

Acknowledgement

This research was supported by a grant from the Wisconsin Electric Utilities Research Foundation.

REFERENCES

1. A. P. Spano, Nuclear Fusion 15, 909, 1975.

2. W. M. Stacey, Jr., C. C. Baker, and M. Roberts, "Tokamak Experimental Power Reactor Studies," Paper IAEA-CN-35/I3-2, presented at the 6th Int. Conf. on Plasma Physics and Controlled Nuclear Fusion Research, Oct. 1976, Berchtesgaden, FRG.

3. The effects of charged particle damage are reviewed elsewhere in this Conference, see paper by R. Behrisch.

4. G. L. Kulcinski, D. G. Doran, and M. A. Abdou, Properties of Reactor Structural Alloys After Neutron Irradiation, ASTM STP-570, Amer. Soc. for Testing and Materials, Philadelphia, 1975, p. 329.

5. L. G. Kirchner, F. A. Smidt, Jr., G. L. Kulcinski, J. A. Sprague, and J. E. Westmorel. NRL-3312, June 1976, p. 3.

6. M. Karenko, to be published.

7. R. Bullough, B. L. Eyre, and K. Krishan, Proc. Roy. Soc. 346, 81 (1975).

8. D. Kramer, K. R. Garr, A. G. Pard, and C. G. Rhodes, AI-AEC-13047, 1972.

9. O. K. Harling, M. T. Thomas, R. L. Bordzinski, and L. A. Rancitelli, "Recent Neutron Sputtering Results and the Status of Neutron Sputtering," to be published, J. Nucl. Mat.

10. Radiation Induced Voids in Metals, J. W. Corbett, L. C. Ianniello, ed., USAEC-Symp. Series 26, April 1972.

11. Physics of Irradiation Produced Voids, R. S. Nelson, ed., AERA-R-7934, 1975.

12. Courtesy of H. R. Brager, HEDL.

13. F. A. Garner, C. L. Guthrie, and T. K. Bierlein, HEDL-TME-75-23, Vol. 2, 1976.

14. Tokamak Experimental Power Reactor Conceptual Design, ANL/CTR-76-3, Vol. 2, 1976.

15. G. B. Engle and W. P. Eatherly, High Temperatures and High Pressures 7, 319, 1969.

16. J. W. Helm, Carbon 3, 493, 1966, see also J. H. Cox and J. W. Helm, Carbon 7, 319, 196

17. W. J. Gray and W. C. Morgan, BNWL-B-288 and 289, June 1973.

18. G. L. Kulcinski, R. W. Conn, and G. Lang, Nuclear Fusion 15, 327, 1975.

19. J. A. Horak and T. H. Blewitt, Phil. Stat. Sol. 9, 721, 1972.

20. W. J. Weber, G. L. Kulcinski, R. G. Lott, P. Wilkes, and H. V. Smith, Jr., p. I-130 in "Radiation Effects and Tritium Technology for Fusion Reactors," CONF-750989, Vol. I, 1976.

21. S. C. Agarwal and A. Taylor, ibid, p. 150.

22. "Fusion Power by Magnetic Confinement - Program Plan," ERDA-76/110/1.

23. A. H. Spano, Nuclear Fusion 15, 909, 1975.

24. "Tokamak Experimental Power Reactor Conceptual Design," Vol. I & II, ANL/CTR-76-3, August 1976.

25. "Experimental Fusion Power Reactor Conceptual Design Study," Vol. I, II, & III, GA-A14000, July 1976.

26. "Oak Ridge Tokamak Experimental Power Reactor Study Reference Design," ORNL-TM-5042, Nov. 1975.

27. D. W. Kearney, private communication, numbers should be considered very temporary as the design is still evolving.

28. D. Steiner, personal communication, numbers should be considered very temporary as the design is still evolving.

29. For a summary of the work on Beryllium, see "Radiation Effects Design Handbook-Section 7-Structural Alloys" by M. Kangilaski, NASA-CR-1873, 1971, and a Radiation Effects Information Center Report, REIC-No. 45 by M. Kangilaski, June 1967.

30. G. J. Thomas, W. Bauer, P. L. Mattern, and B. Granoff, SAND-75-8718, 1975.

31. J. B. Holt et al., to be published, Lawrence Livermore Laboratory.

32. R. J. Price, J. Nucl. Mat. **48**, 47 (1973).

33. R. A. Weiner and A. Boltax, WARD-OX-3045-22, 1976.

34. G. L. Kulcinski et al., "A Tokamak Engineering Test Reactor to Qualify Materials and Blanket Components for Early DT Fusion Power Reactors," UWFDM 171, 1976.

35. C. Taylor et al., Int. Conf. Proc. on Neutron Sources for CTR Surface and Materials Studies, Argonne, Illinois, July 15-18, 1975.

36. J. Davis, J. E. Osher, R. Booth, and C. M. Logan, to be published in the Proc. of the 2nd Topical Meeting on the Technology of Controlled Nuclear Fusion, Richland, Wash., Sept. 1976.

37. C. R. Emigh et al., ibid, Ref. 35.

38. "Accelerator Based Neutron Generator," BNL-20159, 1975.

39. J. Scott, Oak Ridge National Laboratory, private communication.

40. "Proceedings of the International Conference on Radiation Test Facilities for the CTR Surface and Materials Program," ANL/CTR-75-4, 1975.

41. R. E. Schmunk - personal communication.

ENERGY SOURCES AND CONVENTIONAL MAGNETS

Karl I. Selin

AFR Sweden

JET Design Group, Culham Laboratory, Abingdon U.K.

Abstract

A power supply for an EPR toroidal field magnet is described. It is concluded the main network can supply the
necessary power estimated to be of the order of 1 GW. The
power system response to fault conditions in the experiment is
discussed. The utilization of the power dissipated in the magnet
is desirable and is referred to in connection with the cooling
system arrangements.

Introduction

The scenarios of fusion reactor blueprints are that no
commercial reactors will be built for 30 - 40 years and therefore, I would say naturally, they are designed on paper with
superconducting coils. However if we take a look at projects
closeby and consider designs which shall operate within a few
years from now, maybe, some will hesitate. And this might be
the reason why I have been asked to evaluate the possibility of
supplying very large plasma fusion experiments having conventional
magnets with electric power.

The largest U.S/European Plasma experiments, which are
presently near the construction phase, require of the order of
700 MW and 7 GJ peak power and energy.

In principle, a pulse load may either be supplied directly
from the electric power network or indirectly from a motor-
generator-flywheel set, utilizing the kinetic energy released
from it during a speed reduction, [1] . A combination of
these principles can also be used, whereby some loads are
supplied from the mains, others from flywheel generators.
Alternatively a single load is supplied from several convertors
of which some are fed from the mains and others from a generator.
In Fig.1 is shown the probable arrangement to supply power to the
OH coils and the EF amplifiers of JET [2].

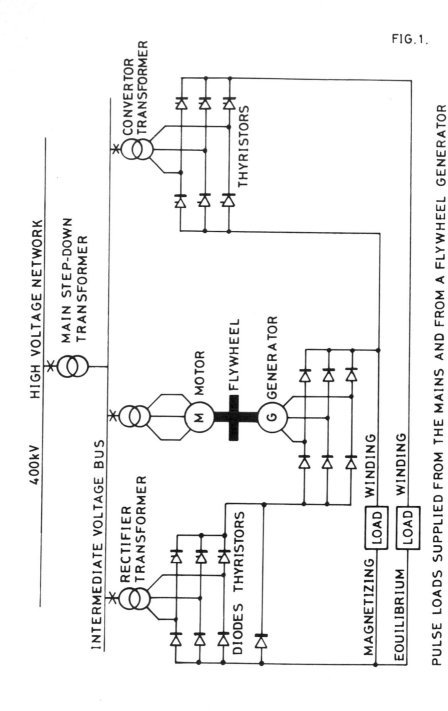

FIG.1.

PULSE LOADS SUPPLIED FROM THE MAINS AND FROM A FLYWHEEL GENERATOR

The attitudes of people concerned with building plasma
experiments are rather diversified stretching from a rather
detached view, that it does not matter how the electricity is
converted and transferred to run the experiment, to a con-
viction that

1.   a supply from the mains is the best not having to bother
     with a local generator plant, or

2.   low reliability thyristors must be used in the rectifier
     of the mains.   Therefore a local generator plant is
     preferable.   Further a network cannot cope with the fast
     power swings of a big experiment, anyway.

As for the JET there has been a considerable amount of
discussion which site is best suited for the electrical supply.
     Our US colleagues may, from experience, add that if the
site of the experiment is not serviced by a high voltage, high
power network, the choice is simple;   local generators are
necessary, and if the generators and loads are arranged with a
common AC-bus system the generator convertors must be fully
equipped with thyristors.

1.   Large Tokamak devices with conventional magnets.
     Is it reasonable from a power supply standpoint to assume
such large magnets as in an EPR to be of a conventional design?
And could such a device be called an EPR?   Are superconducting
coils an essential factor for EPR behaviour evaluation?   Before
you give the answers, let me make some general comments.

First: If superconducting magnets are suitable for and can
meet the operating conditions in a reactor, will not be known
until such magnets are tried out in a large Tokamak device.
Some of the problems and uncertainties involved were presented
by Cornish and Khalafallah [3] at the 8th SOFT Conference.

Second: That a single EPR will be built having the features of a reactor and the possibilities to test all technological aspects of a reactor is not very probable. Much more of a step-by-step procedure can be foreseen with parallel, alternative approaches tried out. And the design of a prototype reactor or any large experiment for that matter within the area of plasma fusion development must depend on whether progress has been made in the technologies involved, e.g. large pulsed field cryogenic magnets for a Tokamak configuration where the magnets are exposed to pulsating loads in an environment where maintenance is difficult and should be kept at a minimum.

Third: When certain conditions are met, the toroidal field power does not constitute a very large portion of the total electric power output of a fusion reactor power station, e.g. < 10% of the electric power produced. To reach such a performance the size of the reactor and also the wall loading would necessarily be very large as shown by Ivanov, Popkov and Strelkov [4].

2. The Oak Ridge Tokamak EPR [5]

The Reference Design has been developed by the Oak Ridge National Laboratory, which is operated by Union Carbide Corporation for the U.S. Energy Research and Development Administration. It is scheduled to start operation in 1985. Consequently the design of the cryogenic magnets should be ready and construction begin in 1979.

The peak power demand of the Reference Design is 300 MW.

It has been suggested to me to consider the Reference Design of the Oak Ridge Tokamak EPR and to evaluate the power demand and a supply source, assuming the reactor magnets to have copper conductors. A further assumption will be that the copper conductors are water cooled.

2.1 The Toroidal field magnet of the Reference Design

The plasma current will be 7.2 MA in a plasma of radius $a = 2250$ mm. The field on torus axis is $B = 4.8$ T. The maximum field $B_{max} = 11$. The number of coils is 20. Each coil is of an oval shape with N turns of mean length $L = 30$ m (approximately). Each coil shall produce $NI = 8.1 \times 10^6$ ampere turns. The coil cross section is shown in Fig.2.

488

3. Aspects on using a conventional magnet for the toroidal field in the EPR.

3.1 Coil cross section A_O

As a first approach, the space occupied by the cryogenic toroidal field coils will be used for a conventional magnet.

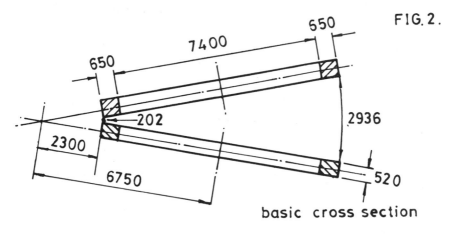

FIG. 2.

basic cross section

Cross section $A_O = 650 \cdot 520 = 338000 /$ Coil

Inner circumference $2\pi\,2300 = 14450$

Inner coil gap $14450/20 - 520 = 202$

The cross section area of each coil including cooling channels etc. is $A_O = 338000$ mm^2. The area of cooling channels and insulation is $(1-f)\,A_O$, where f is the copper filling factor. The area of the cooling channels is assumed to be much larger than the insulation area and approximately equal $(1-f)\,A_O$. The currer ·ensity

$$j_O = \frac{NI}{fA_O}$$

The power loss per unit volume is $\rho j_O{}^2$, where ρ is the copper resistivity. The power loss per coil is $LfA_O\rho j_O{}^2$. If $NI = 8.1$ MAt, $L = 30$ m, $f = 0.8$, $j_O = 30$ A/mm^2 and $\rho = 20$ $\mu\Omega$ mm a power of 2.9 GW is dissipated in the magnet of 20 coils. The copper of the magnet has a volume 20 x 30 x 0.338 f = 203 f (m^3), a weight of 1805 f (ton) and a thermal capacity 0.704 f (GJ/^{0}C).

489

3.2   Statement of problems

In light of these preliminary figures it might be worth-while to consider the following six problem areas:

(A)   Is the necessary amount of power available

(B)   Is the necessary amount of energy available

(C)   Can the magnet be efficiently cooled

(D)   Can the heat dissipated in the magnet be recovered and utilized

(E)   What type of heat exchanger may be used

(E_1) -   in case of a pulsed toroidal field utilizing a primary circuit heat storage of a high capacity.

(E_2) -   in case of a continuously operated experiment

(F)   How much will the experiment cost

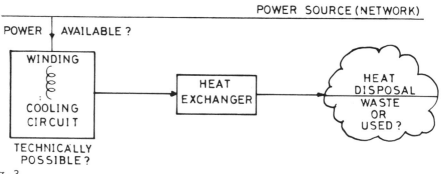

Fig.3.

A power of the order of several Gigawatt can be drawn from either flywheel generators or from a strong network.   From the standpoint of energy the pulse length of EPR operation excludes a local generator plant.

In the U.S. it might very well be feasible to connect an EPR of 2.9 GW load to the mains.   Still, it will be assumed that the EPR, if equipped with conventional magnets, would be re-designed for a resistive load power of 1.2 GW in the toroidal field magnet.   A load power of 1.2 GW is expected to be available at several sites in the US and in Europe.   There might be restrictions in the form of the time of load connection and disconnection, power time derivatives, reactive power swing, etc. In spite of such restrictions an all static power supply from the mains at the most favourable sites should perform satisfactori

3.3 Coil cross section A_1

As compared with the initial assumption A_0 the cross section of each toroidal field coil will have the width increased to 722 and the thickness to 1300 mm, (see Fig.4) and the cross section area A_1 = 9386 cm^2. The number of ampere-turns per coil is increased slightly for the same Shafranov-Kruskal stability factor

$$q = \frac{H_{TOR}}{H_{POL}} \quad \frac{a}{R}$$

$$(NI)_1 = 1.07 \times 8.1 = 8.7 \text{ MAt}$$

The current density

$$j_1 = \frac{(NI)_1}{fA_1} = \frac{9.269}{f} \quad A/mm^2$$

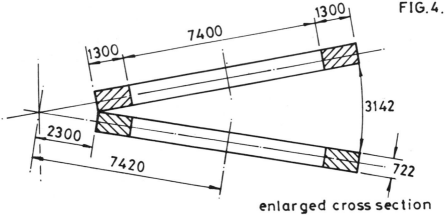

FIG.4.

7400

1300

1300

7420

2300

3142

722

enlarged cross section

Cross section A_1 =1300 × 722 = 938600 mm^2

Mid plane section of coils around torus showing cross section of toroidal field coils, (mm) (mm2)

The average length of one coil turn is still about 30 m and therefore the input power to the magnet

$$2OLfA_1\rho j_1{}^2 = 2OL\rho \frac{(NI_1)^2}{fA_1} = \frac{0.9677}{f} \quad GW$$

The copper of the magnet has a volume 564 f (m^3), a weight of 5012 f (ton) and a thermal capacity 1.95 f (GJ/$^\circ$C).

3.4. Values of voltage/current at the magnet terminals.

Each coil consists of N conductors series connected. Each conductor has a central cooling channel of radius r_1. The relative cross section cooling area is

$$\frac{N\pi r_1^2}{A_1} = 1 - f$$

$$r_1 = \sqrt{\frac{(1-f)A_1}{N\pi}}$$

The resistance of the magnet is

$$R = 20 \frac{\rho L N^2}{f A_1} = \frac{1278}{f} N^2 \quad \mu\Omega$$

the voltage,

$$U = 20 \rho j_1 L N = \frac{111}{f} N$$

and the current

$$I = \frac{8.7 \times 10^6}{N}$$

In order to limit the power to 1.2 GW the f-factor ought not to be less than 0.8. This will be commented upon further on.

3.5   Cost of magnet and magnet power

The installed capacity of the interconnected European Electric Network is about 200 GW. It may be considered reasonable that the society puts aside 1 out of 200 GW to increase the knowledge in this area of a new fuel technology and energy source   The importance of the right perspective on energy R & D should be underlined it does not mean that an EPR of less operating cost would not be preferable.

The electricity bill will no doubt be very big. If the experiment needs 1.2 GW during 3000 hours per year, the energy used (3.6 TWh) is approximately equal to what 300000 persons on an average consume of electricity at home and at work combined. The yearly cost would be about 100×10^6.

The substation and rectifier equipment is estimated to cost about 20×10^6.

The cost of the magnet, in proportion to weight of the JET toroidal field magnet would be approximately $ 60 \times 10^6$.

492

3.6   The load characteristic as compared with other large nuclear
      experiments

The CERN II experiment, which has recently come into operation, has a power demand on the EdF grid of 150 MW loading to 80 MW regeneration, in a periodically repeated sequence.

The JET experimental device requires about 300 MW peak load superimposed on a 300 MW load pulse of about 20 sec, repeated every 10 minutes.   In the early part of the next decade a peak power of about 575 MW can be made available for JET from the grid at the proposed sites in France, Germany, Great Britain and in Italy.

It should be emphasized that the power to an EPR differs in nature from the loads of CERN and JET.   The EPR load on the network would be dominated by the toroidal field load.   In most instances the on-loading and off-loading of the network can be made slow.   Only on rare occasions in connection with a severe fault condition will there be a 100% sudden change of load.

These basic differences relate not only to power demand but also time and energy, which makes in a way an EPR simpler but of course much more expensive than for example JET.

4.   Static power supply for the toroidal field conventional
      magnet.

4.1   Field voltage applied step-by-step

The power supply scheme for the toroidal field load is shown in Fig.5.   The experiment should basically be run as a power reactor and the magnetic field interrupted only to the extent maintenance and experiments require.   This power supply is intended for continuous operation and the toroidal magnetic field is switched on and off using circuit breakers on the primary side of the rectifier transformers.   Oil minimum breaker could be used having a 4 kA continuous current rating, 50 kA fault current interrupting capacity and a load current switching capacity before maintenance of 1000 times.

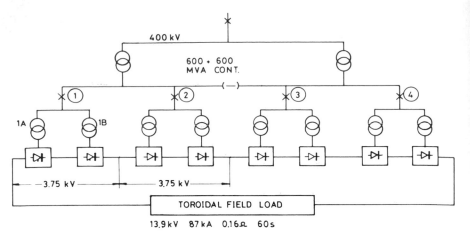

THE TOROIDAL FIELD POWER SUPPLY WITH DOUBLE BUS BARS
CIRCUIT BREAKERS 1.......4. RATED LINE-LINE VOLTAGE 70 kV
    LOAD CURRENT 4 kA, FAULT CURRENT 50 kA
TRANSFORMERS 1A AND 1B ARE CONNECTED Y/Y AND Y/Δ FOR 12 PULSE OPERATION
CONVERTOR 1 IS EQUIPPED WITH THYRISTORS, RECTIFIERS 2.....4 WITH DIODES.

F IG.5.

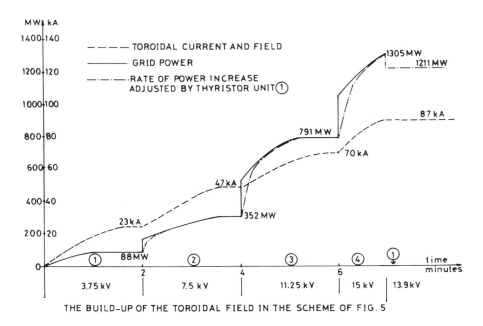

THE BUILD-UP OF THE TOROIDAL FIELD IN THE SCHEME OF FIG. 5

FIG.6.

494

The magnetic field is established as shown in Fig.6 by closing circuit breakers No.1 ... 4 in proper time sequence. The time intervals may be suggested by the responsible power supply authority. Each circuit breaker connects and disconnects two bridges supplied from out of phase transformer secondary voltages. A 12 pulse rectification scheme is hereby secured in all stages of switching, consequently the 5th and 7th harmonics are suppressed and the percentage amount of harmonic currents in the AC supply is constant.

In the scheme shown in Fig.5 each pair of bridges supplies 3.75 kV. Unit number (1) is equipped with thyristors, whereby the output voltage magnitude and polarity can be chosen and controlled by phase delay. Prior to closing C/B (2) ... (4) the rectifiers of same units free-wheel the DC current. As C/B (2) is closed, the power to the magnet would increase in a step-wide manner. If this is objectional to the power supply network the step-wise change can be moderated by voltage inversion in unit (1). By utilizing unit (1) in sequential control with units (2)...(3)...(4) in succession, the power demand can be adjusted as shown in Fig.6 to a power time derivative, which at all times agrees with the power supply network capability. The time scale of switching (every 2 minutes) shown in Fig.6 is not critical and may be changed to fit the planning and load dispatching scheme laid down by the power generation authority.

The control of the power derivative and the final magnet voltage/current control is done using phase delay of the thyristor bridges. The rating of unit (1) is 326 MVA. It would possibly draw the same amount of reactive power, if only for a few seconds, in the process of power derivative control. Unit (1) may be designed in such a way that the reactive power involved in its control action becomes only half the above value but it would necessitate either a further division of unit (1) in twice the number of subunits (diode/thyristor sequential control) or result in a mixed 6-12 pulse reaction on the AC side with more AC current harmonics. An optimum choice can be made to suit local demands and conditions.

495

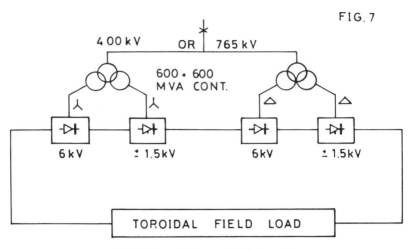

FIG. 7

400 kV   OR   765 kV

600 + 600
MVA CONT.

6 kV   ± 1.5kV   6kV   ± 1.5kV

TOROIDAL FIELD LOAD

13.9 kV, 87 kA, 0.16 Ω, 60 s

ON-OFF SWITCHING OF GRID POWER TO MAGNET.
FIELD ADJUSTMENTS BY 12-PULSE THYRISTOR-CONVERTORS

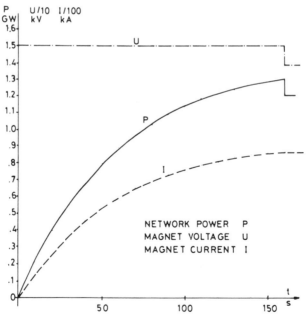

NETWORK POWER   P
MAGNET VOLTAGE   U
MAGNET CURRENT   I

THE BUILD-UP OF THE TOROIDAL FIELD AFTER THE
CLOSING OF THE CIRCUIT BREAKER IN FIG.7.

FIG.8

The reactive power required by transformer reactances and commutation angle phase delay is estimated to be maximum 300 MVAr. Together with the reactive power due to the control actions of unit (1) the reactive power load of the high voltage network may reach values of the order of 600 MVAr. The reactive power swings would not exceed 300 MVAr.

The bridges could be equipped with capacitors on the transformer low voltage side whereby special capacitor circuit breakers or compensators are avoided. At no load a transformer voltage rise would be experienced by such a scheme.

4.2 Field voltage applied by single switching

An alternative power supply scheme is shown in Fig.7. The toroidal field is switched by a single breaker aided by two thyristor bridges. The maximum power derivative, U^2/RT, where U = magnet voltage, R = magnet resistance and T = magnet time constant, is <20 MW/s. If the power system accepts the rate of power change shown in Fig.8 this scheme is less complicated and a less expensive system.

4.3 Fast changes of power input to the toroidal field magnet

Other loads as the ohmic heating coils, the equilibrium coils and the additional heating may be of large amplitudes and power derivatives. During a time interval, which is one or two orders of magnitude less than the magnet time constant, the power input to the magnet may be reduced thereby reducing significantly the overall power consumption and the rate of change of power to ease the stress on the network generators. A diversion of 50% of the power from the toroidal to the poloidal circuit during 2 seconds would only decrease the toroidal field about 1.5%. ,

4.4 Normal and abnormal load conditions

If the load is switched on as described in 4.1 and 4.3 there are no problems of fast load changes. The toroidal field and load would be increased at a tempo which is easy for the generating capacity to respond to. After the magnetic field is established the steady active load is assumed to be 1.21 GW. Most of the time there is a surplus of reactive power available at the voltage level of 400 kV and above and a steady reactive load should be acceptable. When the experiment is to be

497

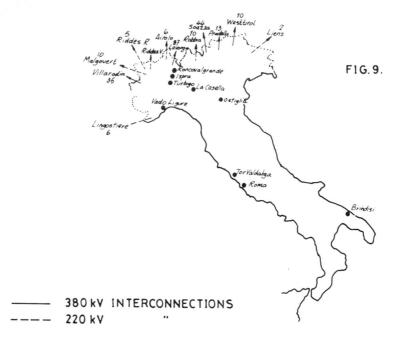

FIG. 9.

—————— 380 kV INTERCONNECTIONS
— — — — 220 kV   "

GEOGRAPHICAL POSITION OF THE MOST IMPORTANT
POWER STATIONS IN THE ISPRA AREA.

INTERCONNECTIONS WITH FOREIGN COUNTRIES
CONSIDERED IN THIS STUDY.

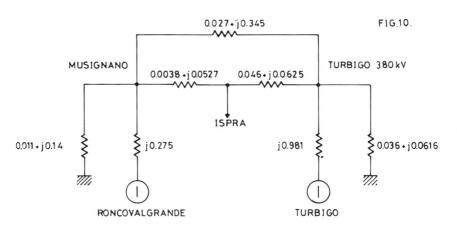

FIG. 10.

EQUIVALENT SCHEME OF NETWORK SEEN FROM ISPRA (OBTAINED FROM
SHORT CIRCUIT IMPEDANCES)
(IMPEDANCES ARE IN PER UNIT OF 1000 MVA REFERRED TO A VOLTAGE OF 400 kV)

interrupted the active power load to the magnet is reduced at
a slow rate causing an orderly reduction of generated power.

A quite different situation appears in case of an operating
condition in which the load has to be suddenly disconnected
following, e.g. a busbar fault in the substation, a step down
transformer fault, a secondary breaker fault or a fault within
the experimental device which would necessitate an opening   of
the primary circuit breaker and a sudden disappearance of full
load.   Then

(1)   suddenly all generators experience a sudden drop in output
      power in proportion to the short circuit power contribution
      of each generator,

(2)   which is followed by power oscillations between the
      generators influencing also the rotating shafts of loads.

(3)   after the oscillations are damped the through-power for the
      majority of the generators is the same as prior to the
      fault incidence;   for a minority of dedicated generators
      the steady state through-power is reduced as much as the
      amount of dropped loading.

4.5   Field tests and computed step power response of the European Network

All power stations in Europe are connected together with
power transmission lines. The UCPTE organization (Unione per il
coordinamento della producione e del trasporto di energia elettrica)
deals with the administrative and technical problems of this very
large network.

ENEL has made computations and field tests of load and
generation changes of the order of 300 MW at Roncovalgrande in
the north of Italy which are illustrative of transient conditions
in a network.

In Fig. 9 is shown some of the power stations and sub-
stations in Italy and also the 220 kV and higher voltage lines
interconnecting Italy with neighbouring countries.

In Fig.10 is shown a very much concentrated equivalent
scheme of Europe as seen from the power stations at Roncovalgrande
and Turbigo.

In Fig. 11 - 12 the computed reaction at various places in
Italy to a sudden load drop at Roncovalgrande is shown.

499

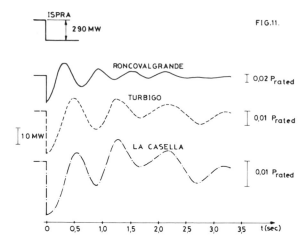

FIG.11.

DIGITAL SIMULATION OF SUDDEN DROP OF ACTIVE LOAD

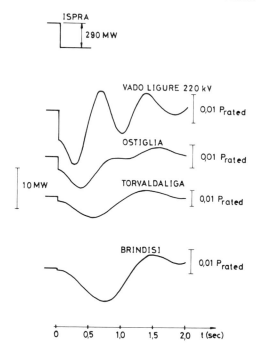

FIG.12.

DIGITAL SIMULATION OF SUDDEN DROP OF ACTIVE LOAD

The first sudden drop is clearly shown, the step depends on the size of generator and the impedance between the place of load disturbance and the generator. Such a step means a sudden change of the torque on the pole faces of the rotor in the generator. In Fig. 13 the network response to the fast active and reactive load swing of a poloidal magnet pulse is shown in an all-static scheme of a large Tokamak experiment.

In Fig.14 the results of field tests are recorded. In these tests a simultaneous change of both the generated active power, ΔP = 244 MW, and the reactive power, ΔQ = 111 MVAR, at Roncovalgrande has been made. The drop in generated reactive power (ΔQ) causes a voltage drop and the active power (ΔP) a frequency drop, which also has been measured in Rome, Fig.15.

These slides all show power oscillations which are damped. But in any network, if the load change is big enough, there will be one or several generators, for which the stability limit is exceeded. It means that during the oscillations the angle between the rotor magnets and the stator rotating field is so large that the syncronising torque is lost. Then, in general, the generator must be disconnected. Eventually the interconnected network has to be broken up by opening transmission line circuit breakers.

If in one area there is an unbalance between generation and load it is necessary to start load shedding in that area of the country to keep up the frequency and voltage. A severe network disturbance has then occurred and it will take some time to recover normal operating conditions again. It is expected that in the future the unit generator will be of the order of 1 GW and disconnection of a load of 1.2 GW is not expected to create a problem as long as the site is properly chosen.

5. Cooling of the toroidal field magnet

In JET the volume of copper is 50 m^3 as compared with 450 m^3 in the redesigned EPR. The ratio of resistive power dissipated in the magnets is about 4. Under these conditions it appears possible to arrange proper cooling, also on a steady state basis. The cooling of the magnet will be done through primary and secondary cooling circuits as shown in Fig. 16.

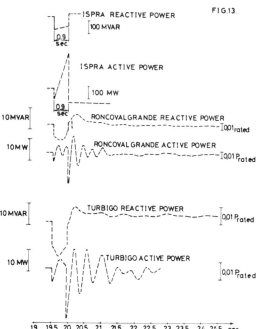

FIG.13.

DIGITAL SIMULATION OF RESPONSE TO COMBINED ACTIVE
AND REACTIVE POWER TRANSIENTS OF A TOKAMAK
POLOIDAL FIELD CIRCUIT LOAD.

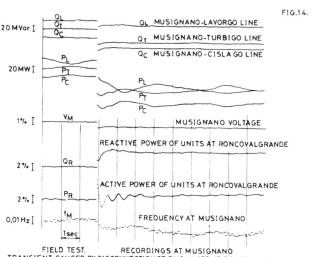

FIG.14.

FIELD TEST.   RECORDINGS AT MUSIGNANO
TRANSIENT, CAUSED BY DISCONNECTION OF TWO UNITS AT RONCOVALGRANDE
$\Delta P = 244\,MW, \Delta Q = 111\,MVAR$

502

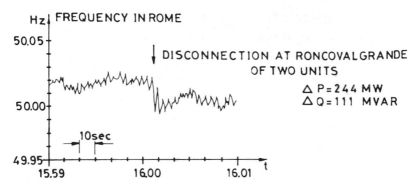

FIELD TEST
TRANSIENT CAUSED BY DISCONNECTION OF TWO UNITS AT
RONCOVALGRANDE.

FIG.15.

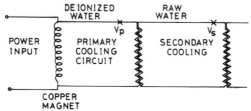

COOLING CIRCUITS FOR TOROIDAL FIELD MAGNET
V_p = VALVE AND PUMPING OF PRIMARY COOLANT
P_s = VALVE AND PUMPING OF SECONDARY COOLANT

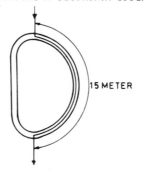

COOLING OF HALF A TURN OF A COIL OF THE TOROIDAL
FIELD MAGNET

FIG.16

503

5.1 Water temperature rise along the conductor

There will be about 6 km of conductor in the magnet. Each cooling water channel will encompass half a turn of coil winding (15 meter). The water velocity is v, density q, thermal capacity c and temperature ϑ. When steady state temperature conditions are reached there will be a temperature differential in the direction of flow (X)

$$\frac{d\vartheta}{dX} = \frac{f}{1-f} \frac{\rho j^2}{vqc}$$

5.2 The copper and cooling water temperature difference

The radial heat flux from a conductor section of length ΔX to the cooling water through a temperature difference of ΔT at heat transfer coefficient h is in steady state

$$\frac{1}{N} fA_1 \Delta X \rho j_1^2 = h \, 2\pi r_1 \Delta X \Delta T$$

$$\Delta T = \frac{469600}{h \, f \, \sqrt{n}\sqrt{1-f}}$$

The temperature difference ΔT will be minimum when the copper area is $f = 2/3$ of the available space A_1. In order to limit the power a second assumption is made about the coil,

$$f = 0.8$$

whereby

$$j = 11.59 \text{ A/mm}^2,$$

the power dissipation in 20 coils

$$P_1 = 1.21 \text{ GW},$$

the resistive voltage of 20 coils

$$U = 139 \text{ N},$$

the cooling channel radius

$$r_1 = \frac{244.4}{\sqrt{N}} \text{ mm},$$

the copper/water contact area, which is the cooling area of the magnet

$$A = 20 \, N2\pi r_1 L$$

$$= 40 \, L \, \sqrt{N\pi(1-f)A_1}$$

$$= 921.5 \, \sqrt{N} \text{ m}^2,$$

and the temperature difference

$$\Delta T = \frac{1\ 313000}{h\ \sqrt{N}}$$

A third assumption is made regarding the coil,

$$N = 100,$$

whereby

$$U = 13.9\ \text{kV}$$
$$I = 87\ \text{kA}$$
$$P_1 = 1.21\ \text{GW}$$
$$r_1 = 24.4\ \text{mm},\ D = 0.049\ \text{m}$$
$$A = 9215\ \text{m}^2$$

A fourth assumption is made regarding the velocity of water through the coil

$$v = 2\ \text{m/s}$$

The heat transfer from the copper to the turbulent flow of water is guided by the Dittus-Boelter relation

$$\frac{hD}{K} = 0.023\ (\text{Re})^{0.8} \left(\frac{c\mu}{K}\right)^{0.4}$$

Reynolds number $\qquad\qquad\qquad \text{Re} = \dfrac{vDq}{\mu}$

Thermal conductivity of water	$K = 0.6\ \text{W/}^0\text{C m}$
Specific heat of water	$c = 4180\ \text{Ws/kg}$
Density of water	$q = 1000\ \text{kg/m}^3$
Dynamic viscosity of water	$\mu = 0.0007\ \text{kg/s m}$
Diameter of cooling channel	$D = 0.049\ \text{m}$
Velocity of water	$v = 2\ \text{m/s}$
Heat transfer coefficient	$h = 6.95\ \text{kW/m}^2$
	$\text{Re} = 140\ 000$
	$\Delta T = 18.9\ \ ^0\text{C}$
	$\dfrac{d\partial}{dX} = 1.285\ ^0\text{C/m}$
	$\Delta \partial = 19.2\ ^0\text{C over 15 m of conductor}$
	length.

5.3 Temperature gradient in conductor

The radial heat flux pattern through the conductor to the cooling water channel is influenced by the form of the conductor. The conductor area is 9386 mm^2 and the copper area is 7509 mm^2. A square conductor would have a side of 97 mm.

For an approximate calculation of the hot spot temperature assume hollow conductor of outer radius $r_2 = 60$ mm. Through unit length of conductor and at radius r, where the radial temperature gradient is $d\partial/dr$, the heat flux

$$\pi(r_2{}^2 - r^2)\rho j_1{}^2 = 2\pi r K_c \frac{d\partial}{dr}$$

The thermal conductivity of copper $K_c = 346$ W/$^\circ$C m.

Integrating

$$\partial = \partial_{r_1} + \frac{\rho j^2 r_2{}^2}{2K_c} \ln\left(\frac{r_2}{r_1} + \frac{r_1{}^2}{2r_2{}^2} - \frac{1}{2}\right)$$

The second term is about 6 $^\circ$C.

5.4 Cooling water flow rate

The cross section of cooling ducts per coil

$$(1-f)\ A_1 = 0.2 \times 938600 = 187720 \text{ mm}^2$$

At a water velocity of v m/s the cooling water flow is $v(1-f)A_1 = 0.37544$ m^3/coil or 15 m^3/s through 4000 cooling channels in 40 half-coils.

The amount of water in the copper coils is

$$20 \times 30 \times 0.18772 = 112.6 \text{ m}^3$$

A similar amount of water may be assumed in the condenser. Extra storage over and above 225 m^3 may be used, if thermal heat storage is found advantageous.

5.5 Utilization of magnet losses

The flow and storage of magnet power losses can be visualized from the electro-thermal equivalent circuit in Fig.17. The copper volumetric power p_1 is equivalent with an input current p_1, temperature with voltage, etc. With V_p in Fig. closed and V_s open the transient temperatures (voltages) are,

$$V_1 = \frac{p_1}{C_1 + C_2}\left[t + \left(\frac{C_2}{y} - \tau\right)\left(1 - e^{-\frac{t}{\tau}}\right)\right]$$

$$V_2 = \frac{p_1}{C_1 + C_2}\left[t - \tau\left(1 - e^{-\frac{t}{\tau}}\right)\right]; \quad \tau = \frac{C_1 C_2}{y(C_1 + C_2)}$$

5.5.1 Thermal energy recovered from the primary cooling loop at reduced power

For energy recovery at a rate less than 1.2 GW a strategy of operation may be to increase the volume of primary loop coolant to last for a length of time acceptable to the mains (not too short).

From the equation of 5.5 it is found that an adiabatic temperature rise (secondary cooling loop open) of $\Delta\vartheta_1 = 70^\circ C$, $\Delta\vartheta_2 = 63\ ^\circ C$ is reached after 140 s using the minimum amount of coolant (225 m^3).

The amount would have to be increased about 80 times to last for one hour pulses. As an EPR in principal should operate continuously any type of pulsed operation would probably be objectional. Also for other devices there are no obvious means to take advantage of pulsed operation apart from the fact that the heat exchanger and the final heat disposal equipment can be made smaller.

5.5.2 Thermal energy continuously extracted

Energy may alternatively be used continuously at fairly high temperature of the raw water in the secondary cooling loop.

Assume a steady state condition of the circuit in Fig.17. If a thermal power (current) of 1.21×10^9 shall be disposed of through a temperature rise of the raw water, $V_3 = \dfrac{P_1}{y_3}$, the flow rate of raw water would be $\dfrac{1.21 \times 10^9}{4180\ \partial_3}$
As an example, $\partial_3 = 20^\circ C$ at a flow rate of 14.5 m^3/s. If $y_2 \approx y_1$ the magnet temperature would be 20 + 19 + 19 degrees above raw cooling water return temperature. With an (average) magnet temperature of 90°C the energy would have to be recovered through a temperature drop from 52°C to 32°C. Such a temperature range is probably too low for the water to be directly used, but its value and possible use for various process industries should be considered because it might eventually influence decisions regarding the use of conventional magnets in still larger Tokamak plasma experiments.

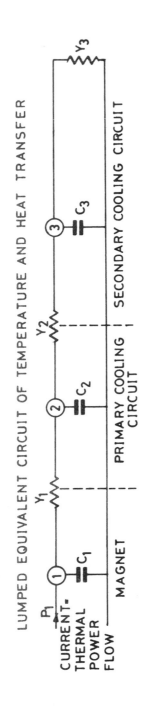

LUMPED EQUIVALENT CIRCUIT OF TEMPERATURE AND HEAT TRANSFER

FIG. 17.

$v_1 = \vartheta_1$ = AVG. MAGNET TEMPERATURE

$v_2 = \vartheta_2$ AVG. DIW TEMPERATURE

$v_3 = \vartheta_3$ AVG. RW TEMPERATURE

$P_1 = 1.21 \times 10^9$

COPPER : $C_1 = C \times MASS = 390 \times 4 \times 10^6 = 1.56 \times 10^9$

DIW   : $C_2 = C \times MASS = 4180 \times 225 \times 10^3 = 0.94 \times 10^9$

RW   : $C_3 = C \times MASS$

C = THERMAL CAPACITY
    PER DEGREE

$Y_1 = h_1 \times A_1 = 6950 \times 9236 = 0.064 \times 10^9$

$Y_2 = h_2 \times A_2$

$Y_3 = C \times (MASS FLOW RATE)$

Y = THERMAL POWER FLOW
    PER DEGREE TEMPERATURE
    DIFFERENCE

508

References

1. K.I. Selin, E. Bertolini: Large Tokamak Power Supplies. A Survey of Problems and Solutions. Proceedings of the 9th Symposium on Fusion Technology, 1976.

2. E. Bertolini et al: The JET Power Supply Comprehensive System
    Proceedings of the 9th Symposium on Fusion Technology, 1976.

3. D.N. Cornish, K. Khalafallah: Superconductors for Tokamak Toroidal Field Coils. Proceedings of the 8th Symposium on Fusion Technology, 1974.

4. D.P. Ivanov, G.N. Popkov, V.S. Strelkov: On the Possibility of using normal conductors for the toroidal magnet system of a Tokamak fusion reactor. Proceedings of the 8th Symposium on Fusion Technology, 1974.

5. Oak Ridge Tokamak Experimental Power Reactor Study, Reference Design. ORNL-TM-5042. Oak Ridge National Laboratory, November 1975.

Lecture 1. on <u>Auxiliary Heating in Breakeven Tokamaks</u>.[†]

by

John Sheffield (EUR-UKAEA)
JET Design Group, Culham Laboratory, Abingdon, Oxon

1.   Introduction

We may identify two main areas where auxiliary heating may be used in breakeven Tokamak reactors.

(1)   In the start up phase of a Tokamak reactor, having a thermal energy distribution, auxiliary heating is required to bridge the gap between the ohmic state where the temperature will probably be ~1 keV, and the ignited condition where T>5 keV. This has been discussed in numerous papers, e.g. Wort[1] Bodin et al[2], Sweetman et al[3], Gibson[4], Sweetman[5], Girard et al[6], Cohn et al[7], Jassby et al[8], etc.

(2)   In two component and three component systems, auxiliary power is required to maintain a non-thermal ion distribution. These systems have been discussed by for example, Dawson et al[9], Furth et al[10], Berk et al[11], Cordey et al[12] Jassby[13], Jassby[14], Perkins[15], Sheffield[16].

I will follow a similar line to that taken in the above papers and will assess the auxiliary heating requirements of a "Breakeven Tokamak" producing ~500 MW of thermal power, using the D-T cycle.   I will consider two and three component systems insofar as they may be used during the start up phase, but not as reactor systems in their own right. In the first lecture I will not discuss details of the heating methods.

Symbols and Units are given in the Appendix.

[†]Work carried out in pursuance of EURATOM JET Design Phase Agreement 30-74-1FUA-C.

II. Power Balance

A simplified electron power balance is given by

$$\frac{\partial T}{\partial t}\left(\frac{3}{2}neT_e\right) = \frac{1}{r}\frac{\partial}{\partial r} \ r\left(neK_e \ \frac{\partial T_e}{\partial r} + \frac{3}{2} \ D_eeT_e \ \frac{\partial n}{\partial r}\right) + P_{ie}$$

$$\qquad\qquad\qquad\qquad\qquad\qquad\qquad\qquad\qquad\text{ion} \rightarrow \text{electron}$$

$$- \ P_b - P_{LR} \quad - \quad P_R \quad - \quad P_c \quad + P_\Omega \ + P_{\alpha e} + P_{ae} \qquad + P_{\alpha ae} \qquad (1)$$

brems	line radn.	recomb. radn.	cyclotron radn.	ohmic	alpha power	auxiliary power	auxiliary alpha power

For the ions

$$\frac{\partial T}{\partial t}\left(\frac{3}{2}neT_i \right) = \frac{1}{r}\frac{\partial}{\partial r} \ r\left(neK_i \ \frac{\partial T_i}{\partial r} + \frac{3}{2} \ D_ieT_i \ \frac{\partial n}{\partial r}\right) + P_{ei}$$

$$- \ P_c \quad + \quad P_{\alpha i} \ + \quad P_{ai} \quad + \quad P_{\alpha ai} \qquad\qquad\qquad (2)$$

charge exchange	alpha power	auxiliary power	auxiliary alpha power

where K and D are the thermal conduction and diffusion coefficients respectively, $p(W cm^{-3})$ are power densities. $P_{ie} = - P_{ei}$ is the ion electron power transfer. $P_{\alpha a}$ is the alpha power, generated by the auxiliary power, which is trapped in the plasma.   It will be assumed that $n_i = n_e$, and that all thermally generated alpha power is trapped in the plasma and transferred immediately to it.

(a) In general a breakeven system will be big enough that the equilibration time of T_e and T_i will be much less than the global energy confinement time τ_E and consequently we set $T_e \approx T_i = T$.

(b) Note also that the α-power is transferred mainly to the electrons as long as $T_e \ll 110$ keV, which means that in practice situations with $T_i > T_e$ can be sustained only by auxiliary heating.

(c) If we now ignore the diffusion loss and the charge exchange loss, since the latter will be confined in large apparatus to the surface, then combining (1) and (2) we obtain with $T_e = T_i = T$

$$\frac{\partial}{\partial t}(3neT) = \frac{1}{r}\frac{\partial}{\partial r}\left(rn \ (K_e + K_i) \ e \ \frac{\partial T}{\partial r} \right) - P_B - P_{LR} - P_R$$

$$- \ P_c + P_\Omega + P_\alpha + P_a + P_{\alpha a} \qquad\qquad (3)$$

Bremsstrahlung

$$P_b = 5.4 \times 10^{-5} \xi \, n^2 Zeff \, T^{\frac{1}{2}} \, W \, cm^{-3} \tag{4}$$

where ξ corrects for e-e collisions and relativistic effects for $1 < T < 50$ keV $\quad 1 < \xi < 1.2$.

For constant density and parabolic temperature $T_p = T_0(1-\frac{r^2}{a^2})$

Gibson[4] gives $\overline{P_b} \approx 5.1 \times 10^{-5} \xi n^2 Zeff \, T^{\frac{1}{2}} \, Wcm^{-3} \tag{5}$

T is taken as the mean temperature.

Line radiation

Meade[17] gives $P_{LR} = 2 \, n\Sigma n_z \, Wcm^{-3} \tag{6}$

where a weak dependence on T has been ignored, n_z is the impurity density of charge z_i

Recombination radiation

Calculations by the TFR group[18] show that for conditions of interest $P_R \lesssim P_{LR} \tag{7}$

Cyclotron radiation

There is no universal formula for this quantity because of the differing behaviour with temperature of the various harmonics, Gibson[4] gives a figure reproduced here in Figure 1, showing various predictions for a particular case. In general the loss is significant at low β and high T and when the walls are not good reflectors. The expression given by Trubrukov[19] is

$$P_c = 0.7 \times 10^{-5} \, n \, B^2 \, Tk_L \quad Wcm^{-3} \tag{8}$$

where k_L is a plasma transparency coefficient lying typically in the range $0.005 \to 0.05$.

Ohmic power

$$\overline{P_\Omega} = C_1 \, I^2 \, \gamma_R \, Zeff/(a^4 \, T^{\frac{3}{2}}) \, Wcm^{-3} \tag{9}$$

$$C_1 \approx 2.8 \times 10^{-7},$$

γ_R is a resistance anomaly, which we will take to be unity.

Alpha-power

For the D-T cycle

$$^2_1D + ^3_1T \to ^4_2He^+ + ^1_0n \tag{10}$$

$$3.52 + 14.06 = 17.58 \; MeV = E_n$$

The fusion power density is

$$P_n = n_D n_T \langle \sigma v \rangle \, E_n \quad W \, cm^{-3} \tag{11}$$

Convenient empirical formulae, for the α-power density are

$$P_\alpha \approx 2 \times 10^2 \, \frac{n_D n_T}{T^{2/3}} \, \exp\left(\frac{-20}{T^{\frac{1}{3}}}\right) \, Wcm^{-3} \quad 2 \lesssim T < \sim 35 \; keV \; + \tag{12}$$

and $P_\alpha \simeq 4\times10^{-7} n_D n_T \, T^4 \, \text{Wcm}^{-3}$ $2 \leq T \leq 10 \text{ keV}$ (13)

For a flat density and parabolic temperature, convenient formulae are

$$\overline{P_\alpha} \simeq 4.5\times10^2 \, \frac{n_D n_T}{T} \cdot \exp\left(\frac{-19.4}{T^{\frac{1}{3}}}\right) \quad \text{W cm}^{-3} \quad 2 \leq T \leq 20 \text{ keV} \quad (14)$$

and $\overline{P_\alpha} \simeq 1.7\times10^{-5} \, n_D n_T \, T^{2.5} \text{Wcm}^{-3}$ $2 \leq T \leq 10 \text{ keV}$ (15)

I will take $n_D = n_T = n/2$.

Fusion power from non-thermal ion distribution

We can expect, as discussed above, that non-thermal ion distribution will only result from the use of auxiliary heating, we therefore define the quantity $Q_a = \dfrac{\text{fusion power produced}}{\text{auxiliary power input}}$

It is important to note that this does not include the efficiency of production of the auxiliary power.

$$P_{\alpha a} = \frac{Q_a}{5} \cdot P_a \, \eta_\alpha \quad (16)$$

where η_α is the fraction of fusion α's which is trapped in the plasma

Conduction loss

The greatest uncertainty lies here, because there is no clear knowledge of this quantity for either electrons or ions in the reactor regime of n and T.   Present experiments have explored the range to $10^{13} \leqslant n \sim 10^{14} \text{cm}^{-3}$ and $T_o \leqslant 2 \text{ keV}$.

(a)   It is convenient to define a global energy confinement time τ_E, as

$$\tau_E = \frac{C_o 2\pi^2 Ra^2 n_e T \left(1 + \dfrac{n_i}{n_e}\right)}{P_L} \quad \text{(s)} \quad (17)$$

where P_L is the total power lost owing to the loss mechanisms considered. $C_o = 2.4 \times 10^{-3}$

We will consider cases where $n_i = n_e = n$, but if there are impurities then strictly $\overline{n_i} < \overline{n_e}$

T refers to the mean temperature.

(b)   Empirical fits have been made to present Tokamak data by for example Daughney[20] and Hugill and Sheffield[21].

In the latter paper, the following formula is given

$$\tau_E = 9.5\times10^{-9} \quad n^{0.84} \, a^{2.13} \, \left(\frac{R}{a}\right)^{0.91} B^{0.49} \, T_{eo}^{0.25} \, A_i^{0.62} \text{(s)} \quad (18)$$

Since the Tokamaks investigated used ohmic heating, a second

relationship between these quantities may be obtained
from (17) with $P_L = P_\Omega$. Combining this power balance with
(18) we may rearrange the formula for τ_E and in fact it
agrees well with that of Daughney. What is not clear is
the correct form for application to higher temperature
regimes $T > 2$ keV. Nevertheless it may be used in the
prediction of the ohmic temperature.

$$T_\Omega \simeq \left[5.6 \times 10^{-13} \ (R/a)^{0.91} \ B^{0.49} \ A_i^{0.62} \ I^2 \ \gamma_R \text{Zeff} / n^{0.16} \ a^{1.87} \right]^{0.44} \quad (19)$$

(c) For higher temperature regimes various predictions have been
made, some representative cases are listed below. T is mean
temperature.

Psuedoclassical loss	$\tau_E = 1.1 \times 10^{-11} \ I^2 T^{0.5} / n \text{Zeff}$ (s)	(20)
bremsstrahlung loss	$\tau_E = 90 \ T^{0.5} / n \text{Zeff}$ (s)	(21)
Bohm loss	$\tau_E = 4 \times 10^{-11} \ a^2 B \ T^{-1}$ (s)	(22)
Plateau loss	$\tau_E = 3.3 \times 10^{-17} R^3 I^2 T^{-1.5} / a^2$ (s)	(23)
Trapped-ion loss	$\tau_E = 5 \times 10^{-18} \ (R/a)^{2.5} \ na^4 B^2 \text{Zeff} \ T^{-3.5}$ (s)	(24)

(d) In order to assess the auxiliary power requirements, I will
denote the losses by

$$P_L = \frac{C_0 \ 4\pi^2 \ Ra^2 \ nT}{\tau_E} + P_B \quad (25)$$

where $\qquad \tau_E = C_3 \ T^\alpha \qquad\qquad\qquad\qquad\qquad (26)$

The bremsstrahlung has been separated out because it is the
inescapable minimum loss.

I will assume that the impurity level and type is such
that we may ignore line radiation and recombination radiation.
Cyclotron radiation becomes important at the higher
temperatures for representative reactor designs, if the
walls are not reflecting. Gibson[4] has discussed this for
a representative case and finds that $P_c < P_b$ for $T < 35$ keV if
the walls are 90% reflecting. In fact the use of controllable
reflectors might allow us to control the loss rate when the
plasma has ignited and thereby prevent temperature runaway
situations. For the discussions below I will neglect
cyclotron radiation.

Thus τ_E represents losses due to conduction.

III. Auxiliary Power Requirements - Thermal Plasma

We may now integrate equation (3) over the plasma volume and we obtain

$$\frac{\partial W}{\partial t} = - C_4 T^{1-\alpha} \quad -C_5 T^{0\cdot5} + C_6 T^{-1\cdot5} + C_7 T^{\gamma} + P_a \qquad (W) \qquad (27)$$

$$\text{conduction} \quad \text{brems} \quad \text{ohmic} \quad \text{Alpha} \quad \text{auxiliary}$$

where $C_4 = 9.5 \times 10^{-2} R\ a\ b\ n/C_3$

$C_5 \simeq 1.0 \times 10^{-3} R\ a\ b\ n^2 Zeff$

$C_6 \simeq 5.5 \times 10^{-6} R\ I^2 \gamma_R Zeff/ab$

$$C_7 = \left. \begin{array}{l} 2.0 \times 10^{-6} \\ 8.4 \times 10^{-5} \end{array} \right\} R\ ab\ n^2 \left\{ \begin{array}{ll} \gamma = 4 & \text{Flat T} \\ \gamma = 2.5 & \text{Parabolic T.} \end{array} \right.$$

The plasma volume has been generalised to include ellipticity.

$$W = 9.5 \times 10^{-2} \ R\ ab\ n\ T \qquad (J) \qquad (28)$$

is the total plasma energy.

The auxiliary power required to maintain the plasma at the temperature T (P_{ao}), is obtained from (27) by getting $\frac{dW}{dt} = 0$. In general there will be two temperatures at which $P_{ao} = 0$; at T_{Ω} when the ohmic power balances the losses; at T_{IG} when the alpha power balances the losses. The ignition temperature. The minimum value is obtained when the losses are solely due to hydrogenic bremsstrahlung and then $(T_{IG})_{min} \sim 4$ keV. Between these two temperatures the auxiliary power required will reach a maximum at T_m, given by

$$\frac{dP_{ao}}{dt} = (1-\alpha)\ C_4 T_m^{-\alpha} + 0.5\ C_5 T_m^{-0\cdot5} + 1.5\ C_5 T_m^{-2\cdot5} - \gamma C_7 T_m^{\gamma\cdot1} = 0 \qquad (29)$$

This is illustrated in figure 2, in the format suggested by Wort[1

Now we note that at T_m, the ohmic power will be small, and because bremsstrahlung has a weak dependence on T, the maximum comes mainly from balancing conduction against α-power which leads to

$$T_m \simeq \left[\frac{(1-\alpha)\ C_4}{\gamma C_7} \right]^{\frac{1}{\alpha-\gamma-1}} \qquad (30)$$

Clearly if $\alpha \gtrsim 0.5$ or conduction is negligible, T_m will be an important factor in determining T_m.

The maximum auxiliary power required to sustain the plasma is

$$P_{am} = C_4 T_m^{1-\alpha} + C_5 T_m^{0\cdot5} - C_6 T_m^{-1\cdot5} - C_7 T_m^{\gamma} \qquad (31)$$

The ignition temperature is given by

$$C_7 T_{IG}{}^\gamma = C_4 T_{IG}{}^{1-\alpha} + C_5 T_{IG}{}^{0.5} = P_{\alpha I} \qquad (32)$$

Consequently $\dfrac{P_{am}}{P_{\alpha I}} = \left[\left(\dfrac{T_m}{T_{IG}}\right)^{1-\alpha} - \left(\dfrac{T_m}{T_{IG}}\right)^\gamma \right] - \dfrac{P_{BI}}{P_{\alpha I}} \left[\left(\dfrac{T_m}{T_{IG}}\right)^{1-\alpha} - \left(\dfrac{T_m}{T_{IG}}\right)^{0.5} \right] \qquad (33)$

If we ignore the bremsstrahlung loss, we find from (30) and (32) that

$$\frac{T_m}{T_{IG}} = \left(\frac{1-\alpha}{\gamma}\right)^{\frac{1}{\gamma+\alpha-1}} \qquad (34)$$

There are essentially two cases

(a) $\gamma > 1 - \alpha$ in which case $\dfrac{P_{am}}{P_{\alpha I}}$ is given as in figure 3

For example for flat profiles $\gamma = 4$, and say $\alpha = -1$ then $\dfrac{1-\alpha}{\gamma} = 0.5$ and $\dfrac{P_{am}}{P_{\alpha I}} = 0.25$.

For the ignition systems being considered where the thermal power P_{th} is in the range ~ 500 MW, we have $P_{\alpha I} \sim 100$ MW, and taking the upper level as being realistic for support of the plasma against the expected losses we have $P_{am} = 25$ MW.

(b) $\gamma \leqslant 1 - \alpha$ in which case the α-power apparently would exceed the losses at all temperatures, and in this case the neglect of bremsstrahlung would be incorrect because it would be the dominant loss for determining P_{am}.

In figures 2, 8 and 9 a plot is given of the various powers for a representative case.

Unfortunately P_{am} is only the power required to maintain the plasma at T_m, and in fact we must also supply power to raise the plasma temperature, equation (27). The total auxiliary power is set at $P_a > P_{am}$ $\qquad (35)$

We now approximate P_{ao} as a parabola going from $T=0 \rightarrow T=T_{IG}$

$$P_{ao} = P_{am}\, 4 \left(\frac{T}{T_{IG}} - \frac{T^2}{T_{IG}{}^2}\right) \qquad (36)$$

This is a reasonable approach since the time consuming part of the heating occurs in the region of T_m and it is excess power $P_a - P_{am}$ that determines the heating time.

$$C_8 \frac{dT_o}{dt} = P_a - P_{am}\, 4 \left(\frac{T}{T_{IG}} - \frac{T^2}{T_{IG}{}^2}\right) \qquad (37)$$

where $C_8 = 9.5 \times 10^{-2} R\ ab\ n$.

Let the time to raise the temperature from $T=0 \rightarrow T=T_{IG}$ be t_{IG}, solving (37)

$$t_{IG} = \frac{C\ T_{IG}}{P_{am}\ (y-1)^{\frac{1}{2}}}\ \text{arc tan}\ \left(\frac{1}{(y-1)^{\frac{1}{2}}}\right) \tag{38}$$

where $y = \dfrac{P_a}{P_{am}}$ is a measure of how much we must exceed P_{am} in order that the time to ignition is acceptable, by this we essentially mean $t_{IG} \lesssim 10s$. In figure 4 we plot

$$\frac{t_{IG}}{\dfrac{C_8 T_{IG}}{P_{am}}} \quad \text{vs} \quad \frac{P_a}{P_{am}}$$

For example let t_{IG} = 10 s, for a machine of R = 600 cm, a = 200 cm, b = 300 cm, n = 1.0 x 10^{14} and flat T, where T_{IG} = 10.0 keV and P_{am} = 25 MW. then $\dfrac{P_a}{P_m} \simeq 2.1$ e.g. we need

$P_a \simeq 53$ MW.

Even if P_{ao} = 0, we require $P_a \simeq 34$ MW in order to raise the plasma temperature to ignition in 10 s.

IV. Auxiliary Power Requirements - Non-thermal Plasma

(a) In situations where the ion energy distribution is not thermal we may have an enhanced fusion reactivity. For example the application of ion-cyclotron heating might be used to provide a high energy tail to the ions Jassby[14], Perkins[15].

In addition there are of course the two component and three component beam injection systems. Dawson et al[9] Cordey et al[12], Jassby[13].

In these cases the fusion is enhanced and if we allow that the thermal fusion production remains unchanged we may add a new contribution[16]

$$P_{\alpha a} = \frac{Q_a\ \eta_\alpha\ P_a}{5} \tag{39}$$

(b) Now for a two component system with a deuterium beam of energy in the range 100 $< \varepsilon_o \lesssim 150$ keV we have Cordey et al[23]

$$Q_a \sim 0.23\ T_e, \quad \text{and}\quad \eta_\alpha \sim 1 \tag{40}$$

see figure 5. These energy beams may however only be appropriate to expanding plasma scenarios. Girard et al[6]. Sweetman et al,[3].

(c) For a three component system if we can arrange for the colliding beams to be in a region where $n_b \sim n_e$, we have $Q_a \sim 0.5$ Te, see figure 6. However such a system is probably possible only in an expanding plasma, Sheffield[16], with the beams on the outer edge and then $\eta_\alpha \sim 0.5$. Thus for reacting beam systems from (39) with Q_a and η_a given as above

$$P_a + P_a \sim P_a (1 + 0.05 \text{ Te}) \tag{41}$$

Thus we may use a lower value of P_a, call it $P_{aQ} = P_a/(1+0.05 \text{ Te})$. This could be an important contribution if T_m in ≥ 5 keV.

(d) In some plasma start up scenarios, adiabatic compression has been proposed, Jassby et al[8], Furth et al[10], Berk et al[11]. This adiabatic compression has two features; first it increases the temperature, for major radius compression $\frac{T_2}{T_1} = \left(\frac{R_1}{R_2} \right)^{\frac{4}{3}}$; second if injected beams are present, the beam ions may be held at high energy during the compression, "energy damping", enhancing the fusion reactivity.

   Three important points about compression should be noted.

Green et al[24], (1) in general adiabatic compression is useful only if the initial plasma has sizeable auxiliary heating, (2) for some scaling laws compression is not effective. Fortunately it is effective for radiation losses and trapped particle losses. (3) the compression time τ_C must be significantly less than τ_E for the process to be adiabatic we require $\tau_C \lesssim \left(\frac{C^{1\cdot33}-1}{C^{1\cdot33}} \right) \tau_E$.

V.   Representative Breakeven Tokamak

(a) Consider the representative Tokamak, figure 7, whose parameters are R = 600 cm, a = 200 cm, b = 300 cm, B=50 kG, I = 8.0 MA, Shield thickness on inner part of torus 75 cm, shield + blanket on outer part of torus 125 cm, Wedged copper toroidal coils of radial thickness 80 cm, coil stress 7.5 kG/ mm^2, toroidal coil power 500 MW, Volt seconds = 100.

(b) We take the plasma parameters $\overline{A_i}$ = 2.5, n = 1.0×10^{14} cm^{-3} T_{IG} = 10 keV, Zeff = 2, γ_R = 1, effective circular radius (ab)$^{0\cdot5}$ = 245 cm.

519

$$\beta_J = \frac{2 \times 10^7 \text{ ab n T}}{I^2} = 1.9$$

$$P_\alpha = 72 \text{ MW flat T profile}$$

$$P_\alpha = 95 \text{ MW parabolic T profile}$$

At the operating temperature we require

$$\tau_E = 4.9 \text{ s flat T}$$

$$\tau_E = 3.4 \text{ s parabolic T.}$$

Now we have $\tau_E = C_3 T^\alpha$, and in table 1 we compare values of C_3 required for different scalings to balance $P_\alpha - P_B$ at the ignition temperature with the nominal values eqs.(20→25

Table 1. Compression of required factors C_3 with nominal values
(parabolic T cases in brackets)

	nominal C_3	α	required C_3
Psuedoclassical	35.0	0.5	2.2 (1.5)
Bremsstrahlung	4.5	0.5	2.2 (1.5)
Bohm	0.12	-1.0	72 (50)
plateau	11.0	-1.5	230 (160)
Trapped-ion	6.2×10^3	-3.5	2.2×10^4 (1.5×10^4)

Thus we see that we do not require a loss as favourable as psuedoclassical or as bremsstrahlung (Zeff = 2), however we do require a conduction loss, an order of magnitude better than plateau or trapped-ion scaling, and two or more orders better than Bohm.

In table 2, the values of T_Ω(19), T_m(29), P_{am}(31), and the additional power P_a needed for the achievement of ignitio in a time $t_{IG} = 10$ s (38), and P_{aQ} for a non-thermal state with $T_e = T_m$ (41).

Note $T_{IG} = 10$ keV.

The power balance for flat and parabolic T profiles is illustrated in respectively figures 8 and 9.   Note the similarity of the empirical scaling (19) and plateau scaling (Wesson (25)).

Table 2. Characteristics of a Breakeven Tokamak for various loss Scalings. (Parabolic T cases in brackets)

	$^\phi T_{\Omega_0}$ (keV)	T_m (keV)	P_{am} (MW)	P_a (MW)	P_{aQ} (MW)
Psuedoclassical	1.0 keV	5.5 (4.0)	48 (52)	71 (74)	56 (62)
Bohm	1.0 keV	7.0 (5.0)	31 (17)	56 (45)	41 (36)
Plateau	1.0 keV	6.5 (5.0)	24 (11)	54 (37)	41 (30)
Trapped-ion	1.0 keV	5.0 (3.0)	14 (7)*	42 (35)	34 (30)

ϕ T_Ω is for empirical scaling

*   Ignition occurs at 5.5 keV because $(1-\alpha) > \gamma$

Conclusions.

(1)   We see that it will be necessary to provide ~ 50 MW of auxiliary power to ignite the plasma in a "reasonable" time. This is irrespective of the scaling of the dominant conduction loss with T, assuming of course that $P_L = P_\alpha$ at the operating temperature. (P_α = 70 →100 MW).

(2)   The plasma behaviour is affected by the profile of T and the way in which it evolves.   Note that the profile which governs α and γ depends upon the relationship of $1 - \alpha$ to γ. if $1 - \alpha > \gamma$, then the profiles will probably be flat. if $1 - \alpha < \gamma$, they may peak.   However if the temperature peaks this may well lead to enhanced losses, acting to flatten the T profile.   Clearly self-consistent calculations of various representative cases should be made to give a better understanding of these effects.

(3)   As a general comment, I would point out that the α-power is not transferred immediately to the plasma, nor is it necessarily deposited where it is created, see Tsuji et al[26]. This correction should also be included.

(4)   The power requirement may well be less in systems which involve expanding plasmas with a propagating burn, that is if ignition can occur in a smaller radius plasma, see Girard et al[6], Jassby et al[8] and Sheffield[16].   If plasmas of the scale discussed here are needed for ignition then the minimum power required will be ~50 MW, however in bigger devices it should then not be necessary to put even

higher auxiliary power because with a propagating burn they should ignite when the plasma reaches the scale of the breakeven plasma.

(5) For some of the loss scalings $P_\alpha > P_L$ when $T > T_{IG}$ and consequently the temperature will tend to rise in an uncontrolled fashion, the problem is reduced because of point (3), but nevertheless some control on the P_L is desirable. One possibility might be to control the reflection and therefore reabsorption of the cyclotron radiation. Since multiple reflections occur when the plasm is transparent, a small change in wall reflectivity will have a large effect.

(6) A rough estimate of the cost of the representative Break-even Tokamak has been made using a combination of the costing formulae of Sheffield and Gibson[27] and Hancox[28]. These costs are given in table 3. The auxiliary power is set at $P_a = 50$ MW and is costed at 1.5$ US per watt delivere to the plasma. This is consistent with present estimates for the large tokamaks.

The costs of Diagnostics, Data acquisition, Remote handling and Buildings Design costs and Labour costs are not included.

Table 3. Cost of Representative Tokamak ⚡US dollars - April 1975

Toroidal Coils	48
Toroidal Power	20
Poloidal Coils	14
Poloidal Power	27
Vacuum System	9
Structure	6
Blanket and Shield	57
Auxiliaries	5
Auxiliary Heating	75
	270

We see that auxiliary heating represents about 25% of the total hardware costs.

Acknowledgements

I have appreciated valuable discussions with Dr. A. Gibson and other members of the JET Design Group and of the Associated Laboratories.

Symbols, Units

R (cm), a (cm) major and minor toroidal radii; b/a ellipticity;
B (G) toroidal magnetic field; I(A) plasma current; $n(\times 10^{13} cm^{-3})$
plasma density; $n_2(\times 10^{13} cm^{-3})$ impurity density; Z_i ionic charge.
$Z_{eff} = \sum_i n_i Z_i^2/n_e$; $p(W\ cm^{-3})$ power density; P (W) power
W (J) plasma energy; T (keV) temperature; T_0 (keV) peak temperature.
K $(cm^2 s^{-1})$, D $(cm^2 s^{-1})$; thermal conduction and diffusion coefficients; A_i, atomic mass number: $\tau_E(s)$ energy confinement
time. Subscripts, e, i, z, electrons, ions and impurities
respectively: subscript a, α, Ω, auxiliary heating, alpha and
ohmic terms.

REFERENCES

(1) Wort D. (1969) Proc. B.N.E.S. Conf. on Nuclear Fusion Reactors (Culham) 517.

(2) Bodin, H. et al (1974) Proc. IAEA Workshop Culham, Nuclear Fusion. Special Supplement 455.

(3) Sweetman D, et al (1971) CLM-R112, Culham, England.

(4) Gibson, A (1972) Lecture Ial. Erice School on Fusion Reactor Technology.

(5) Sweetman D. (1973) Nuclear Fusion $\underline{13}$, 157.

(6) Girard J.P. et al (1973) Nuclear Fusion $\underline{13}$, 685.

(7) Cohn D. et al (1976) Nuclear Fusion $\underline{16}$, 31.

(8) Jassby D. et al (1975) Trans. A.N.S. $\underline{22}$, 72

(9) Dawson J. et al (1971) Phys. Rev. Letts. $\underline{26}$, 1156.

(10) Furth H. et al (1974) Phys. Rev. Letts. $\underline{32}$, 1176.

(11) Berk H. et al (1975) Tokyo conference IAEA CN-33/G2-3.

(12) Cordey J.G. et al (1975) Nuclear Fusion $\underline{15}$, 710 and (1976) Nature $\underline{259}$, 526.

(13) Jassby D. et al (1975) VIIth E.P.S. Conference Lausanne and (1976) Nuclear Fusion $\underline{16}$, 15.

(14) Jassby D. (1974) 2nd Conf. on RF Plasma Heating Lubbock, E-10.

(15) Perkins F (1976) IIIrd Int. Conf. on Heating of Toroidal Plasmas, Grenoble.

(16) Sheffield J. (1976) IIIrd Int. Conf. on Heating of Toroidal Plasmas, Grenoble.

(17) Meade D. (1974) Nuclear Fusion $\underline{14}$, 289.

(18) Breton C. et al (1976) Fontenay-aux-Roses. EUR-CEA-FC-822.

(19) Trubnikov B. et al (1958) 2nd Int. Conf. on Peaceful uses of Atomic Energy, Geneva vol.31, 93.

(20) Daughney C. (1975) Nuclear Fusion $\underline{15}$, 967.

(21) Hugill J. et al (1976) submitted to Nuclear Fusion.

(22) Kadomstev B. et al (1967) Rev. of Plasma Physics, Vol.5., p. 321. Consultants Bureau, New York.

(23) Cordey J. et al (1975) Nuclear Fusion $\underline{15}$, 755.

(24) Green B. et al (1975) Phys. of Fluids $\underline{13}$, 2593.

(25) Wesson J. (1976) Culham. Private communication.

(26) Tsuji H. et al (1976) Nuclear Fusion $\underline{16}$, 287.

(27) Sheffield J. et al (1975) Nuclear Fusion $\underline{15}$, 677.

(28) Hancox R (1975) Culham. Private communication.

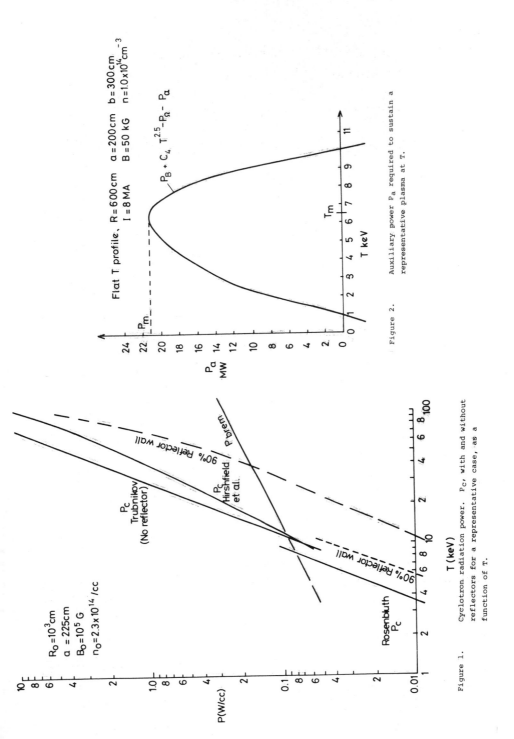

Figure 1. Cyclotron radiation power. P_c, with and without reflectors for a representative case, as a function of T.

Figure 2. Auxiliary power P_a required to sustain a representative plasma at T.

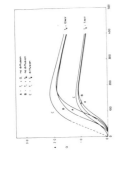

Figure 5. Q values for a D beam (energy ε_0 keV) injected into a tritium plasma including effects of finite T_i(B) and energy diffusion (C).

$$P_L = C_4 T^\alpha \qquad P = C_7 T^\gamma$$

$$\frac{T_m}{T_{IG}} = \left(\frac{1-\alpha}{\gamma}\right)^{\frac{1}{\gamma + \alpha - 1}}$$

Figure 3. Ratio of the maximum power deficit P_{am} to the alpha power at ignition vs $x = \frac{1-\alpha}{\gamma}$.

Figure 6. Q values for a colliding beam D-T system, as a function of deuterium energy (tritium energy is 1.5 x greater) and ratio of beam density to electron density. (Excluding the effect of energy diffusion).

Figure 4. Relation to determine by how much P_a must exceed

526

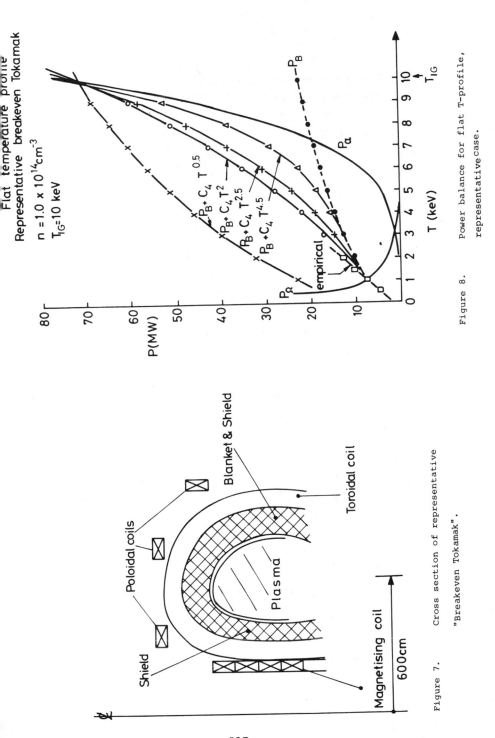

Flat temperature profile
Representative breakeven Tokamak

$n = 1.0 \times 10^{14} \text{cm}^{-3}$
$T_{IG} = 10 \text{ keV}$

Figure 8.   Power balance for flat T-profile,
            representative case.

Figure 7.   Cross section of representative
            "Breakeven Tokamak".

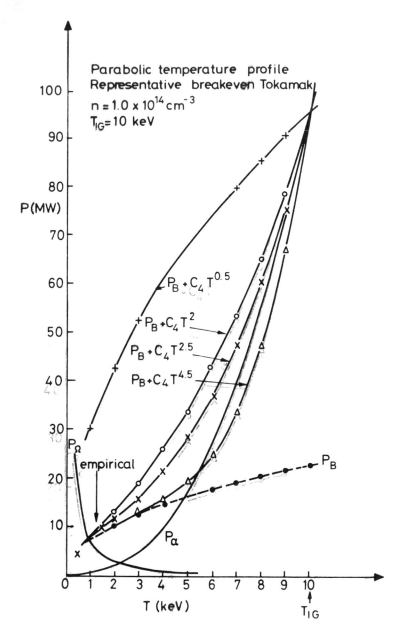

Figure 9.   Power balance for parabolic T-profile, representative case.

"AUXILIARY HEATING IN BREAKEVEN TOKAMAKS"[(+)]

Summary of Varenna Meeting (September 1976) **Lecture 2**

John Sheffield (EUR-UKAEA)

JET Design Group, Culham Laboratory, Abingdon, Oxon (U.K.)

I - INTRODUCTION

The reader is referred to the Varenna Symposium Proceedings for a full account of the meeting, for graphs, figures and authors. Individual contributions will not be named in this review. References given below and symbols and units may be found in Lecture 1.

a) There are numerous problem areas in the development of a Tokamak fusion reactor, among them, at least three relate to start -up and ignition. Good confinement of the plasma is clearly essential but in addition it is necessary to raise both the temper-

---

[(+)]Work carried out in pursuance of EURATOM JET Design Phase Agreement 30-74-IFUA-C

ature and density to the required ignition level. In Lecture a discussion is given of the heating requirement for a break-even Tokamak with the assumption that confinement is adequate at the ignition condition. Ignition and operation are taken to be at $n = 1.10^{14}$ cm^{-3} and $T = 10$ keV (mean values). The required auxiliary power for a wide range of loss scalings with start up in <u>full aperture</u> at full density was in the range 50 ± 20 MW ($\sim 500 \pm 200$ MJ energy).

b) It must, however, be emphasized that the start-up requirement is ~ 50 MW heating + some method of raising the density. In other words the start-up scenario is also an important factor

c) It is convenient to define two types of heating, "high grade, which heats to the centre of the plasma", and "low grade, which heats the plasma edge". The latter may have a value in coping with the thermal capacity of the system during density built up (i.e., be heating new plasma). The former is needed to raise the central region of the plasma to the ignition temperature.

d) It is important also to distinguish between "short term" and "long term" heating methods. There is a world of difference between the requirements of respectively laboratory experiment breakeven Tokamaks and reactors.

II - STATE OF THE ART. FUTURE APPLICATIONS, AND APPLICATION TO A BREAKEVEN TOKAMAK

II.1 - Ohmic heating

The initial heating of a Tokamak is provided by the axial current $P_\Omega \propto \dfrac{I^2 \gamma_R \cdot z_{eff}}{T_e^{1.5}}$. The maximum value of the __mean__ temperature achieved in present Tokamaks is $T \lesssim 1$ keV.

The use of ohmic heating to raise $T \longrightarrow$ ignition was discussed, to improve T over the present values it would, in essence, by necessary to raise the current. Now

$$I = \frac{0.5}{q_a} \cdot \frac{a^2}{R} \quad B \tag{A}$$

Given that R/a will not change significantly, increasing I means, increasing B and/or decreasing q_a (e.g. to $B \gtrsim 100$ kG $q_a \lesssim 2$).

Three factors must be appreciated.

1. At a minimum losses will be due to bremsstrahlung. Balancing $P_\Omega = P_b$ shows that typically the plasma is heated until $\beta_p \lesssim 0.5$ by this technique, when $I \gtrsim 1.5$ MA. If this is the operating condition, it is a wasteful use of the toroidal magnetic energy, and the whole question of the economics of such an approach should be investigated.

2. At large electron drift velocities, microinstabilities will be driven which are known to heat the ions. In general this occurs when $h \leq 10^{13}$ cm^{-3} and

the problem of raising the density still remains.
In addition the microinstabilities may enhance
the losses.

3. Losses in present Tokamaks are more severe than
bremsstrahlung, and simple empirical scalings
suggest that one would apparently have to raise
the toroidal field to unacceptable levels in
order to achieve ignition. To summarize, the use
of ohmic heating alone to reach ignition would
require the availability of very large magnetic
fields, low q_a and, probably, microinstabilities
of a type which raise the resistivity without
umpairing the confinement.

II.2 - Adiabatic compression

Major radius compression has been used successfully
in ATC to raise the temperature and density, some
typical figures are.

$$R_1 = 88 \quad a_1 = 17 \quad I_1 = 8.10^{14} \quad \hat{T}_{e1} \sim 1.0, \quad \hat{T}_{i1} \sim 0.2 \, , \quad n_1 \sim 2.10^{13}$$

$$R_2 = 35 \quad a_2 = 11 \quad I_2 = 2.10^{5} \quad \hat{T}_{e2} \sim 2.5 \, , \quad \hat{T}_{i2} \sim 0.8 \quad n_2 \sim 10^{14}$$

The weak features of compression are the relatively
inefficient use of magnetic energy, since the whole
chamber is not used, and the requirement, given
practical limitations on the compression ratio and
time, $C = R_1/R_2$ and τ_c, that the initial temperature

should be hot. For example $C = 1.4$ and $\frac{T_2}{T_1} \sim 1.7$
and we require $\tau_c \lesssim 0.4\, \tau_E$. To reach say $T_2 \sim 7$ keV
we must have $T_1 \sim 4$ keV, and from the previous
discussion we can expect that this will require some
auxiliary heating.

Finally it must be remembered that the effectiveness
of compression depends upon the loss scaling.
Fortunately it is effective for radiation losses and
trapped particles losses.

The good features of compression are that it is
efficient, it heats the whole plasma (high grade
heating), it raises the density, and it minimizes
the requirement, for other kinds of auxiliary heating.
In addition during compression it removes the plasma
from the limiter, and if high energy ions are present,
it may either clamp or even their energy, thereby
enhancing the beam fusion release.

An attractive scenario has been suggested [8,10,11],
which involves establishing and preheating a plasma
in the outer region of the torus. This plasma is
rapidly compressed to the centre of the torus, and
it is then, surrounded by a gas blanket.

The compression causes the centre to ignite and then
the burn region should propagate out through the gas.

II.3 - High Frequency Heating

The technique involves driving oscillating currents
in the plasma at a frequency for which the plasma
impedance is high and the applied fields do not
disturb the confinement. Systems operating at various
frequencies in the range 30 kHz $\rightarrow$ 300 GHz are being
considered. Each system consists of an oscillator
and feed (cables or waveguides) going to a launching
structure (coils or waveguides) which are close to the
plasma. There are a number of stages involved.

A) Generation: i.e. can one produce the power efficiently at the desired frequency? At the present
time sources exist with $\gtrsim$ 50% efficiency for
for frequencies up to a few GHz. For higher
frequency, up to 300 GHz, sources are being studied
(the gyratron) and there is considerable optimism
that efficient supplies, operating at powers in
the range 100 kW $\rightarrow$ 1 MW will be developed.

B) Trasmission-coupling: of the H.F. power to the
plasma, i.e. can the launching structure be
arranged so that the power couples efficiently
(i.e. penetrates and is absorbed) to the plasma?
In this context it should be noted that treating
the system, launching structure, plasma, torus
separately and in infinite slab geometry is in
general not correct, and the real finite geometry
situation must be treated.

534

There are two important situations found as the wave travels in the plasma.

a) The phase velocity $\frac{\omega}{k} \rightarrow \infty$, and the wave is reflected $(\omega = 2\pi\nu$, $K = 2\pi\lambda$) this is called a CUT-OFF. In the absence of a magnetic field it occurs when $\omega = \omega_{pe}$ (the electron plasma frequency). At higher frequencies the electrons inertia prevents them from shorting out the incident field. When a magnetic field is present, penetration can occur when $E_{i\perp}$ B, because the charges are tied to the field in the perpendicular direction and in general cannot short out the field.

b) To obtain good absorption we require that the phase velocity of the incident wave should match the thermal velocity of either electrons or ions i.e. $\omega/K \rightarrow \nu_q$ (q = e or i) this is called a RESONANCE. More generally $(\omega - m\omega_{cq}) / K \rightarrow \nu_q)$ where m is an integer and ω_{cq} is the cyclotron frequency. The actual absorption and thermalization of the energy taken from the wave may be by collisions, by Landau damping, and at sufficiently high power densities by turbulent waves driven by the plasma. The heating methods are described below in order of increasing frequency. The illustrative parameters are for deuterium plasma and realistic n , T and B values.

II.3.1 - Transit Time Magnetic Pumping

TTMP ($m = 0$) and $\omega = K_{||} V_{AF}$.

The absorption is by Landau damping but some "colli-sions" are required to thermalize the energy taken from the wave.

a) Ion TTMP $\nu = V_i/\lambda_{||}$ $= 3.5 \times 10^9 T_i^{\frac{1}{2}}/\lambda_{||}$ Hz

$30 < \nu < 150$ Hz, $\lambda = c/\nu \sim 10 \to 2$ (km).

The parallel wavelength is relevant to this re-sonance because the ions are tied to the mag-netic field. The coils are wound around the torus in the poloidal direction and are separated in the toroidal direction by $\lambda_{||}/2$. Alternate coils are driven 180° out of phase.

Advantages (1) The heating goes directly to $T_{i||}$, this is good because a too rapid increase in $T_{i\perp}$ leads to ion loss disc in velocity space. (2) The whole plasma is heated. (3) Simple power supplies.

Problems. (1) In experiments on the Proto-Cleo and WII stellarators there was severe plasma pumpout. It is suggested that this effect was due to parametric instabilities.

Theoretically, in larger devices, at higher B, it should be possible to operate with fluctuating field $\widetilde{B}$ at a level low enough that the parametric effects will not occur.

Further investigations, under far more satisfactory conditions, will begin soon on the PETULA Tokamak

536

at Grenoble. (2) A major technological problem
to be faced in the application to breakeven
system is that the coils __must__ be close to the
plasma.

b) __Electron TTMP__ $\nu = \overline{v}e/\lambda_{||} = 2\times10^{11} \, T_e^{\frac{1}{2}}/\lambda_{||}$ Hz,

   $2 \lesssim \nu \lesssim 10$ mHz, $\lambda \sim 150 \rightarrow 30$ (m).

The electron TTMP modes and magnetosonic resonances
extend in the frequency range up to the ion cyclotron resonant frequency range. Heating in this
range has been applied successfully in the USSR.

c) __Shear Alfvèn wave heating__ $\omega = k_{||} \, V_{aF}$, $\nu \lesssim 1$ MHz.

A different coil arrangment could be used to drive
shear Alfvèn waves. Theoretically these should
convert at the resonant layer to kinetic Alfvèn
waves and these in turn should heat to the middle
of the plasma. An advantage over ion TTMP is that
the plasma is not compressed.

__Problem.__ All these low frequency methods require
coils close to the plasma, it is not clear whether
in a breakeven Tokamak, given the problems expected
with the first wall, that such methods will be
acceptable.

II.3.2. - __Ion Cyclotron Resonant Frequencies__

   I.C.R.F. $m \simeq 1 \rightarrow 2$

   I.C.R.F. heating has been investigated at low power
in T.F.R. (Fontenay), at high power in ST and ATC

(Princeton), and at the Kurchatov Institute. In all cases the observed absorption, determined from the damping length around the torus, is greater than that predicted by simple linear theory.

In general the heating should occur along a vertical chord where B has the chosen resonant value. The drift orbits of the heated charges trasmit the power to the rest of the plasma.

$\underline{m = 1 \quad \nu = \nu_{ci}} = 7.6 \times 10^2 xB$ Hz (deuterium),

$15 < \nu < 75$ MHz $\lambda = 20 \to 4$ (m).

$\underline{\omega < \omega_{ci}}$ (hydrogen) . The expected ion cyclotron damping is too weak to explain the observations and the wave damping is believed to be due to the presence of a singular layer, i.e. a region where

$$n_{\parallel}^2 - 1 + \mathcal{E}_i \left(\omega_{pi}^2 / (\omega^2 - \omega_{ci}^2) \right) = 0$$

$\underline{\omega_{ci} < \omega < 2\omega_{ci}}$ (deuterium). Theoretically there should be electron Landau and transit time damping. In TFR stronger damping is observed. There is at present no explanation but the present theory may be too simple.

In this region there are a large number of toroidal eigenmodes and the coupling goes into and out of resonance as the density changes. One interesting observation is that this occurs even when $\bar{n}$ = constant, because of internal density fluctuations. Feedback is needed to keep the system locked onto a resonance. Preliminary work on this at low power has been successful. Theoretically if electron damping occurs it should heat the body of the plasma, with a preference for the centre because the E-field is higher there.

<u>$m = 2$ $\nu = 2\nu_{ci}$, $30 < \nu < 150$ MHz, $\lambda = 10 \rightarrow 2$ (m).</u>

<u>$\omega = 2\omega_{ci}$ (deuterium).</u> Theoretically there should be harmonic ion cyclotron damping, and the method should be particularly effective at high n and T_i. Two problem have been observed; (1) The damping is stronger than expected. This is believed to be due to contamination by a small quantity of hydrogen. This can lead to an ion-ion hybrid resonance or, since $\omega_{cH} = 2\omega_{cD}$, to hydrogen ion cyclotron damping, though the latter seems too weak to explain the observations. (2) At high power in ST and ATC, the energy was transferred to perpendicular ion motion. These fast ions then escaped via the loss disc in velocity space.

It was necessary to limit the applied power and energy input because the ions generated impurities which caused disruptions.

<u>ST</u> $\Delta T_{\parallel}$ = 120 eV, with P = 480 kW, $\omega_{max} \sim 300$ J at $\omega = 2\omega_{cD} \sim 20\%$ of the power was transferred to the plasma (at $\omega = \omega_{cD} \sim 4\%$ was transferred and there was strong surface heating).

<u>ATC</u> ΔT_i = 130 eV with P ~ 75 kW, $\omega_{max} \sim 1500$ J at $\omega = 2\omega_{cD}$. The improvment occurred because the

539

ions were contained better than in ST. In PLT the
problem should be even less because the ions are
extremely well contained.

Future Applications

a) TFR-600. The present diagnostic work will be
   extend to high power operation at 500 kW, with
   coil launchers, at the end of 1977.

b) PLT. It is proposed to apply up to 5 MW of
   heating at 55MHz during 1979.

c) JET and TFTR. ICRF heating is being actively
   studied.

d) Further in the future, suggestions have been
   made to use ICRF to produce a Hot Ion Tokamak
   (one with $T_i \gg T_e$) and to generate a two component ion distribution (i.e. one with a large
   high energy tail).

Breakeven Systems

Advantages (1) The power supplies are straight-
            forward.

(2) The body of the plasma should be
    heated (high grade heating), and in
    some modes the ions are preferentially
    heated.

Problems. (1) Present theory does not account
           clearly for the observed damping

(2) Application via coils or side loops
    close to the plasma may present

difficulties even in a breakeven
system and almost certainly in a
reactor.

The application via loaded waveguides is possible
in a large apparatus at $\omega = 2\omega_{cD}$ and this is a
more attractive alternative.

II.3.3 – <u>Lower Hybrid Resonant Heating</u> L.H.R.H. $m \gg 1$

This uses a natural resonance of the plasma, it
has the angular frequency

$$\omega_{LH} = \omega_{pi} / ((\ 1 + \omega_{pe}^2)/\omega_{ce}^2)^{\frac{1}{2}} \quad \text{rads s}^{-1}.$$

If $\omega_{pe}/\omega_{ce} \ll 1$, $\nu = 1.6 \times 10^2\ (n\ (cm^{-3}))^{\frac{1}{2}}$ Hz.

500 MHz $< \nu <$ 2.5 GHz, $\lambda = 60 \rightarrow 12$ cm.

A) One problem with the method is that the incident
wave must have a finite wavenumber along the
magnetic field if it is to have access to the
resonant zone. This $K_{\parallel}$ is provided by using a
set of phased waveguides (the Brambilla-Grill).

B) It was also previously a matter for concern as
to whether there would be good coupling of power
to the plasma. Measurements at low power, with a
double waveguide launcher, on the H1 linear
plasma at Princeton, of coupling as a function
of phase angle agreed well with the theory of
Brambilla. Results ATC at higher power were not
in such good agreement, with theory, however the

maximum coupling $\sim$80 → 90% was obtained with
proper phasing and without special matching. On
the WEGA Tokamak (Grenoble) the launching structure consist of two loops. The coupling coefficient without plasma was $\sim$ 0, with plasma it
was $\gtrsim$ 90% with careful turning. On Alcator
(M.I.T.) a single waveguide was used and then
$\sim$ 70% coupling obtained.

C) Experiments at high power have been made on ATC
(up to 160 kW at 800 MHz), on Alcator and on
WEGA (up to 80 kW at 500 MHz). The following
observations were made.

1) On ATC additional frequencies were generated
in the plasma near zero frequency and near
ω_{LH} (sidebands).
These sidebands had an onset at $\sim$10 kW and
their amplitude increased with increasing
power. On WEGA, the threshold was lower $\sim$1kW.
The ATC sidebands decreased as the plasma
density increased, and for conditions where
$\omega > 2 \omega_{LH}$ no sidebands were observed, the
coupling remained good but there was no ion
heating.

2) In ATC and WEGA ion temperature were measured
with charge-exchange (C.X.) detectors, and from
the doppler broadening of the OVII more. In ATC
$T_{i\perp}$ rose from 170→230 eV, in WEGA $\Delta T_{c\perp}$ was

542

$\sim$ 20 $\rightarrow$30% above 200 eV. In both cases a high
energy ion tail was observed with equivalent
temperature $\sim$1$\rightarrow$2 keV. In both cases the high
energy tail dissappeared rapidly when the H.F.
was switched off ($\lesssim$0.1 ms). However while in
ATC the body temperature also decayed rapidly
in Wega is decayed in a few millisecond ($\sim\tau_E$).
In Wega it is estimated that the body of the
ions received $\sim$ 10 kW of the input 80 kW.

3) In both experiments Thomson scattering indicated
no electron body heating, however soft x-rays up
to 20 keV increased in intensity, suggesting
that power was transferred to an electron high
energy tail. On ATC it was estimated that this
accounted for $\lesssim$20% of the input power.

4) On Alcator at low densities $\sim$ few x10^{13}cm^{-3},
hard X-ray emission suggested that the coupling
was to runaway electrons. At high densities
$\sim 10^{14}$ cm^{-3}, the fast ion signal seen on C-X
detectors suggested that only the edge was heated.

5) On ATC with compression, the ion temperature rose
significantly above the pure ohmic heating level.

6) A neutral beam at 26 keV was injected tangent to
the major axis in ATC, and at the same time L.H.R.
was applied. Ions were seen at 15 keV above the
injection energy, indicating that H.F. penetrates
to the plasma centre.

Summary 1) Coupling good ~80 →90%.

2) There is at present no measurement indicating where the power is lost. The explanation is probably that ions are given perpendicular energy and are then rapidly lost through the loss disc.

3) Theories were presented which indicated that parametric instabilities were probably responsible for effects, such as the generation of high energy ions and electrons and the behaviour of the sidebands. The theories agreed with the idea that power was deposited at radii outside the resonant zone. It was clear that more measurements were urgently needed to assist the theoretical investigations.

D)   Future Applications

For the future there was optimism that some of the problem such as the edge heating would be overcome, possibly with the aid of more complex launching structures with different values of $h_{\parallel}$ and of frequency. The heating of the ions on axis was considered in this respect to be an encouraging result. It was not clear whether the optimum system would heat electrons or ions. The presently envisaged experimental programme is as follows.

544

a) H1, continue physics tests, in particular of
   electron Landau heating (1977)

b) Wega, try a bigger range of plasma parameters,
   apply more diagnostics, particularly to the
   edge. Raise power to $\sim$ 400 kW, (1977) possibly
   try more than one frequency

c) Try two waveguide grill, 2 MWatt at 1 GHz
   for ~ 1 μs, on Petula, (1977), (Grenoble)

d) Doublet IIA (G.A.) 200 kW of 800 MHz for electron profile heating (1977)

e) JFT 2 (Japan) 4 waveguide tests (1977)
   PDX possibly try ~ 1 MW (1979)

f) FT (Frascati) LHRH is under investigation.

g) JET, the application of LHRH is being actively,
   studied, the power supplies are compatible
   (1981 $\rightarrow$). JT-60 (Japan) Klystrons at
   (1 $\rightarrow$ 1.8 GHz) are under development.

E) Application to a Breakeven Tokamak

The method is attractive because it uses small
waveguides which may be bent to shield the insulating vacuum barrier. Thus all the sensitive
components are shielded from radiation. However
if the present penetration problems are not
overcome it can be considered only as a "low grade"
heating system. If the efficiency is not improved
it cannot be used.

545

II.3.4 - <u>Electron Resonance Heating</u>

Heating near the electron plasma frequency,

$$\nu_{pe} = 9\text{x}10^3 \; (n \; (cm^{-3}))^{\frac{1}{2}} \; Hz, \quad 25 \; GHz < \nu < 250 \; GHz$$

$$\lambda = 1.2 \rightarrow 0.12 \; cm.$$

Heating near the electron cyclotron frequency

(E.C.R.H.)

$$\nu_{ce} = 2.8\text{x}10^6 \; B \; Hz, \quad 15 \; GHz < \nu < 300 \; GHz \quad \lambda = 2 \rightarrow 0.1 \; cm.$$

These are attractive possibilities because small

waveguides may be used. Power supplies are being

developed.

The electron temperature has been raised using E.C.R.H.

on the TM 3 Tokamak (Kurchatov).

The method will be tried on the ELMO bumping torus

and ISX torus at ORNL and possibly on T-10 and P.D.X.

II.4 NEUTRAL INJECTION HEATING

II.4.1 - <u>Introduction</u>. Since it will be necessary to raise

the density above the initial filling density (this

typically yields $\bar{n} \lesssim$ few x 10^{13} cm^{-3}), there will

certainly be neutral injection of deuterium and

tritium in the general sense. We may divide this

generalized neutral injection into 5 types.

a) <u>Gas Puffing</u>. This has been used successfully on

Alcator, Pulsator, ATC, TFR and Ormak... to raise

the density. The limitations on gas puffing occur

when the edge cools because the heating cannot

cope with the gas influx, and the plasma disrupts.

On Ormak and ATC it has been shown that with
additional heating higher densities may be
achieved. Thus gas puffing + some "low grade"
heating of the edge region may be an attractive
part of a start-up scenario.

b) Pellets: an extreme form of gas puffing, initial
tests on Ormak look good.

c) Clusters: not only fuel but also give some energy
to the plasma $\sim 1 \rightarrow 10$ keV. This technique will
be tried on TFR and ASDEX.

d) Neutral injection "low energy" $\epsilon \lesssim A_b \cdot 100$ keV
$A_b = 1, 2, 3$ for H, D, T. At lower energies in
this range e.g. $\epsilon < 100$ keV neutral injection
may be used for fuelling.

e) Neutral injection "high energy" $\epsilon > A_b \times 100$ keV.
For permitted power levels the current is in
general not large enough to give total fuelling.

II.4.2 - Penetration

a) The rapid penetration of ions added by gas puffing
in present experiments is much faster than expected. In general however this must be classified
as a poor penetration, fuelling technique.

b) Pellets and clusters should have theoretically
better penetration than atoms of the equivalent
energy but they have not been tried yet.

547

c) Neutral injection: for injection of hydrogen into a hydrogen plasma the number of trapping lengths to the plasma centre is

$$F = \frac{a}{\lambda} \simeq \frac{0.18 \, n \, a}{\sin \theta} \, \frac{A_b}{\epsilon} \quad Z_{eff} \quad where$$

$\epsilon \gtrsim A_b \times 40 \text{ keV}$ is the injection energy (keV). θ is the mean injection angle $\theta = 90°$ perpendicular injection, $\theta = \left(\frac{a}{4R}\right)^{\frac{1}{2}}$ for injection tangent to the plasma axis.

Computed deposition profiles indicate that for conventional body heating it is desirable to have $F \lesssim 3$. Thus for heating in full aperture with a deuterium beam ($A_b = 2$), $n = 10 \times 10^{13} \text{ cm}^{-3}$, $a = 200$ cm, $\theta = 90°$ we need $\epsilon > 240 \times Z_{eff}$ keV. Now for a breakeven Tokamak high Z_{eff} would not be acceptable, but it might well be necessary to live with $Z_{eff} \sim 2$. Thus high grade NI requires $\epsilon > 240$ keV. In practice the problem is made worse in positive ion systems by the presence of 1/2 and 1/3 energy species.

Lower energy injection must be classified for this application as "low grade" heating, but it may play a valuable role in expanding plasma density build up scenarios [6] [16]. Alternatively ripple injection may be used to get low energy beams in raising them to the "high grade" category (Jassby, 1976) (Grenoble Conference).

Positive ion systems. At low energies $\epsilon < A_b \times 100 keV$, such systems are relatively efficient, though at the higher part of the range the efficiency drops to $\sim 10\%$. Direct recovery systems have been proposed to improve the overall efficiency. Preliminary experiments at F.A.R. and L.L.L. have been encouraging. It must be noted, however, that practically $\sim 5\%$ of the ion beam is lost in the accelerator grids. Conceivably another 5% is lost in the neutralizer and recovery system, in addition large powers are used in the arc source. Consequently even with a perfect recovery system acceptable overall efficiencies, say $> 25\%$? Can only be obtained for $\epsilon < A_b \times 100$ keV. Systems for recirculating the ions through a neutralizer are being studied but it is not clear whether such systems will be workable.

Negative ion systems. For $\epsilon \gtrsim A_b \times 100$ keV, we must turn to negative ions from the penetration point of view they have the advantage that there is only one species. These systems are being studied actively at L.L.L. There are four systems for negative ion production.

1) Direct extraction from a plasma, an example is the Dimov, negative ion magnetron source (Novosibirsk).

2) Attachment in a cooling plasma.

3) Surface production.

4) Double charge exchange of positive ions ($D^+-2e \rightarrow D^-$)
in an exchange cell. Measurements at L.L.L.
using a Cs cell show $\sim 20\%$ efficiency at
~ 1 keV.

4) Pumping. A simple formula was presented which
relates the cryopumping panel area to the beam
parameters.

$$A = \frac{I}{5.6 \; Q \; \eta_g \; FP} \quad cm^2 .$$

I = neutral beam current, Q = pumping speed =
= 9 L/cm^2 s for D2.

η_g = source gas efficiency $\quad \begin{matrix} \sim 0.2 \rightarrow 0.5 \; \text{positive ions} \\ \sim 0.01 \quad\quad \text{negative ions} \end{matrix}$

P = pressure Torr, F = conversion efficiency for
beam. $\eta_{gF} = 0.01$, P = 10^{-4} A/I $\sim 2 \times 10^3$ cm^2/amp
equivalent.

II.4.2 - State of the Art

Neutral injection has been used successfully first
at low power levels in CLEO (Culham), ATC (PPL),
ORMAK (ORNL), T-11 (Kurchatov), during 1973/1975,
and subsequently at higher power levels in ATC,
ORMAK, and TFR. The general conclusions from this
work are:

a) Essentially all the observed effects may be explained by classical theory. Detailed comparisons
have been made of ion energy spectra obtained

550

with C-X detectors with the prediction of Fokker-
-Planck code calculations. The slowing down of
the fast ions, their scattering and the transfer
of power to the plasma ions agrees with theory.
Some effects observed such as the distortion of
the plasma ion energy distribution by the beam
and the acceleration by electrons of the injected
ions to greater than ϵ have subsequently been
derived from the classical theory.

b) There is no indication that injection directly
impairs the confinement properties.

c) The bulk of the ions are heated.

d) Studies of the C-X spectra obtained when neutral
beams are injected are now being used as a diagnostic to measure for example, Z_{eff}, the current
profile and the ion temperature on axis.

Some recent results at high power wave reported.

Ormak. (Tangential injection 35 keV D)

a) The measured charge in T_i is consistent with a
neoclassical conductionloss . The ion temperature
rises linearly with injection power and there
is no sign of saturation.

b) The mean ion temperature has been raised from
$0.37 \longrightarrow 1.03$ keV the peak ion temperature to
1.5 keV.

c) The maximum injected power was 360 kW of which
160 kW went to electron and to ions and 40 kW

was lost directly to the wall by injected ions on uncontained orbits.

d) The mean electron temperature reminded uncharged, but with large counterinjection power, the electron temperature profile, became hollow in the middle and T_e $(0) \sim 1$ keV $< T_i$ $(0) \sim 1.5$ keV. The only factor that can reproduce this phenomenon in code calculations is the buildup of a large impurity concentration on the axis. If fact Z_{eff} is observed to increase and this is probably caused by the large number of counter injection ions which are on uncontained orbits.

T.F.R. (injection at $80°$, and 0, $\pm 9°$, $\pm 18°$, in the vertical plane).

a) The maximum injected power was 570 kW, and $\sim 20-30\%$ was lost, in reasobable agreement with calculations.

b) The peak ion temperature when a 455 kW $D°$ beam was injected into a D plasma was 1.9 keV, with $\bar{n} = 3 \times 10^{13}$ cm^{-3}. $(\Delta T_i / T_i)_0 = 0.96$, and $\Delta T_i / T_i = 0.86$, $\Delta T_i \alpha P_{bi} / \bar{n}$.

In this case Z_{eff} increased from $3 \rightarrow 5$ during injection.

c) A diagnostic neutral beam was used to calibrate T_i (0), and confirmed previous estimates that the C-X detector was underestimating the central T_i.

d) The neutron output during D → D operation was consistent with the plasma-beam conditions.

e) The electron temperature rose only slightly (∼ 10%) because the electrons receive beam powers mainly at the plasma edge and lose it rapidly by radiation. In the centre the power goes mainly to the ions. No hollow electron profiles occurred and T_e (0) $\gtrsim T_i$ ( ).

f) Under most conditions the ion temperature rise is consistent with neoclassical conductivity losses, but at low collisionality the losses appear to be 2 → 4X higher than the neoclassical value. The ion energy confinement time seems to scale as $\tau_{Ei} \propto n^\epsilon \; T_i^\delta$ where $\epsilon \lesssim 1$ and $\delta \sim 0$, further $\tau_{Ei} \sim \tau_{Ee}$.

g) Both ion cyclotron frequencies and low frequency noise are generated during injection.

II.4.3 - <u>Future Applications</u> (some of the proposed systems, numerous other machines ω VII, ASDEX, TEXTOR etc. will also use injection).

MACHINE	PROBE LENGTH (s)	TOTAL POWER (MW)	NEUTRAL POWER (MW)	E (max) (keV)	SPECIES	NUMBER of Sources	DATE
TFR 600	$\gtrsim$ 0.05	3.5 → 5	1 → 1.5	35 → 50	D	10	77
DITE	$\lesssim$ 0.2	2.5	1.6	35	H/D	4	77
ORMAK Upgrade	0.25	10	4.0	$\gtrsim$ 40	H/D	4	78
PLT	0.25	10	3.0	40	H/D	4	77
TFTR	$\gtrsim$ 0.5	80	20/15 [*]	120	D	12	81
JET	$\lesssim$ 5 → 10	25 → 62.5	10 → 25	80 160	H D	20 → 50	81 83
JT 60	$\sim$ 1	72	15 → 20	75	H	16	80+
m_x mirror	> 0.5	160	50	80	D	24	> 80?

[*] Low Energy Species.

a) In these experiments $P_b > P_r$, and it should be possible to determine the limits of beam pressure/plasma pressure, and the limits imposed by impurity generation. One of the mean concerns, based upon code predictions, is that at the higher energies where $F \propto Z_{eff}$, impurity blocking can lead to an unstable situation in which the beam power is deposited nearer and nearer the edge as the impurities build up.

b) On the development side vigorous programmes are underway to develop the 80 keV (H) $\rightarrow$ 160 keV (D) beams for JET, TFTR and JT 60.

On the positive ion side, attempts are being made to improve the atomic species content in the source, and some success has been achieved at ORNL by using a hot gas feed. On the negative ion side, both the Dimov source and double exchange systems are being studied. At L.L.L. the initial goal is to produce 1A of D^- at 100 keV for 10 ms, at present they have 0.1 A of D^- at 1 keV.

In the longer term the goal is 5 A D° at 200 keV for 1 sec.

II.4.4 - Application to a Breakeven Tokamak

a) Low energy beams $E \lesssim 100$ keV, may be used for fuelling and as a "low grade" heating system in

the conventional modes. With the ripple injection technique they become a "high grade" system.

b) High grade beams in the reference design essentially means $\leftrightarrow$ 240 keV. Such beam can only be
produced efficently from negative ions.

c) The question of operating injectors in the hostile
of a reacting plasma has not been studied in
detail. There is concern about the effects of
radiation on the source for line of sight injectors. It is not clear whether it is practicable
to bend the in beam before neutralization.
Either way in both this case and for the cryopumps
which also do not have to see the plasma, we must
not forget that neutrons can scatter, so that it
will not be possible to eliminate them "completely".
Estimates of the acceptable level are being made.

d) While complex systems have a place during the
research phase, and may be used (reluctantly) on
a breakeven system, on a reactor it appear that
system must be simple and reliable as well as
efficient and inexpensive.

II.5 - <u>SUMMARY</u>

SYSTEM	GRADE OF HEATING	RAISES DENSITY	TECHNOLOGY	BREAK-EVEN	REACTOR	STATUS
Gas Puffing	-	✓	✓	✓	✓	Tested, good
Pellets	-	✓	Developing	✓?	✓ ?	Preliminary, good
Clusters	low	✓	Developing	✓ ?	✓ ?	Untested
NI $\lesssim$ 200 keV D positive ion	low	✓	Developing	✓	✓ ?	Good at $\lesssim$ 40 keV
NI $\gtrsim$ 200 keV D negative ion	high	poor	Long term development	✓	✓ ?	Untested
Ohmic	high	no	✓	✓	✓	Tested but limited to lowest temperatures
Adiabatic compression	high	(✓ ?)	✓	✓	✓	Good
TTMP	high?	no	?	?	?	Open
ICRF	high?	no	developing?	✓?	✓?	Some good some bad
L.H.R.	?	no	✓	✓?	✓?	Some good some bad
E.C.R.H.	high?	no	?✓	✓?	✓?	Some good

This table indicates that for start up in a breakeven system one may well need a range of methods. For purely illustrative purposes one might consider the following scenario,

1) Plasma expansion and build up using low energy neutral beams and ohmic heating$^{(6), (16)}$ with a plasma against on the outer wall of the torus. (Beams of 40 MW at say 50 keV would raise the density to 10^{14} cm^{-3} in a plasma of size a =160, b = 240, R = 600 in about 10 seconds).

2) At large radius, we add a "high grade" heating system with $P \sim 10$ MW for example an HF system, to make the central temperature rise well above the level given by edge heating.

3) Before the plasma reaches full aperture we compress it in major radius to the centre of the torus ($c \sim 1.2$ for example). The centre ignites.

4) We add a gas blanket and there is a propagating burn from the centre to the edge[8].

THE IMPACT OF SERVICING REQUIREMENTS ON TOKAMAK

FUSION REACTOR DESIGN

J T D Mitchell

Culham Laboratory, Abingdon, Oxon OX14 3DB

(Euratom-UKAEA Fusion Association)

INTRODUCTION

There is one design requirement which appears to be common to all D-T fusion reactors, irrespective of the plasma confinement principle embodied in the design. It must be practicable to repair the structure or "first wall" nearest the plasma. A repair can include a range of operations between finding and sealing a pinhole leak in a weld to complete replacement of the whole structure of the breeding blanket. The two reasons why these repairs are necessary are:

(1) The variability and defects in generally perfect products, such as those caused by errors of human performance or by flaws in raw materials.

(2) The extreme severity of the environment of the first wall, which causes the deterioration of the properties of materials situated there. This deterioration is the result of neutron irradiation and transmutation in the materials under conditions of stress and high temperature and of surface corrosion and erosion.

Repair or servicing of a fusion reactor blanket is complicated because its proximity to the plasma means that the blanket is also furthest from the 'normal engineering environment' where the servicing operation starts. This is illustrated by the early 'onion-skin' reactor models as used in many blanket and shielding neutronic and activation studies. Servicing is also complicated further by the interconnections through the various 'skins' required for operational access - i.e. vacuum ports, coolant ducts, mechanical supports etc.

This paper discusses briefly some of the major servicing functions essential to fusion reactor repair and maintenance: it examines the servicing operations proposed in some published fusion reactor designs. A new concept for a tokamak reactor is described and the principle blanket servicing operations are outlined.

By utilising replacement or exchange servicing of segments of the blanket, the more difficult and time consuming blanket maintenance is carried out on spare units while the reactor itself is fully operational. To incorporate these facilities imposes new requirements on the design of some sub-systems of a Tokamak fusion reactor. It will be necessary to incorporate similar concepts in Experimental Power Reactors.

SERVICING FUNCTIONS AND CONSTRAINTS

All operations of servicing - or maintenance - other than control adjustment - involve breaching the boundary of outer regions, component disconnection, service or replacement and reconnection, after which the breach or breaches must be closed; this must be done within the two constraints (a) of practicability and (b) of time.

Practicability of the operation includes:

(1)   questions of scale, that the components be neither too large or too small;

(2)   simplicity of the actual operation, that motions should be linear and not compound;

(3)   inclusion of effective inspection and testing;

(4)   number of operations, and;

(5)   reliability of the necessary servicing sub-systems, including facilities for maintenance of the servicing sub-systems themselves and training and relaxation for the service personnel.

The constraint on time means that plant and equipment should be shut down for the shortest possible time. All shutdowns should occur on dates and at times selected to coincide with periods of lower total equipment utilisation i.e. low product demand. The Central Electricity Generating Board of the United Kingdom places a very high premium on the availability of new and efficient generation plant, especially base load plant, in January and February of each year - aiming for 100%. This premium is considerably reduced during the late summer months.

SOME FUSION REACTOR SERVICING SCHEMES

Methods of blanket servicing and replacement have been examined in a number of reactor studies and at fusion reactor design workshops [1]. One of the more novel solutions is that incorporated

560

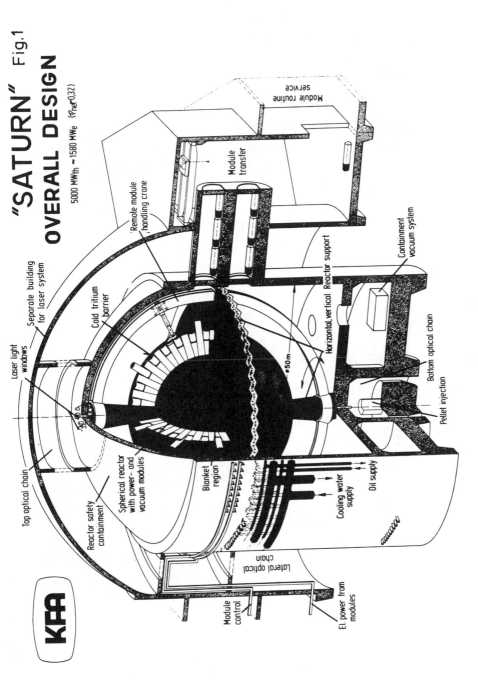

"SATURN" Fig.1

OVERALL DESIGN

5000 MWth ~ 1580 MWe (η_{net}=0,32)

Separate building for laser system

Laser light windows

Remote module handling crane

Cold tritium barrier

Top optical chain

Reactor safety containment

Spherical reactor with power- and vacuum modules

Blanket region

Module control

Lateral optical chain

El. power from modules

Cooling water supply

Oil supply

Horizontal, vertical Reactor support

∅50m

Containment vacuum system

Bottom optical chain

Pellet injection

Module transfer

Module routine service

KFA

Figure 1   "SATURN" 5 GW(t) Laser Fusion Reactor overall design.

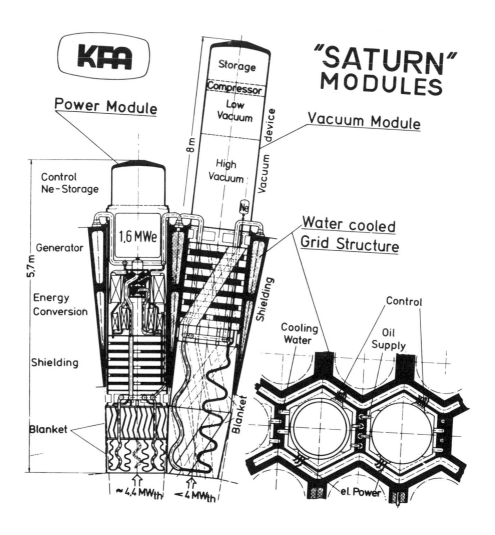

Figure 2   "SATURN" 5 GW(t) Laser Fusion Reactor.
            General Arrangement of Power and Vacuum Modules.

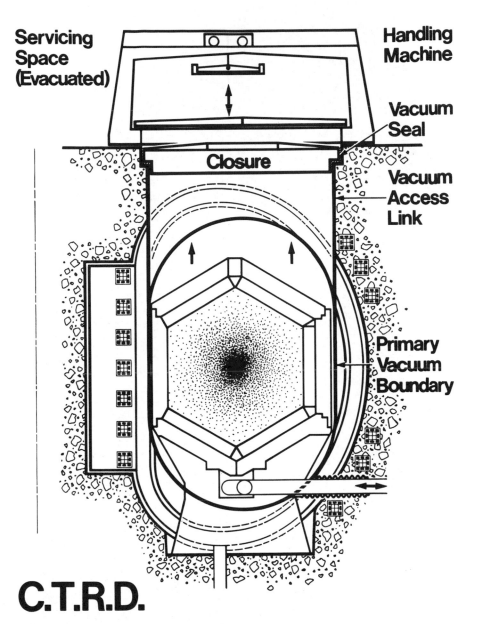

Servicing Space (Evacuated)

Handling Machine

Vacuum Seal

Closure

Vacuum Access Link

Primary Vacuum Boundary

C.T.R.D.

CONCEPTUAL BLANKET SERVICING SYSTEM

Figure 3   Collaborative Tokamak Reactor Design
Showing conceptual blanket servicing arrangement.

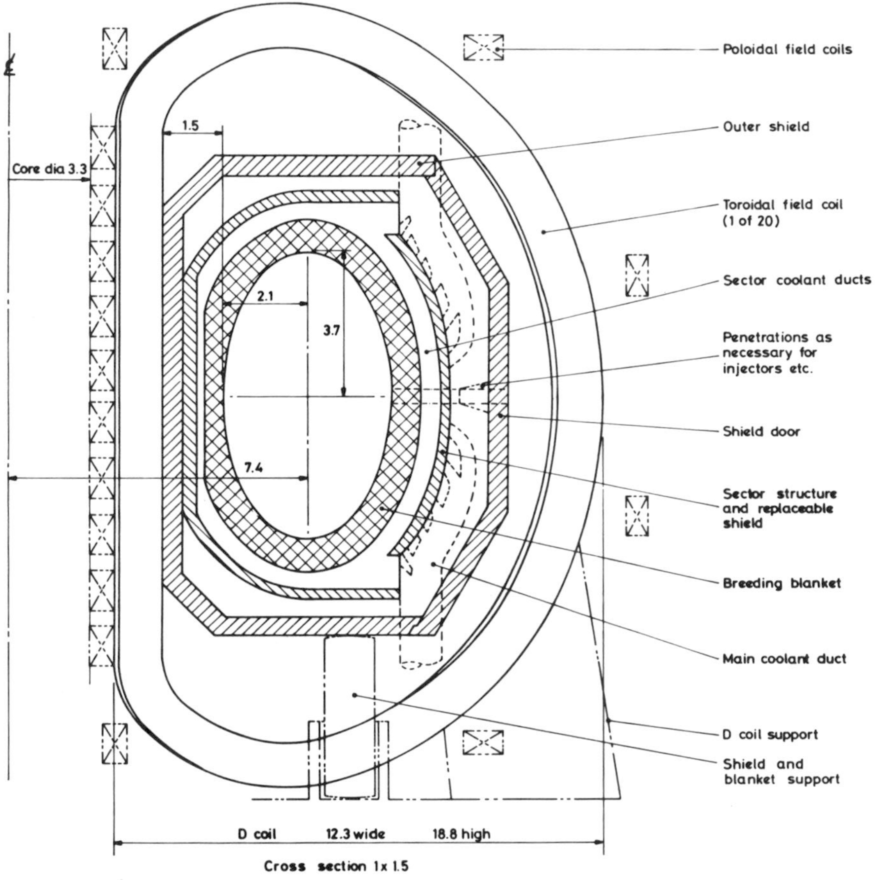

REACTOR PARAMETERS

Electric power output	MWe =	2000
Wall loading	MW/m^2	6.7
Major radius	m	7.4
Plasma semi-axis (horiz)	m	2.1
ellipticity	—	1.75
aspect ratio	—	3.5
Magnetic field peak on axis	T	8.0
	T	4.1
Core field peak	T	4.3
Plasma current	MA	11.6
temp – Te = Ti	keV	12
density – n	10^{14}	3.5
pressure ratio $\bar{\beta}_t$	%	9.3
β_p	%	1.9 ($=\sqrt{R/a}$)
Safety factor q$_{(surface)}$	—	2.5 ($q_0 = 1$)

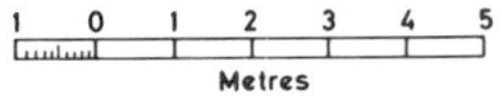

Metres
0 1 2 3 4 5

— Poloidal field coils

Outer shield

Toroidal field coil (1 of 20)

Sector coolant ducts

Penetrations as necessary for injectors etc.

Shield door

Sector structure and replaceable shield

Breeding blanket

Main coolant duct

D coil support

Shield and blanket support

3.7

2.1

7.4

1.5

Core dia 3.3

D coil   12.3 wide   18.8 high

Cross section 1 x 1.5

Figure 4   Culham Conceptual Tokamak Reactor Mark II
Principal reactor dimensions and parameters.

in the Saturn 5 GW(t) Laser Fusion Reactor {2}. This reactor has an output of 50 GJ per pulse, 100 pulses per second and a mean wall loading of 4 MW/m^2 (Figure 1.) The spherical plasma chamber is formed by about 1100 circular cross section power modules comprising blanket, shield and a 1.6 MW(e) Brayton cycle gas-turbine-alternator (Figure 2). These are mounted in a hexagonal gridded and cooled structure which also ducts all services to individual modules. All the service disconnections operate at normal ambient temperature and in low radiation field, and should perform reliably. Vacuum exhaust is produced by 70 similar self-contained pumping modules which also have normal temperature service connections. Remotely controlled machines are arranged to disconnect, remove and replace the modules for servicing, which includes tritium recovery and first wall replacement. The authors suggest a cycle time of between 3 to 6 months, a shut down of 10 minutes to exchange one module, and even postulate the possibility of "onload" module exchange. This design shows how simple shape plus the absence of magnet coils can improve the overall concept of fusion reactor servicing.

UWMAK I and II {3,4} are the most detailed Tokamak reactor designs produced. In UWMAK I, 30^0 sector sub-assemblies of the reactor comprising blanket, shield, toroidal magnet and divertor components, in all some 3,500 tonnes each, are removed to servicing bays. The operation will be slowed down by the extreme size and weight and the many service disconnections: re-connection must also be very reliable to avoid extensive fault-finding during reactor start-up. In UWMAK II, after shut down and disconnecting major services such as divertor vacuum lines and helium cooling circuits, the outer poloidal field coils are moved, allowing hinged doors in the shield to be swung open in between the toroidal field coils. This gives access to the rear of the outer blanket structures for cutting out and removing the blanket components and sub-assemblies. Although a low power density design like UWMAK I, the total weight of material to be handled in UWMAK II is reduced by more than 90%. But large and heavy machines and complex remote handling operations are necessary and the reactor must be shut down whilst all the detailed work is done. The authors have estimated that using 3 parallel shift operations, the whole blanket rebuild could be completed in a six-week shutdown. This programme would require a large and extended peak of effort every few years; such well separated peaks are not conducive to reliable human performance. Extensive sub-assembly testing in-situ and the necessary remedial repairs will be required during the rebuild. This is not mentioned in the study and it is likely that the estimated 6-week shutdown is optimistic.

THE CULHAM CONCEPTUAL TOKAMAK REACTOR MK II

In reactor designs at Culham, we have been concerned with various forms of subdivided blanket designs and with different concepts for replacing the blanket, either as cells or modular sub-assemblies {5}. In the Collaborative Tokamak Reactor Design Study {6}, starting from fission reactor experience, we examined carefully a vertical access concept for a tokamak (Figure 3), but no easy solution appeared because of limitations on access and the complex nature of the blanket structure. Resulting from an economic study of tokamak reactors, Hancox recently proposed new parameters and dimensions for a 5 GW(t) reactor (Figure 4), now the Culham Conceptual Tokamak Reactor Mk II (CCTRII) {7}. The small aspect ratio and high wall loading are important features of the design and derive from the elliptical plasma cross section and from the saturation of the central core of the pulsed field transformer. To allow sufficient space for ducting and access, the horizontal opening of the D shaped B_ϕ coils was increased by 2.5m, without altering the plasma dimensions etc: the resulting cost increase is only marginal. With such a high wall loading, the blanket first wall life will be short - perhaps between 1-3 years. Therefore the engineering concept must incorporate facility for easy blanket servicing - see Figure (5) and reference {8}.

An important feature of the CCTRII concept arises from the fact that there can be no neutron flux levelling in a fusion reactor blanket and shield. Though the life of the front of the blanket may be only a year or so, radiation damage of the outer region of the shield is many orders of magnitude less, indicating a potential structure life > 30 years, more than the working life of the reactor. Thus the outer region of the shield can be designed as a permanent primary vacuum wall, and the support structure for the inner shield and blanket. In between each toroidal coil, the outer shield has a vacuum tight door, extending for the full height and width of the blanket segment inside (Figure 6). Thus removal of the door gives access to cut the cooling ducts, disconnect any instrumentation and remove the blanket segment. After replacing auxiliary shield components, the procedure is reversed to instal a replacement segment. From the figure it will be seen that the outer poloidal coils and toroidal coil struts, plus any injectors, have also first to be moved and subsequently replaced. Whilst doing so simplifies initial assembly, the poloidal and vertical field coils have all been located outside the toroidal field structure especially so the largest coils can be moved for access to the shield doors. However this will increase the coil and energy supply system costs for the poloidal/vertical field system, in addition to the increase of D coil cost mentioned above.

In these operations all the principal load motions are horizontal, straight and few. The only material moved is that which must be replaced, and components blocking access to the interior of the reactor. The replacement blanket/shield segment is a pre-tested assembly, leaving only the primary cooling ducts, tritium recovery and instrumentation connections to be joined and tested before replacing the outer shield door. Doors can also be repaired or renewed if necessary.

Building layouts for the reactor hall and servicing workshops have been developed and the operations involved in replacing a blanket segment have been scheduled. About 10 transporters will be required in the reactor hall for moving doors and segments and positioning the specialised remote controlled machines for duct cutting and welding etc. The active workshops for sector repair must be equipped to dismantle the blanket structure partially or completely, repair and/or replace the components, dispose of waste material and finally to test the repaired unit.

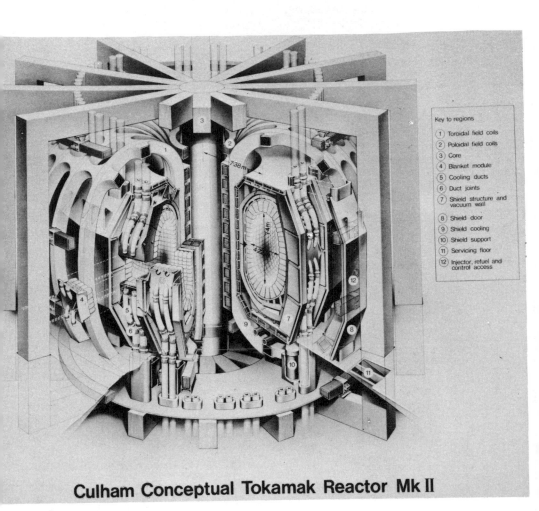

Culham Conceptual Tokamak Reactor Mk II

Key to regions
1. Toroidal field coils
2. Poloidal field coils
3. Core
4. Blanket module
5. Cooling ducts
6. Duct joints
7. Shield structure and vacuum wall
8. Shield door
9. Shield cooling
10. Shield support
11. Servicing floor
12. Injector, refuel and control access

Figure 5   Artist's impression of Culham Conceptual Tokamak
            Reactor Mark II.

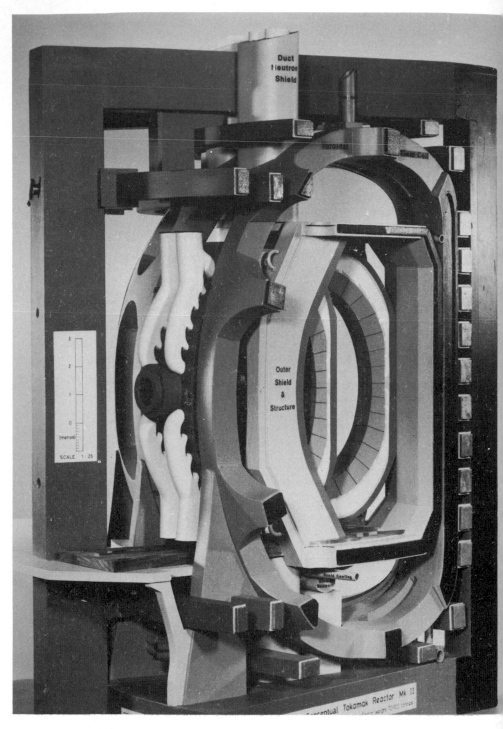

Figure 6   Model of Culham Conceptual Tokamak Reactor showing
shield door removed and blanket segment partly withdrawn onto transporter.

It is a most important integral feature of the concept that the detailed blanket repair work is done with the reactor fully operational and that the reactor is shut down only for the essential replacement operations. It seems also preferable for replacements to be made singly every few weeks: capital investment is reduced, the work load is levelled and flexibility increased because the work sequence can be adjusted to permit early exchange of any blanket segment found faulty. Using the blanket segment exchange operations schedule, target times for individual operations have been derived. To achieve 0.9 generation plant availability, including turbine maintenance but not breakdowns, requires that the segment cooling ducts be joined and tested in a 4 hour operation. This has been assessed as the longest and most difficult single operation and the practicability of performing it in the allotted time is now being studied.

DISCUSSION

Fusion reactors and their associated generating plant will represent a large investment of human effort and it will always be important to reduce outage time to relatively short and selected periods in the reactor service life. The complex nature of fusion reactor concepts and the high activation levels resulting from using high wall loadings requires that the engineering design of the reactor incorporate facilities for remote servicing. The most advanced remote machine technology will have to be employed, incorporating also any useful experience from fission, space and undersea exploration projects.

The importance of remote handling capability using sophisticated equipment such as the Mascot servo manipulator {9} has been disclosed during the engineering design of JET. The present suggestions are that between 5-10% of the total capital cost will be required for provision of remote handling facilities, if the experiment is completely successful and is ultimately operated on D-T mixtures.

An example of a practical design for remote handling is shown in Figure 7. It is a development of particle accelerator magnet design by the staff of CERN, Geneva. The new magnet can be inserted into position in the beam line by the simple act of lowering it onto a locating framework whose position can be adjusted by small motors to align the magnet. The lowering action also completes all service connections to the magnet, i.e. heavy and light electrical circuits and water cooling connections. To complete the operation a servo manipulator can be used to connect in the two ends of the beam vacuum tube, using special single fastening vacuum joints.

Design for 'minimum disturbance access' will also help to reduce reactor outage time, first by reducing the total number of operations and secondly by reducing the probability of re-assembly faults. An example is incorporated into the Mirror Reactor proposal by Carlson and Moir {10}. The whole reactor is divided horizontally into two parts, the one floating in a tank and the other supported from a bridge structure above. By changing the water level in the tank, the two halves can be separated: the

primary vacuum joint is the only disconnection required to gain
access to the blanket, providing of course that the service
connections to the lower half have sufficient flexibility.

There are other examples: as CCTRII design progresses, the removable supports between the toroidal coils might be eliminated,
(Figure 6). Regrettably the outer poloidal coils have to be
moved, but it may be quite practicable to incorporate flexible
electrical cables and coolant pipes eliminating these disconnections. Furthermore, if redundant injectors or refuellers
can be accommodated, reactor start-up could be achieved without
replacing the specific injector in the segment position just
serviced. This final operation could then proceed more leisurely
but with the reactor operational.

Design is a matter of compromise. Thus in a sophisticated
concept such as a fusion reactor, each element and sub-system
has to function off-optimum for the optimum performance of the
complete reactor system. A desirable optimisation would be that
any faulty component could be quickly changed for a serviceable
replacement. While it seems inescapable that quick blanket
replacement is a necessity, in any present Tokamak reactor, a
long and difficult operation would be required to replace a
toroidal field coil because of the complex magnet "structure"
and activation of the shield.

The designer could be guided by failure probability assessment.
However at the present stage of large superconducting magnet
development, this is not possible and engineering judgment must
be used instead, Electrical equipment generally fails because of
insulation breakdown. In a superconducting coil, the environment
is benign (4K and inert) and radiation damage and mechanical
movement must be closely controlled to avoid quenching. Stringent
quality control at marginal increase of cost would also improve
the reliability of the coils. Terminations are highly probable
locations for electrical failures: it seems feasible to make all
terminations accessible for repair without removing the coils.
Whilst an early coil failure resulting in a very costly repair
or a prematurely decommissioned reactor cannot be ruled out it
might nevertheless be practicable to manufacture toroidal field
coil systems to the same degree of reliability and long life
expectation as for example a fission reactor pressure vessel. It
will be some time before this question can be decided on facts
but clearly it has a very significant impact on fusion reactor
concepts. This can be seen by comparing CCTRII with an alternative
tokamak reactor concept which has more easily replaced magnet
coils {11}. In the latter case, both coils and energy supply
systems would have to be very much larger and possibly too
expensive altogether.

There is no divertor in the present CCTRII concept partly
to simplify the initial studies but also because of general
ignorance of a practical burn cycle for a Tokamak reactor, i.e.
how to exhaust - and refuel - the plasma. A bundle divertor
cannot be used for exhaust because of its high electrical power
requirements {12} and if one is required, the only alternative
is an axisymmetrical divertor. Neither single-null median plane

or conventional 2-null divertors are compatible with the blanket and shield structure of CCTRII.

A possible alternative, which also retains the high wall loading of CCTRII is suggested in Figure 8: there is a single null divertor above the elliptical plasma- based on the UWMAK II coil configuration - and it is suggested that plasma equilibrium might be provided by difference current coils below the shield structure. The neutron shielding is extended upwards to enclose the divertor coils, their local shielding and structure, the trapping region and the main vacuum exhaust line to the cryopumps etc. Doors are also included giving access for trapping surface maintenance. The divertor coils must be made very reliable like the toroidal field coils- but in this case high reliability will be difficult to achieve because the divertor coils will have to be constructed inside the toroidal coils during final site erection.

The blanket segment is a unit structure on a flat base and can be replaced in the same way proposed for the divertorless CCTRII concept. An extra operation sequence will be required to replace the separate divertor section of the blanket: since this unit is relatively small and low weight and the movements are not complex, the exchange should not take very long. Comparing this concept with the original CCTRII shows the extensive change in the engineering concept resulting from incorporating a divertor. It does seem vitally important for any power reactor application of the Tokamak confinement concept to initiate study of possible burn cycles and clarify requirements and practicability of divertor and refuelling concepts.

CONCLUSIONS

Fusion reactors will have to incorporate facilities to meet the operational requirements of the electricity supply system for efficient periodic replacement of the blanket structure. Some increase of the size and cost of both the reactor and its supporting facilities will be required to provide adequate access space for the remote handling operations. Designers of Experimental Power Reactors will have to include in their designs philosophies and techniques similar to those outlined and of course compatible with the confinement geometry chosen for their particular reactor. It would seem very desirable to explore thoroughly these additional constraints and requirements in any future experimental equipment design.

ACKNOWLEDGEMENTS

I acknowledge with thanks the assistance of many colleagues at Culham who participated in the conceptual reactor studies and also at AERE Harwell who contributed to the overall engineering design and layout - also permission from KFA Jülich to use Figures 1 and 2 of the SATURN reactor and from CERN Geneva to use Figure 7 of an accelerator magnet.

571

REFERENCES

1. **Fusion** Reactor Design Problems. Proceedings of an IAEA Workshop, Culham. Nuclear Fusion Supplement 1974.

2. Bohn F H et al. Some design aspects of inertially confined fusion reactors. Fifth Symposium on Engineering Problems of Fusion Research, Princeton 1973.

3. University of Wisconsin Report, UWFDM-68 Vol 1 (1974).

4. University of Wisconsin Report UWFDM-112 (1976).

5. George M W. Structure renewal and maintenance requirements in a fusion reactor. Fifth Symposium on Engineering Problems of Fusion Research, Princeton 1973. ·

6. Collaborative Tokamak Reactor Design - Euratom Associated Laboratories, October 1975.

7. Hancox R and Mitchell J T D. Reactor costs and maintenance with reference to the Culham Mk II Conceptual Tokamak Reactor Design - to be presented at the Sixth IAEA Conference on Plasma Physics and Controlled Nuclear Fusion Research Berchtesgaden 1976.

8. Mitchell J T D and Hollis A. A Tokamak reactor with servicing capability. Ninth Symposium on Fusion Technology, Garmisch-Partenkirchen (1976) - to be published.

9. Galbiati L et al. A compact and flexible servo-system for master-slave electric manipulators. Proceedings 12th Conference on Remote Systems Technology (ANS) 1964.

10. Carlson G A and Moir R W. Mirror fusion reactor study. American Nuclear Society, Winter Meeting, San Francisco, California November 1975 (UCRL Preprint 76985).

11. Challender R S and Reynolds P. Accessibility and replacement as prime constraints in the design of large Tokamak experiments. Ninth Symposium on Fusion Technology, Garmisch-Partenkirchen (1976) - to be published.

12. Crawley H J. The practical feasibility of a bundle divertor for a Tokamak power reactor. Ninth Symposium on Fusion Technology, Garmisch-Partenkirchen (1976) - to be published.

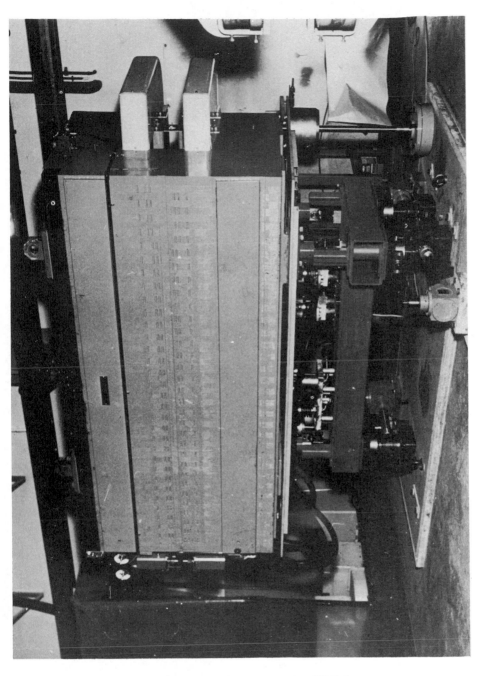

Figure 7   Accelerator beam magnet developed by CERN Geneva
            showing special mounting designed for remote handling.

573

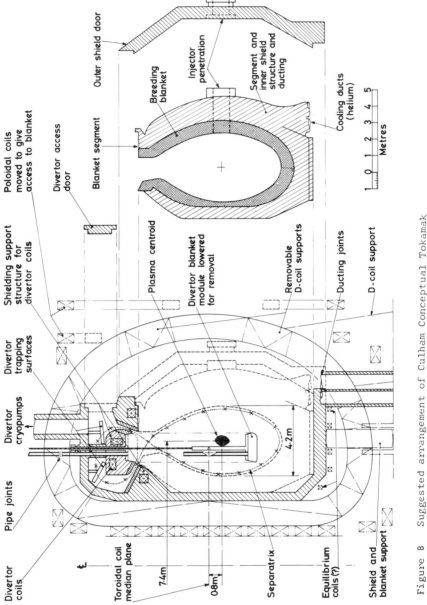

Divertor coils

Pipe joints

Divertor cryopumps

Divertor trapping surfaces

Shielding support structure for divertor coils

Poloidal coils moved to give access to blanket

Divertor access door

Blanket segment

Outer shield door

Breeding blanket

Injector penetration

Segment and inner shield structure and ducting

Cooling ducts (helium)

Metres

Toroidal coil median plane

7·4m

0·8m

Plasma centroid

Divertor blanket module lowered for removal

Separatrix

Equilibrium coils (?)

Removable D-coil supports

Ducting joints

D-coil support

Shield and blanket support

4·2m

Figure 8   Suggested arrangement of Culham Conceptual Tokamak Reactor incorporating a single null divertor and showing blanket component assembly.

574

STRATEGY

FUSION POWER BY MAGNETIC CONFINEMENT - PROGRAM PLAN

Stephen O. Dean
U.S. Energy Research and Development Administration

Abstract

This Fusion Power Program Plan treats the technical, schedular and budgetary projections for the development of fusion power using magnetic confinement. It was prepared on the basis of current technical status and program perspective. A broad overview of the probable facilities requirements and optional possible technical paths to a demonstration reactor is presented, as well as a more detailed plan for the R&D program for the next five years. The "plan" is not a roadmap to be followed blindly to the end goal. Rather it is a tool of management, a dynamic and living document which will change and evolve as scientific, engineering/technology and commercial/economic/environmental analyses and progress proceeds. The use of plans such as this one in technically complex development programs requires judgment and flexibility as new insights into the nature of the task evolve.

The presently-established program goal of the fusion program is to DEVELOP AND DEMONSTRATE PURE FUSION CENTRAL ELECTRIC POWER STATIONS FOR COMMERCIAL APPLICATIONS. Actual commercialization of fusion reactors will occur through a developing fusion vendor industry working with Government, national laboratories and the electric utilities. Short term objectives of the program center around establishing the technical feasibility of the more promising concepts which could best lead to commercial power systems. Key to success in this effort is a cooperative effort in the R&D phase among government, national laboratories, utilities and industry.

There exist potential applications of fusion systems other than central station electric plants. These include direct production of hydrogen gas and/or synthetic fuels; direct energy production for chemical processing; fissile fuel production; fission product waste disposal; and fusion-fission hybrid reactors. Efforts are in progress to evaluate these applications; the present and planned programs will permit timely information on which decisions can be made to pursue these goals.

The pace of the fusion program is determined by both policy variables and technical variables. A multiplicity of plans, referred to as Program Logics, are outlined. These range from "level of effort research" to "maximum effective effort" and are primarily describable by the presumed level of funding. Within these program logics there are many optional technical paths. A few of the potential paths or options are outlined.

577

The tokamak is currently the most promising approach to fusion and is closer to achieving a demonstration reactor for commercial application than other fusion concepts; but active programs in other concepts are pursued. The plan permits changes to alternate concepts on a timely basis as the physics and engineering/technology studies evolve.

The total cost to develop fusion power from FY 1978 through the date of operation of the first demonstration reactor is found to be roughly $15 billion dollars in constant FY 1978 dollars. With such funding a demonstration reactor could operate in the time frame 1993 to 2005 depending on near-term funding profiles and progress. A reference case (called Logic III) which aims at a demonstration reactor in 1998 is treated in detail.

The Fusion Power Program Plan consists of five documents as follows:

ERDA 76-110/0	Executive Summary
ERDA 76-110/1	Volume I: Summary
ERDA 76-110/2	Volume II: Long Range Planning Projections
ERDA 76-110/3	Volume III: Five Year Plan
ERDA 76-110/4	Volume IV: Five Year Budget and Milestone Summaries

ii

I. INTRODUCTION

A. GOALS

The presently-established program goal of the fusion program is
to DEVELOP AND DEMONSTRATE PURE FUSION CENTRAL ELECTRIC POWER
STATIONS FOR COMMERCIAL APPLICATIONS. The program is based upon
the assumption that it is in the national interest to demonstrate
safe, reliable, environmentally acceptable and economically competitive production of fusion power in a Demonstration Reactor that
extrapolates readily to commercial reactors. Actual commercialization of fusion reactors is assumed to occur primarily through a
developing fusion vendor industry working with Government, national
laboratories and the electric utilities. Hence it is also an
objective of the fusion program to develop sufficient data that
utilities and industry can address all critical issues (e.g.,
capital and operating costs, reliability, safety, etc.) involved
in arriving at power plant purchase decisions.

Short term objectives of the program center around establishing
the technical feasibility of the more promising concepts which
could best lead to commercial power systems. Key to success in
this effort is a cooperative effort in the R&D phase among
Government, national laboratories, utilities and industry.

There exist potential applications of fusion systems other than
central station electric plants. These include:

1

- Direct production of hydrogen gas and/or synthetic fuels
- Direct energy production for chemical processing
- Fissile fuel production
- Fission product waste disposal
- Fusion-fission hybrid reactors

These applications hold the possibility of increasing the overall impact of fusion power and of hastening its commercial application.

The physical and economic characteristics of these potential applications have been analyzed only partially. Efforts are currently in progress to further evaluate the advantages and disadvantages of these applications; the present and planned programs will provide timely information on which decisions can be made to pursue these goals.

B.   FUSION ADVANTAGES

The potential advantages of commercial fusion reactors as power producers would be:

- An effectively inexhaustible supply of fuel -- at essentially zero cost on an energy production scale;
- A fuel supply that is available from the oceans to all countries and therefore cannot be interrupted by other nations;
- No possibility of nuclear runaway;
- No chemical combustion products as effluents;
- No afterheat cooling problem in case of an accidental loss of coolant;
- No use of weapons grade nuclear materials; thus no possibility of diversion for purposes of blackmail or sabotage;
- Low amount of radioactive by-products with significantly shorter half-life relative to fission reactors.

C. FUEL CYCLES

First generation fusion reactors are expected to use deuterium and tritium as fuel. Several environmental drawbacks are, however, commonly attributed to DT fusion power. First, it produces substantial amounts of neutrons that result in induced radioactivity within the reactor structure, and it requires the handling of the radioisotope tritium. Second, only about 20% of the fusion energy yield appears in the form of charged particles, which limits the extent to which direct energy conversion techniques might be applied. Finally, the use of DT fusion power depends on lithium resources, which are less abundant than deuterium resources.

These drawbacks of DT fusion power have led to the proposal of alternatives for longer term application -- for example, fusion power reactors based only on deuterium. Such systems are expected to (1) reduce the production of high energy neutrons and also the need to handle tritium; (2) produce more fusion power in the form of charged particles; and (3) be independent of lithium resources for tritium breeding.

It has also been suggested that materials with slightly higher atomic numbers (like lithium, beryllium, and boron) be used as fusion fuels to provide power that is essentially free of neutrons and tritium and that release all of their energy in the form of charged particles.

Although such alternatives to DT fusion power are attractive, there is an important scientific caveat. To derive useful amounts of power from nuclear fusion, it will be necessary to confine a suitably dense plasma at fusion temperatures (10^8 $^{\circ}$K) for a specific length of time. This fundamental aspect of fusion power is expressible in terms of the product of the plasma density, n, and the energy confinement time, τ, required for fusion power breakeven (i.e., the condition for which the fusion power release equals the power input necessary to heat and confine the plasma). The required product, nτ, depends on the fusion fuel and is primarily a function of the plasma temperature. Of all the fusion fuels under current consideration, the deuterium–tritium fuel mixture requires the lowest value of nτ by at least an order of magnitude and the lowest fusion temperatures by at least a factor of 5. When the plasma requirements for significant power generation are compared with the anticipated plasma performance of current approaches to fusion power, it is apparent that fusion power must initially be based on a deuterium-tritium fuel economy. However, the eventual use of alternate fuel cycles remains an important ultimate goal and consequently attention will be given to identifying concepts which may permit their ultimate use.

D. FOREIGN EFFORTS

The United States fusion effort is a part of a much larger world effort. At present the U.S. effort is estimated to be about one-third of the world effort as measured by total man-years expended. Extensive collaboration exists among all nations of the world active in fusion R&D. This is effected through bilateral arrangements, both formal and informal, for the exchange of information and manpower and through multilateral arrangements facilitated by the International Atomic Energy Agency and the International Energy Agency. A particularly close collaboration between the

U.S. and the U.S.S.R. has developed during the past three years. This collaboration is supervised by a sixteen member group called the Joint (U.S.-U.S.S.R.) Fusion Power Coordinating Committee (JFPCC).

The program presented in this plan will permit the U.S. to achieve the desired end goal independent of any activity in another nation's program. However, coordination is maintained with other nations which insures that the steps taken in each country are complementary rather than redundant. This procedure leads to a reduction in the risk of failure in the overall effort. Examples of complementary large devices being built or considered in various nations at this time are the Tokamak Fusion Test Reactor (TFTR) in the U.S., the Joint European Tokamak (JET) in Europe, the Japanese Tokamak-60 (JT-60) in Japan, and the Tokamaks T-10M and T-20 in the U.S.S.R. Differences among the devices can be noted, for example, in Figures V-3 and V-4 of Volume II.

E. STRUCTURE

The entire Fusion Power Program Plan consists of five documents as follows:

> ERDA 76-110/0, Executive Summary
> ERDA 76-110/1, Volume I: Summary
> ERDA 76-110/2, Volume II: Long Range Planning Projections
> ERDA 76-110/3, Volume III: Five Year Plan
> ERDA 76-110/4, Volume IV: Five Year Budget and Milestone
> Summaries

In Section II of this volume, the contents of the Long Range Planning Projections are summarized. The Long Range Planning Projections treat the R&D program of the 1980's and 1990's and, in particular, consider the range of optional paths to a Demonstration Reactor that may exist for commercial application. The relationship among funding patterns, physics and engineering progress, and the date of achievement of the end goal is described. In Section III of this volume, the contents of the Five Year Plan are summarized.

Data such as that presented in this plan can be used in the performance of cost/benefit analyses which will provide quantitative supporting information to help decide whether increased funding at some level is desirable. The basic framework for performing such cost/benefit analyses is being constructed by ERDA contractors.

II. LONG RANGE PLANNING PROJECTIONS

A. PROGRAM LOGICS

The most significant policy and technical variables that affect the pace of the fusion program are:

Policy Variables:

- The perceived NEED for fusion power
- The nation's INTENT (what is expected by when? What priority does the program have?)
- FUNDING

Technical Variables:

- PHYSICS RESULTS
- ENGINEERING RESULTS

Because NEED, INTENT, and FUNDING are finally decided by others, the fusion program requires a number of plans by which the program can be conducted. The following plans, referred to as LOGICS, are considered.

LOGIC I. LEVEL OF EFFORT RESEARCH

Research and development are supported at an arbitrary level in order to develop basic understanding. (If this pace were continued, a practical fusion power system might never be built.)

LOGIC II. MODERATELY EXPANDING, SEQUENTIAL

Funds are expanding but technical progress is limited by the availability of funds. Established commitments are given funding priority but new projects are not started until funds are available. In spite of limited funding a number of problems are addressed concurrently. (At this rate, a fusion demonstration reactor might operate in the early 21st century.)

LOGIC III: AGGRESSIVE

The levels of effort in physics and engineering are expanded according to programmatic need, assuming that adequate progress is evident. New projects are undertaken when they are scientifically justified. Many problems are addressed concurrently. Funding is ample but reasonably limited. (This program would be aimed at an operating demonstration reactor in the late 1990's.)

LOGIC IV: ACCELERATED

A great many problems are addressed in parallel and new projects are started when their need is defined. Fabrication and construction are carried out on a normal basis with enough priority to minimize delays. The availability of funds is still limited but a secondary factor in program planning and implementation. (This approach would be aimed at demonstration reactor operation in the early to mid-1990's.)

LOGIC V: MAXIMUM EFFECTIVE EFFORT

Manpower, facilities and funds are made available on a priority basis; all reasonable requests are honored immediately. Fabrication and construction are expedited on a priority basis so that completion times for major facilities are reduced to a practical minimum. (An operating demonstration plant around 1990 would be the program goal.)

Although the five Logics are most easily distinguished by costs and end-goal dates, it should also be noted that the degree of risk varies among Logics. Risk can increase under faster-paced Logics. On the other hand, risk can decrease with higher budgets due to increased effort and partial overlapping of facility goals. It is not possible to quantify the net change in risk among the Logics in general; it is necessary, however, to assess the risk at every point along the way.

The interplay among policy variables, technical variables and program Logics is shown in Figure II-1. Real world requirements as perceived by the Division of Magnetic Fusion Energy (DMFE) determine the program goals and objectives and, as perceived by ERDA, OMB, and Congress, fix the policy variables. An interaction takes place between the Division and ERDA, OMB and Congress; eventually ERDA, OMB and the Congress determine which LOGIC the program is to follow. The goals and objectives, as modified by the policy variables, prescribe the R&D program scope. The choice of LOGIC influences the activity within the program scope and a specific path (called a Logic Option) emerges. The results from following that option constitute the technical variables which the Division evaluates in the process of proposing the program goals and adjusting objectives.

The Logics, numbered I through V, are differentiated grossly according to funding levels of the operations budget in Figure II-2. The funding is such that the funding level for Logic I will result in a DEMO far out in time, while the funding level for Logic V will result in a DEMO as soon as is practically possible. It should be noted that the degree of "pessimism" or "optimism" that one assumes

587

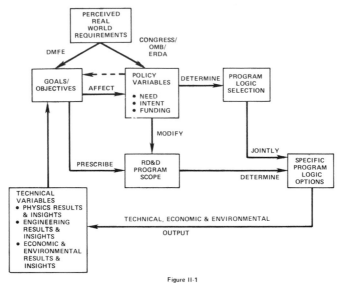

Figure II-1
DMFE PLANNING METHODOLOGY

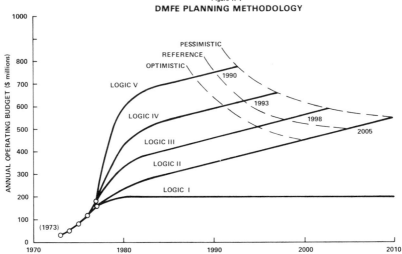

FUSION R&D PROGRAM OPERATING BUDGET AND LOCI OF DEMO OPERATING DATES FOR LOGIC I THRU V

Figure II-2

588

substantially affects the projected date for operation of the DEMO. The projected operating date for a DEMO will also be affected by the degree of "risk" the program is willing to accept in moving from one step to the next. Clearly it is possible to aim at the same dates with lower funding, or earlier dates with the same funding if higher risks are taken, i.e., if less R&D and fewer demonstrated results are required to justify succeeding steps. The projected total annual budgets required for the five Logics are shown in Figure II-3.

B. LOGIC III OPTIONS

Reference Option

The Logic III Reference Option is shown in Figure II-4. The devices listed in Figure II-4 are the following: Demonstration Reactor (DEMO), Experimental Power Reactor (EPR), Prototype Experimental Power Reactor or Ignition Test Reactor (PEPR/ITR), Tokamak Fusion Test Reactor (TFTR), Doublet III (D-III), Poloidal Divertor Experiment (PDX), Princeton Large Torus (PLT), Fusion Engineering Research Facility or Engineering Test Reactor (FERF/ETR), Large Mirror Experiment (MX), Baseball Mirror Device (BB), 2X Mirror Device (2X), Staged Scyllac (SS). The characteristics of the major new facilities (DEMO, EPR, PEPR/ITR, and TFTR) are given in Volume II, Section IV. For planning and costing purposes, it is assumed that a selection process takes place among the various concepts so that EPR's for two concepts and a DEMO for one concept result. Under the Logic III Reference Option a DEMO would operate in 1998.

- Tokamak Assumptions

   Four major devices are postulated beyond Doublet III; namely TFTR, PEPR/ITR, EPR and DEMO. In addition a major engineering facility (FERF/ETR), which may be a tokamak, is constructed.

589

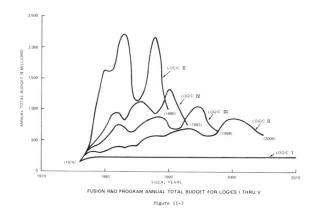

FUSION R&D PROGRAM ANNUAL TOTAL BUDGET FOR LOGICS I THRU V

Figure II-3

LOGIC III REFERENCE OPTION

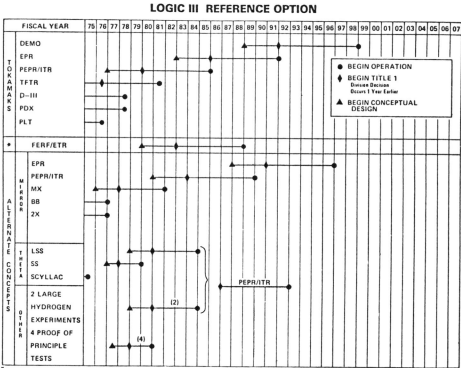

*MAJOR ENGINEERING FACILITY

Figure II-4

590

- Alternate Concepts Assumptions

  Mirror Assumptions
  Three major devices past 2X/BB are postulated; namely
  MX, PEPR/ITR and EPR. The FERF/ETR could be a mirror.
  This Logic could result in a mirror DEMO (not shown
  in Figure II-4 but see Figure II-5) by 2004.

  Theta Pinch and Other Alternate Concept Assumptions
  Five concepts, including Staged Scyllac, are examined
  in parallel on a moderate scale for proof-of-principle
  tests. Once a proof-of-principle has been established
  the most promising concepts are evaluated in large
  hydrogen experiments (LHX). Three LHX's, including
  Large Staged Scyllac, are assumed. After operation
  of the LHX's, one concept is selected for a PEPR/ITR.
  This Logic could result in a Theta Pinch or Other
  Alternate Concept DEMO (not shown in Figure II-4, but
  see Figure II-5) by 2007.

- Logic III Alternate Paths

  As decision dates occur for major facilities, it is
  possible that the decision will be to wait for
  further information. This is shown in Figure II-5,
  in which the program path alternatives for the
  Logic III Reference Option are presented. The
  circles represent facility operation dates and the
  diamonds indicate the initiation of Title I funding
  based upon a decision made the previous year. Note,
  for example, the first decision along the PEPR/ITR
  tokamak line. The result of this decision will be
  to either construct a tokamak PEPR/ITR or delay until
  more information becomes available for both tokamaks
  and mirrors. Assuming that the result of the decision
  is to wait, the next identified decision point is
  along the mirror PEPR/ITR line. This decision can
  result in three alternatives: (1) construct a
  tokamak PEPR/ITR; (2) construct a mirror PEPR/ITR;
  (3) delay until more information becomes available
  for all three approaches to magnetic fusion.

  A decision made for one confinement concept (say
  tokamak PEPR/ITR) will not prevent a second
  decision, at a later point in time, for a second
  confinement method (say mirror or advanced concept
  PEPR/ITR).

591

Figure II-5

LOGIC Ⅲ REFERENCE OPTION PROGRAM PATH ALTERNATIVES

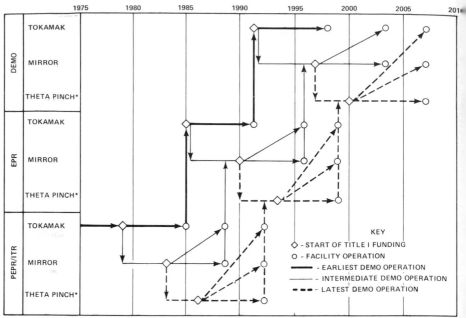

KEY

◇ - START OF TITLE I FUNDING
○ - FACILITY OPERATION
▬▬▬ - EARLIEST DEMO OPERATION
───── - INTERMEDIATE DEMO OPERATION
▬ ▬ ▬ - LATEST DEMO OPERATION

* OTHER ALTERNATE CONCEPTS WOULD FOLLOW THE THETA PINCH PATH

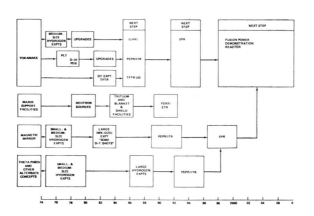

General Features of the Logic III Refernence Option

Figure II-6

> The figure shows that the earliest possible Fusion
> Power Demonstration Reactor would be a tokamak
> operating around 1998. Note that this operation
> date could be delayed until 2004 (and could be a
> mirror) if the 1979 decision resulted in a decision
> to delay selection of the PEPR/ITR until 1983. This
> delay would be reduced if one were willing to permit
> a "continuum" of decision points between 1979 and
> 1983.

The general features of the Logic III Reference Option are
summarized in Figure II-6. The present experimental program
consists of several small and medium-sized hydrogen experiments
(most notably the ORMAK and Alcator Tokamaks, the 2XIIB Mirror,
and the Scyllac Theta Pinch) and the larger PLT at Princeton
which came into operation in December 1975. Two other large
tokamaks, Doublet III at General Atomic and PDX at Princeton,
are in fabrication and scheduled to operate in early 1978.
The first DT burning tokamak, the Tokamak Fusion Test Reactor
(TFTR), is scheduled to operate in mid-1981. A large mirror
experiment, called MX, has been proposed for operation in 1981.

Under a Logic III program each of these devices would be
upgraded, primarily by adding more auxiliary heating power, to
test physics scaling laws at higher temperatures and higher
power density. In the mid- to late-1980's, large device(s)
would be built assuming good results are obtained on earlier
facilities. The next step in the tokamak line is assumed to
be either a Prototype Experimental Power Reactor or an Ignition
Test Reactor (PEPR/ITR). TFTR would be upgraded and possibly
another large hydrogen experiment might be required. An
engineering test reactor (FERF/ETR) is assumed, which could
be a tokamak. By the early 1990's an Experimental Power
Reactor (EPR) would be built, which makes net electrical
power with high reliability. This device would be followed
by the Fusion Power Demonstration Reactor in 1998.

593

The Magnetic Mirror Program is assumed to evolve from the present small- and medium-size experiments, most notably 2XII at Livermore, to a larger device in which a limited number of DT shots would be possible. A major objective of this device would be to test confinement scaling for longer times, and to test methods for improving power balance, a prerequisite to the feasibility of a pure fusion mirror reactor. This would be followed by a PEPR device in the late 1980's. The FERF/ETR could be a mirror. This could be followed by an EPR operating in 1996 and a DEMO around 2004.

For the other Alternate Concepts, larger hydrogen experiments, such as the Large Staged Scyllac, are assumed to operate in the mid-1980's, followed by a PEPR/ITR in the early 1990's. Next could come an EPR in the late 1990's and a DEMO around 2007.

For costing purposes of Logic III, 3 PEPR/ITR devices, 1 FERF/ETR, 2 EPR's, and 1 DEMO are assumed. Depending on progress and periodic assessments, not all the devices and facilities projected in this plan would necessarily be built.

Critical Parameter Assessments

Decisions to move ahead, mark time, retreat, change approach, etc., are based on assessments of the status of the physics and engineering/technology at a given point in time. These assessments include a prognosis on the implications of our current understanding for future commercial fusion reactors. These assessments are called "critical parameter assessments" and currently are scheduled to take place in 1979 for tokamaks, in 1982 for magnetic mirrors and in 1985 for the toroidal theta pinch and other alternate concepts.

Although these assessments clearly involve complex scientific/technical issues, the projected results of these assessments are described herein in simpler terms. Each of the critical parameters is assessed by assigning a "good", "fair", or "poor" rating to that part of the assessment and an overall rating of "good", "fair", or "poor" is then assigned to the physics and engineering/technology parts of the assessment separately. These latter ratings are used in deciding the nature of the best next step in the program.

The definitions of the critical parameters and the proposed definitions of the "good", "fair", or "poor" ratings are given in detail in Volume II, Section III.

Options

In the planning process, assumptions must be made on the range of possible physics and engineering/technology results and the time at which these results will be forthcoming. This gives rise to a multiplicity of potential paths for each approach to fusion power, called "Options". Analysis shows that many of these results lead to decisions to build large devices which are similar in general character, although they may differ in timing and in physics and engineering detail. Consequently the different options are characterized primarily by the nature of the next major facility to which the particular option path leads.

A matrix of possible tokamak options is shown in Figure II-7. In Column (3), "Critical Parameter Results by CY 1979", the assessment is shown by giving a good to poor rating for the physics and engineering/technology assessment discussed above.

Logic Option Matrix for Tokamaks

Option	Description	Results of Critical Parameter Assessment in 1979 Physics	Eng. Techn.	Best Next Step	Completion Date	Best Next Step	I/C	Best Next Step	I/C
1. Reference	D-Shaped	G	F	PEPR/ITR FERF/ETR	79/85 82/88	EPR	85/91	DEMO	91/98
2. Optimistic	Doublet	G	G	EPR-I	79/85	EPR-II	84/90	DEMO	88/95
3. Pessimistic	Circular	F	F	LHX PEPR/ITR	79/84 85/91	EPR	91/97	DEMO	98/05
4. Reassessment	Any	P		Reassess in 1982 based upon further results from upgrades and TFTR					
5.	High Field	G	G	PEPR/ITR	79/85	EPR	85/91	DEMO	91/91
6.	High Field	G	F	Reassess					
7.	Doublet	G	F	PEPR/ITR	79/85	EPR	85/91	DEMO	91/98

Figure II-7

596

A matrix of possible options for mirrors and toroidal theta
pinches is shown in Figure II-8.

Alternate concepts are pursued if they offer potential physics,
engineering/technology or economic advantages. An aggressive
but sequential alternate concepts program is maintained in
Logic III to examine all of the potentially promising confinement
approaches at least to the point of "proof-of-principle" tests
(see Figure II-9). In particular about six proof-of-principle
experiments would be completed by 1980-82. Large hydrogen
experiments for the two most promising concepts would then be
initiated in the early 1980's and completed in FY 1984-86.
One PEPR/ITR would be initiated in 1986 and completed in 1992.
One EPR could be initiated in 1993 and completed in 1999. A
DEMO could be initiated in 2000 and completed in 2007.

Supporting Engineering Facilities are required. The principal
ones envisioned are shown in Figure II-10 and described in
Volume II, Section IV. The engineering and materials test
reactor (FERF/ETR) is the most costly of the supporting
facilities.

C. ROLL-BACK PLANNING

In the preceeding sections the primary planning approach may
be described as "roll-forward", i.e., the current program is
considered and, from that consideration, the nature and timing
of the next step is determined. A successful fusion power R&D
program requires, in addition, a "roll-back" approach in which

597

Logic III: Option Matrix for Mirror and Toroidal Theta Pinch
Alternate Concepts

Option	Physics Prototype	I/C	Results of Critical Parameter Assessment-1982		Best Next Step	Initiation/ Completion Date	Best Next Step	I/C	Best Next Step	I/C
			Physics	Eng/Tech.						
Mirror										
1. Reference	MX		G	F	PEPR	83/89	EPR	90/96	DEMO	97/04
2. Optimistic	MX		G	G	PEPR	82/88	EPR	88/94	DEMO	93/01
3. Pessimistic	MX	77/81	F	F	LHX PEPR	82/87 88/94	EPR	95/01	DEMO	02/09
4. Reassess			P	P	-					
5. Fusion/Fission	MX	77/81	F	F	F/F PEPR	83/89	F/F DEMO	90/96		
6. FERF/ETR	MX	77/81	F	F	FERF/ETR	82/88	-			
			Assessment-1985							
Toroidal θ-Pinch										
1. Reference	Large		G	F	PEPR/ITR	86/92	EPR	93/99	DEMO	00/07
2. Optimistic	Staged	80/84	G	G	PEPR/ITR	86/89	EPR	90/96	DEMO	97/04
3. Pessimistic	Scyllac		F	F	LHX PEPR/ITR	85/90 91/97	EPR	98/04	DEMO	05/12
4. Reassess			P	P	-					

Figure II-8

Logic III: Option Matrix for Other Alternate Concepts

Option	Physics Prototype	I/C	Results of Critical Parameter Assessment-1985		Best Next Step	I/C	Best Next Step	I/C	Best Next Step	I/C
			Physics G	Eng/Techn. F						
Reference	LHX	81/85			PEPR/ITR	86/92	EPR	93/99	DEMO	00/07

Concept	Principal	I/C
EBT	EBT-II	78/80
TORMAC	TORMAC VI	78/80
ZT	ZT-II	79/81
Linear	Scylla IV-P	74/76
	Long Linear Expt.	78/82
Liner	Linus I	78/80

LHX	PEPR/ITR	EPR	DEMO
Up to two large hydrogen experiments would be fabricated based on the most promising concepts. Initiation would occur in FY80-82 with completion in FY84-86.	One PEPR/ITR among theta pinch and other alts. would be fabricated based on '85 assessment of critical parameters	One EPR from all alternate concepts could be fabricated based on '89, '92 assessments of critical parameters.	One DEMO from among all fusion approaches could be fabricated based on '90, '96, '99 assessments of critical parameters.

Figure II-9

599

the nature of the desired end-product, a Fusion Power Demonstration
Reactor that extrapolates readily to commercial reactors, is defined
in detail and in which the physics and engineering tests required
for a DEMO are identified and programs established to provide the
required tests. This "roll-back" approach is discussed in Section V
of Volume II . Clearly "roll-forward" and "roll-back" approaches
must both be used and be complementary for a successful fusion R&D
program.

In order to build a Fusion Power Demonstration Reactor of any type,
certain physics understanding must be demonstrated and certain
technological subsystems must be developed. These activities may
be categorized as "Major Program Elements". Figure II- 11 lists
twenty-one Major Program Elements identifiable at this time.
Inspection of Figure II-11 suggests that there are two basic
classes of Major Program Elements; physics and engineering/techno-
logy. Elements I-IV are basically Physics Elements and the
remainder are basically Engineering/Technology Elements. There
are both explicit and implicit relationships among these "Elements".
Overall technological and economic outlook is determined by the
interrelated progress of each Element towards meeting the needs
of a fusion DEMO. Tests of the critical physics and/or the
technology of the Elements may be made individually in small
test facilities and/or collectively in larger facilities. These
tests can be described as falling into four classes of tests
as follows:

       1.   Early Tests

       2.   High Confidence Level Tests

       3.   Definitive Tests

       4.   Full Scale DEMO Prototype Tests

ENGINEERING FACILITIES (LOGIC III)

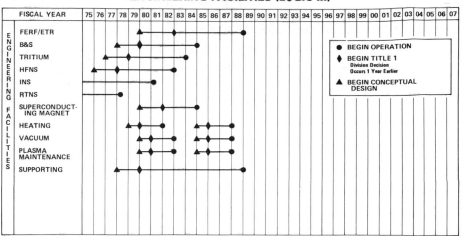

Figure II-10

Major Program Elements
for all Concepts

Physics

    I.   Scaling

    II.   Impurity Control

    III.   Beta Limits

    IV.   DT Burn Dynamics

Engineering/Technology

    V.   Plasma Maintenance and Control

    VI.   Heating Technology

    VII.   Superconducting Magnets

    VIII.   Pulsed Energy Systems

    IX.   Blanket and Shield

    X.   Tritium Processing and Control

    XI.   Electrical Subsystems

    XII.   Power Handling

    XIII.   Plant Availability

    XIV.   Instrumentation and Control

    XV.   Plant Maintenance

    XVI.   Vacuum Technology

    XVII.   Materials

    XVIII.   Balance of Plant

    XIX.   Systems Integration

    XX.   Environment and Safety

    XXI.   Economics

Figure II-11

601

Gross program progress may be measured and described in the above terms for each fusion concept. Early tests along with theoretical models provide the definition of the problems for the progress of each program element. High Confidence Level tests are conducted via model and machine experiments. Definitive tests provide the understanding of scaling laws necessary for the DEMO design. Full scale DEMO prototype tests demonstrate the readiness for DEMO application.

Major facilities are justified in part, by stating the level of test they will provide for each Major Program Element. Major Program Element tests, at different levels of confidence, are performed at different times depending on the option taken and the fusion concept assumed for DEMO. Figure II-12 is a flow chart showing the times at which various classes of tests are expected in the areas of the twenty-one program elements for the Logic III Reference Option program for the tokamak concept.

Each of the horizontal arrows represents the progress of a Major Program Element. The numbers indicate the various classes of tests expected.

The twenty-one Major Program Elements are discussed in some detail in Volume III, Section V to provide further insight into this planning method. The scope of the discussion covers all the fusion concepts, but more information is provided on the tokamak concept because of the current preeminence of this approach. The elements are general enough to cover all the fusion concepts.

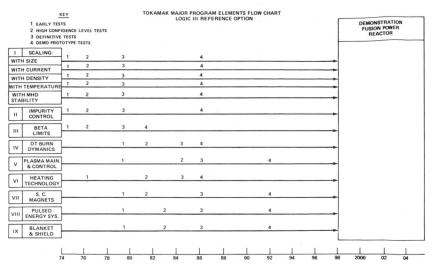

Figure II-12

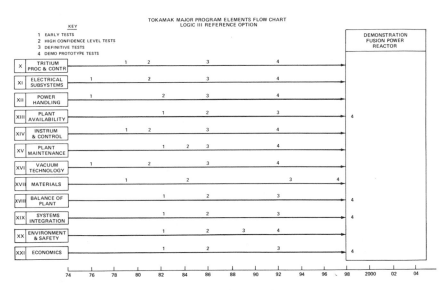

Figure II-12 - continued

603

D. BUDGET SUMMARY

The total integrated program costs from FY 1978 to the date of
initial operation of the DEMO are shown below. All costs are
in constant FY 1978 dollars. Details are presented in Figures
II-13, 14, 15, 16.

LOGIC	I	II	III	IV	V
TOKAMAK PACE	Indeterminate	2630	2630	2630	4140
ENG. FAC. PACE		875	875	1050	1710
ALT. CONC. PACE		1600	2000	2000	4940
OPERATIONS		10120	9017	8260	8490
EQUIPMENT		1013	992	826	849
TOTAL		16238	15514	14766	20129

Figure II-13 PROGRAM COSTS BY YEAR FOR LOGIC II REFERENCE OPTION ($M)

	FY76	FY77	FY78	FY79	FY80	FY81	FY82	FY83	FY84	FY85	FY86	FY87	FY88	FY89	FY90	FY91
TOKAMAK PACE *	20	80	95	35	15	35	35	15	40	60	100	100	60	40	0	40
TFTR	20	80	95	35	15	35	35	15	0	0	0	0	0	0	0	0
PEPR/ITR	0	0	0	0	0	0	0	0	40	60	100	100	60	40	0	0
EPR	0	0	0	0	0	0	0	0	0	0	0	0	0	0	0	40
DEMO	0	0	0	0	0	0	0	0	0	0	0	0	0	0	0	0
ENG. FAC. PACE	2	18	10	10	20	35	56	66	54	25	20	16	44	88	168	163
FERF/ETR	0	0	0	0	0	0	0	0	0	0	0	0	25	75	150	150
HFNS	0	0	0	0	0	0	20	10	0	0	0	0	0	0	0	0
HTTF	0	0	0	10	15	0	20	6	9	0	0	0	0	0	0	0
TF	0	0	0	0	0	10	10	10	5	0	0	0	0	0	0	0
B&S	0	3	0	0	5	10	0	20	10	0	0	0	0	0	0	0
RTNS	2	15	0	0	0	5	0	0	0	0	0	0	0	0	0	0
INS	0	0	10	0	0	0	0	5	0	0	0	0	0	0	0	0
PMCTF	0	0	0	0	0	0	3	5	0	0	0	0	0	0	0	0
VTF	0	0	0	0	0	0	3	0	0	0	0	0	0	3	4	3
SMTF	0	0	0	0	0	10	0	10	10	10	10	10	10	0	0	0
Eng. Test. Fac.	0	0	0	0	0	0	0	0	20	15	10	6	9	10	14	10
ALT. CONC. PACE	0	0	0	15	35	35	15	0	0	0	45	105	105	65	60	120
LHX																
MX	0	0	0	15	35	35	15	0	0	0	0	0	0	0	0	0
LSS	0	0	0	0	0	0	0	0	0	0	15	35	35	15	0	0
#	0	0	0	0	0	0	0	0	0	0	15	35	35	15	0	0
#4	0	0	0	0	0	0	0	0	0	0	15	35	35	15	0	0
PEPR	0	0	0	0	0	0	0	0	0	0	0	0	0	20	60	120
EPR	0	0	0	0	0	0	0	0	0	0	0	0	0	0	0	0
TOTAL PACE	22	98	105	60	70	105	106	81	94	85	165	221	209	193	228	323
OPERATIONS	120	156	180	210	235	255	270	280	290	300	310	320	330	340	350	360
EQUIPMENT	17	20	18	21	24	26	27	28	29	30	31	32	33	34	35	36
TOTAL PROGRAM	159	274	303	291	329	386	403	389	413	415	506	573	572	567	613	719

*PACE: Plant and Capital Equipment line item construction projects.

605

Figure II-13 - continued

	FY92	FY93	FY94	FY95	FY96	FY97	FY98	FY99	FY2000	FY2001	FY2002	FY2003	FY2004	FY2005	28yr. Total FY78-2005
TOKAMAK PACE*	80	160	240	160	80	40	50	100	200	250	250	200	100	50	2630
TPFR	0	0	0	0	0	0	0	0	0	0	0	0	0	0	230
PEPR/ITR	0	0	0	0	0	0	0	0	0	0	0	0	0	0	400
EPR	80	160	240	160	80	40	0	0	0	0	0	0	0	0	800
DEMO	0	0	0	0	0	0	50	100	200	250	250	200	100	50	1200
ENG. FAC. PACE	75	25	0	0	0	0	0	0	0	0	0	0	0	0	875
FGRF/ETR	75	25	0	0	0	0	0	0	0	0	0	0	0	0	500
HFNS	0	0	0	0	0	0	0	0	0	0	0	0	0	0	75
HTTF	0	0	0	0	0	0	0	0	0	0	0	0	0	0	30
TF	0	0	0	0	0	0	0	0	0	0	0	0	0	0	50
B&S	0	0	0	0	0	0	0	0	0	0	0	0	0	0	50
RTNS	0	0	0	0	0	0	0	0	0	0	0	0	0	0	0
INS	0	0	0	0	0	0	0	0	0	0	0	0	0	0	10
PMCTF	0	0	0	0	0	0	0	0	0	0	0	0	0	0	15
VTF	0	0	0	0	0	0	0	0	0	0	0	0	0	0	15
SMTF	0	0	0	0	0	0	0	0	0	0	0	0	0	0	30
Eng. Test. Fac.	0	0	0	0	0	0	0	0	0	0	0	0	0	0	100
ALT. CONC. PACE	120	60	20	0	40	80	160	240	160	80	40	0	0	0	1600
LHX	0	0	0	0	0	0	0	0	0	0	0	0	0	0	100
MX	0	0	0	0	0	0	0	0	0	0	0	0	0	0	100
LSS	0	0	0	0	0	0	0	0	0	0	0	0	0	0	0
#3	0	0	0	0	0	0	0	0	0	0	0	0	0	0	0
#4	0	0	0	0	0	0	0	0	0	0	0	0	0	0	0
PEPR	120	60	20	0	0	0	0	0	0	0	0	0	0	0	400
EPR	0	0	0	0	40	80	160	240	160	80	40	0	0	0	800
TOTAL PACE	275	245	260	160	120	120	210	340	360	330	290	200	100	50	5105
OPERATIONS	370	380	390	400	410	420	430	440	450	460	470	480	490	500	10120
EQUIPMENT	37	38	39	40	41	42	43	44	45	46	47	48	49	50	1013
TOTAL PROGRAM	682	663	689	600	571	582	683	824	855	836	807	728	639	600	16238

*PACE: Plant and Capital Equipment line item construction projects.

Figure II-14 PROGRAM COSTS BY YEAR FOR THE LOGIC III REFERENCE OPTION ($M)

	FY76	FY77	FY78	FY79	FY80	FY81	FY82	FY83	FY84	FY85	FY86	FY87	FY88	FY89	FY90
TOKAMAK PACE*	20	80	95	50	65	115	140	105	45	15	40	120	240	240	120
TFTR	20	80	95	35	15	35	35	15	0	0	0	0	0	0	0
PEPR/ITR Fac.	0	0	0	15	35	35	15	90	45	15	0	0	0	0	0
PEPR/ITR Dev.	0	0	0	0	15	45	90	0	0	0	0	0	0	0	0
EPR	0	0	0	0	0	0	0	0	0	0	40	120	240	240	120
DEMO	0	0	0	0	0	0	0	0	0	0	0	0	0	0	0
ENG. FAC. PACE	2	18	20	20	45	82	79	60	100	160	172	102	35	0	0
FERF/ETR	0	0	0	0	0	0	0	25	75	150	150	75	25	0	0
HFNS	0	0	10	15	20	20	10	0	0	0	0	0	0	0	0
HTTF	0	0	0	0	0	6	9	0	0	0	6	9	0	0	0
TF	0	0	0	5	10	10	10	5	5	0	0	0	0	0	0
B&S	2	3	0	0	5	20	20	10	0	0	0	0	0	0	0
RTNS	0	15	10	0	0	0	0	0	0	0	0	0	0	0	0
INS	0	0	0	0	0	3	5	0	0	0	0	0	0	0	0
PMCTF	0	0	0	0	0	3	5	0	0	0	3	4	0	0	0
VTF	0	0	0	0	0	0	10	10	10	0	3	4	0	0	0
SMTF	0	0	0	0	0	0	10	10	10	0	0	0	0	0	0
Eng. Test. Fac.	0	0	0	0	10	20	0	0	0	10	10	10	10	0	0
ALT. CONC. PACE	0	0	15	35	35	60	105	105	65	60	120	140	120	140	120
LHX	0	0	15	35	35	15	0	0	0	0	0	0	0	0	0
MX	0	0	0	0	0	15	35	35	15	0	0	0	0	0	0
LSS	0	0	0	0	0	15	35	35	15	0	0	0	0	0	0
#3	0	0	0	0	0	15	35	35	15	0	0	0	0	0	0
#4	0	0	0	0	0	0	0	0	20	60	120	120	60	0	0
M-PEPR 1	0	0	0	0	0	0	0	0	0	0	0	20	60	20	0
A-PEPR 2	0	0	0	0	0	0	0	0	0	0	0	0	0	120	120
EPR	0	0	0	0	0	0	0	0	0	0	0	0	0	0	0
TOTAL PACE	22	98	130	105	145	257	324	270	210	235	332	362	395	380	240
OPERATIONS	120	183	248	280	321	346	376	390	400	410	420	430	440	450	460
EQUIPMENT	17	23	32	45	55	45	55	55	40	41	42	43	44	45	46
TOTAL PROGRAM	159	304	410	430	521	648	755	715	650	686	794	835	879	875	746

*PACE: Plant and Capital Equipment line item construction projects.

Figure II-14 - continued

	FY91	FY92	FY93	FY94	FY95	FY96	FY97	FY98	21yr. Total FY78-98
TOKAMAK PACE									
TFTR	40	60	120	240	360	240	120	60	2630
	0	0	0	0	0	0	0	0	230
PEPR/ITR Fac.	0	0	0	0	0	0	0	0	100
PEPR/ITR Dev.	0	0	0	0	0	0	0	0	300
EPR	40	0	0	0	0	0	0	0	800
DEMO	0	60	120	240	360	240	120	60	1200
ENG. FAC. PACE									
FERF/ETR	0	0	0	0	0	0	0	0	875
	0	0	0	0	0	0	0	0	500
HFNS	0	0	0	0	0	0	0	0	75
HTTF	0	0	0	0	0	0	0	0	30
TF	0	0	0	0	0	0	0	0	50
B&S	0	0	0	0	0	0	0	0	50
TRNS	0	0	0	0	0	0	0	0	0
INS	0	0	0	0	0	0	0	0	10
PMCTF	0	0	0	0	0	0	0	0	15
VTF	0	0	0	0	0	0	0	0	15
SMTF	0	0	0	0	0	0	0	0	30
Eng. Test Fac.	0	0	0	0	0	0	0	0	100
ALT. CONC. PACE	100	140	240	240	120	40	0	0	2000
LHX									
MX	0	0	0	0	0	0	0	0	100
LSS	0	0	0	0	0	0	0	0	100
#3	0	0	0	0	0	0	0	0	100
#4	0	0	0	0	0	0	0	0	100
M-PEPR 1	0	0	0	0	0	0	0	0	400
A-PEPR 2	60	20	0	0	0	0	0	0	400
EPR	40	120	240	240	120	40	0	0	800
TOTAL PACE	140	200	360	480	480	280	120	60	5505
OPERATIONS	470	480	490	500	510	520	530	540	9017
EQUIPMENT	47	48	49	50	51	52	53	54	992
TOTAL PROGRAM	657	728	899	1030	1041	852	703	654	15514

608

Figure II-15   PROGRAM COSTS BY YEAR FOR LOGIC IV REFERENCE OPTION ($M)

	FY76	FY77	FY78	FY79	FY80	FY81	FY82	FY83	FY84	FY85	FY86	FY87	FY88	FY89	FY90	FY91	FY92	FY93	15yr. Total FY78-93
TOKAMAK PACE *	20	80	95	35	55	115	195	95	120	180	280	180	140	180	360	360	180	60	2630
TFTR	20	80	95	35	15	35	35	15	0	0	0	0	0	0	0	0	0	0	230
PEPR/ITR	0	0	0	0	40	80	160	80	40	0	0	0	0	0	0	0	0	0	400
EPR	0	0	0	0	0	0	0	0	80	180	280	180	80	0	0	0	0	0	800
DEMO	0	0	0	0	0	0	0	0	0	0	0	0	60	180	360	360	180	60	1200
ENG. FAC. PACE	2	18	25	50	102	129	125	140	230	132	87	20	10	0	0	0	0	0	1050
PERF/ETR	0	0	0	0	0	0	50	100	200	100	50	0	0	0	0	0	0	0	500
HFNS	0	0	15	30	40	40	25	0	0	0	0	0	0	0	0	0	0	0	150
HTTF	0	0	0	0	6	9	0	0	0	6	9	0	0	0	0	0	0	0	30
TF	0	0	0	5	10	20	10	5	0	0	0	0	0	0	0	0	0	0	50
B&S	2	3	0	5	10	20	10	5	0	0	0	0	0	0	0	0	0	0	50
RTNS	0	15	0	0	0	0	0	0	0	0	0	0	0	0	0	0	0	0	0
INS	0	0	0	0	3	0	0	0	0	3	4	0	0	0	0	0	0	0	10
PMCTF	0	0	0	0	5	3	0	0	0	3	4	0	0	0	0	0	0	0	15
VTF	0	0	0	0	8	7	0	0	0	0	0	0	0	0	0	0	0	0	15
SMTF	0	0	0	10	20	0	0	0	0	0	0	0	0	0	0	0	0	0	30
Eng. Test. Fac.	0	0	10	0	0	30	30	30	30	20	20	20	10	0	0	0	0	0	200
ALT. CONC. PACE	0	0	20	50	90	150	90	40	80	200	160	200	160	200	320	160	80	0	2000
LHX																			
MX	0	0	20	50	30	0	0	0	0	0	0	0	0	0	0	0	0	0	100
LSS	0	0	0	0	20	50	30	0	0	0	0	0	0	0	0	0	0	0	100
#3	0	0	0	0	20	50	30	0	0	0	0	0	0	0	0	0	0	0	100
#4	0	0	0	0	20	50	30	0	0	0	0	0	0	0	0	0	0	0	100
M-PEPR	0	0	0	0	0	0	0	40	80	160	80	40	0	0	0	0	0	0	400
A-PEPR	0	0	0	0	0	0	0	0	0	40	80	160	80	40	0	0	0	0	400
EPR	0	0	0	0	0	0	0	0	0	0	0	0	80	160	320	160	80	0	800
TOTAL PACE	22	98	140	135	247	394	410	275	430	512	527	400	310	380	680	520	260	60	5680
OPERATIONS	120	200	270	350	420	470	490	510	530	540	550	560	570	580	590	600	610	620	8260
EQUIPMENT	17	31	27	35	42	47	49	51	53	54	55	56	57	58	59	60	61	62	826
TOTAL PROGRAM	159	329	437	520	709	911	949	836	1013	1106	1132	1016	937	1018	1329	1180	931	742	14766

*PACE: Plant and Capital Equipment line item construction projects.

Figure II-16 PROGRAM COSTS BY YEAR FOR LOGIC V REFERENCE OPTION ($M)

	FY76	FY77	FY78	FY79	FY80	FY81	FY82	FY83	FY84	FY85	FY86	FY87	FY88	FY89	FY90	13yr. Total FY78-90
TOKAMAK PACE*	20	90	110	230	465	485	385	385	370	150	160	320	600	320	160	4140
TFTR	20	90	110	30	15	35	35	15	0	0	0	0	0	0	0	240
PEPR/ITR	0	0	0	80	180	180	80	0	0	0	0	0	0	0	0	520
EPR I	0	0	0	120	270	270	120	0	0	0	0	0	0	0	0	780
EPR II	0	0	0	0	0	0	150	370	370	150	0	0	0	0	0	1040
DEMO	0	0	0	0	0	0	0	0	0	0	160	320	600	320	160	1560
ENG. FAC. PACE	2	18	40	130	190	260	385	335	180	70	50	50	20	0	0	1710
FERF/ETR	0	0	0	0	0	100	225	225	100	0	0	0	0	0	0	650
HFNS	2	18	30	70	80	60	40	20	0	0	0	0	0	0	0	300
HTTF	0	0	0	10	10	0	10	0	10	0	0	0	0	0	0	40
TF	0	0	0	10	20	20	15	0	0	0	0	0	0	0	0	65
B&S	0	0	10	20	20	15	0	0	0	0	0	0	0	0	0	65
RTNS	0	0	0	0	0	0	0	0	0	0	0	0	0	0	0	0
INS	0	0	0	0	0	0	5	5	0	0	0	0	0	0	0	10
PMCTF	0	0	0	0	0	10	0	0	0	10	0	0	0	0	0	20
VTF	0	0	0	0	10	0	10	0	0	0	0	0	0	0	0	20
SMTF	0	0	0	0	10	5	0	5	10	10	0	0	0	0	0	40
Eng. Test. Fac.	0	0	0	20	40	50	80	80	60	50	50	50	20	0	0	500
ALT. CONC. PACE	0	0	104	364	312	156	520	728	520	156	312	728	728	312	0	4940
LHX	0	0	52	78	0	0	0	0	0	0	0	0	0	0	0	130
MX	0	0	52	78	0	0	0	0	0	0	0	0	0	0	0	130
LSS	0	0	0	52	78	0	0	0	0	0	0	0	0	0	0	130
#3	0	0	0	52	78	0	0	0	0	0	0	0	0	0	0	130
#4	0	0	0	52	78	0	0	0	0	0	0	0	0	0	0	130
#5	0	0	0	52	78	0	0	0	0	0	0	0	0	0	0	130
#6	0	0	0	0	0	0	0	0	0	0	0	0	0	0	0	0
M-PEPR	0	0	0	0	0	78	182	182	78	0	0	0	0	0	0	520
M-PEPR	0	0	0	0	0	78	182	182	78	0	0	0	0	0	0	520
A-PEPR	0	0	0	0	0	0	78	182	182	78	0	0	0	0	0	520
A-PEPR	0	0	0	0	0	0	78	182	182	78	0	0	0	0	0	520
EPR	0	0	0	0	0	0	0	0	0	0	156	364	364	156	0	1040
EPR	0	0	0	0	0	0	0	0	0	0	156	364	364	156	0	1040
TOTAL PACE	22	108	254	724	967	901	1290	1448	1070	376	522	1098	1348	632	160	10790
OPERATIONS	120	240	360	510	600	640	660	680	690	700	710	720	730	740	750	8490
EQUIPMENT	17	31	36	51	60	64	66	68	69	70	71	72	73	74	75	849
TOTAL PROGRAM	159	379	650	1285	1627	1605	2016	2196	1829	1146	1303	1890	2151	1446	985	20129

*PACE: Plant and Capital Equipment line item construction projects.

Possible Options and Strategy of the European Tokamak Programme

G. Grieger

Commission of the European Communities,
Fusion Programme, Brussels (Belgium)

In this paper the European fusion programme and its probable evolution
will be analysed and explained. First, what are the options for such
a programme, or better, what potential is available in the European
Community as basis for the future evolution of the programme? The
answers fall into three categories:

(1) Nearly all of the nine member states of the Community are running
fusion labs, some of them being rather large. The total number of
professionals working in these labs is 750. In addition, Sweden has
just joined the Community's fusion programme and Switzerland will
probably follow soon.

(2) Concerning the future reactor technology programme, an even larger
number of experienced staff exists in the large fission centres of
the Community which, at least in principle, would be very appropriate
to take care of many of the technological questions simultaneously
with transferring their resources to the fusion programme, and indeed,
such a development has started already; and

(3) Europe's industry has proved its strength and potential often enough.

Therefore the conclusion is that there is a large potential of highly
qualified staff, sufficient to tackle all essential fusion problems;
but the question remains to which extent this potential can be made
available for the benefit of the fusion programme, or in other words
which amount of money one is willing to spend for the development of
a fusion reactor, and at which rate. This, however, is a purely
political question dependent on which position fusion has in a much
more general energy R and D policy. Nobody would doubt that the
Community as a whole would be capable of bearing the required expenses,
provided fusion ranked high enough in energy R and D planning. This,
however, is still a matter of dispute, and as yet Europe has not put
itself in a position to follow the rather aggressive US strategy.

In such a situation it is the task of Europe's scientific fusion
community to provide the decision-making bodies with all the necessary
information, and to push towards early decisions. One of the related

questions is Europe's need for fusion power compared to that of the US or the USSR, for instance. Such a comparison yields that

(1) the Community has much smaller resources of conveniently available fossil fuels such as oil and gas,

(2) the Community has also poor resources of fissile fuel, so that its present dependence on energy import could not be significantly reduced by this means, and

(3) the Community has a lower and much more variable sunshine intensity, making its use rather difficult by requiring either large storage capacities or back-up power plants.

But, on the other hand, the Community has

(4) a population density larger by a factor of 7—8 than that of the US, which requires particular attention in selecting the proper energy-producing system and in its public acceptance;

(5) a high degree of industrialization and a power consumption density already four times as large as that of the US. (In this connection it is worth noting that the consumption of primary energy already exceeds the geothermal flux in Europe by a factor of six, which gives some measure for the importance of the other natural energy sources).

(6) The Community is, and will remain, very much dependent on the ability to export highly industrialized goods, perhaps including even fusion reactors.

These arguments unavoidably lead to the conclusion that, provided fusion can prove itself to be a powerful energy source, Europe should have an even greater interest in seeing it fully developed than, for example, the US. It shares this need with Japan. It would be wrong, though, to advertise fusion as the only meaningful candidate, but it should have a strong position in the future energy scene.

It is more difficult to argue for the target date of its introduction. Energy independence seems to be a wish impossible to satisfy for Europe during the next decades. But already from the limited resources of fossil fuel there seems to develop a need for an additional energy source with the beginning of the next century. The selection of such a target date is strongly favoured also by a comparison with the US fusion programme, which aims at the year 2000 for the start of operation of a demonstration reactor.

612

A much slower European programme arriving at a fusion reactor let us say 20 years later than the US is perhaps not worth the money. But assuming a time scale similar to the US offers large advantages through cost- and risk-reducing cooperations. The Community should therefore select the same date for achieving its target.

The Community's Tokamak Programme

The Community's fusion programme is run as a series of five year programmes. The present one covers the years 76—80 and comprises 415 Mua not including the costs for the next step, the jointly built and operated large toroidal experiment JET.

The Community's programme is widely coordinated, and the means to achieve this became available by the introduction of preferential support for such new investments which rank high in priority in the programme. These experiments obtain a financial contribution by the Commission of 45% of their investment costs, whereas others get only up to 25%.

It is perhaps not fully recognized that the Community has concentrated its fusion activity to 85% on toroidal magnetic confinement, 65% alone in the tokamak field. These figures hold true if one relates them to the distribution of the professionals among the different lines and includes all the preparatory work for injection and heating.

According to the underlying strategy for the Community's tokamak programme for the near future JET will be its central part. But it is nevertheless felt that a number of special and even basic questions are better to be investigated in somewhat smaller experiments than JET. This allows to use more flexible machines, to study these questions in parallel and to make the machines available to the different groups of experts. JET will then extend their findings closer to the reactor regime.

In fig. 1 the following problem areas are listed on the left-hand side:

(1) How do the plasma parameters scale with increasing size of the devices, particularly when entering new reactor relevant regimes with particle mean free paths long compared to typical connection lengths?

613

(2) What are the attainable limits of q and Beta with respect to plasma
     stability? Are there possibilities for improvement by optimization
     of the confinement configurations, e.g. by shaping of the plasma
     cross-section?

(3) Are there significant advantages of high magnetic field and, conse-
     quently, high plasma current density systems compensating the higher
     technical burden of such arrangements?

(4) and (5) How can the introduction of impurities be minimized

     (4a) by divertors, how can they be implemented in the tokamak scheme?

     (4b) by a cold gas blanket; is this concept feasible?

     (5) what are the mechanisms and implications of plasma wall interact
          and how can the interaction be minimized?

(6) How can the plasma state and its radial variation be effectively
     controlled and optimized?

(7) How can the electron run-away development be minimized?

(8) What are the most powerful and effective methods of additional
     heating, and how do they physically interact with the confined plasma.

(9) What are the best methods for refuelling?

(10) Demonstration of nuclear heating.

(11) Development and demonstration of the feasibility of a superconducting
     magnetic field assembly able to be operated under tokamak conditions.

(12) Build-up of experience in tritium handling and induced radioactivity.

     On the right-hand side devices either existing or being built or
     planned are listed, and the connections between left and right indica
     which are the major questions tackled by them. In this connection it
     is understood that a clear separation of problems is usually impossib

     It is believed that these more basic and goal-oriented investigations
     lead to a more concentrated operation of JET and to a better definiti
     of its successor, whose definition certainly has to start before the
     experimental information from JET is fully available.

     As far as the technology argument is concerned, we have two conflict
     arguments: on the one hand it is true that we are still facing quite
     considerable plasma physics problems which require attack on a broad
     front, and the results of these investigations might lead to such

changes of the picture that any technology programme one might select now could be strongly affected. On the other hand it is also convincing that technology will finally have a decisive influence on the properties of a fusion reactor and that one cannot start early enough to assess its possibilities. Up to now the approach of the Community to this problem was more pragmatic than systematic, and only now activity on long term planning is starting. This was stimulated by a growing confidence in fusion, and a growing support for a gradual transition to a technology programme. This already finds its ex – pression in the decision of the European Council of Ministers that during this five year programme about 15–20% of the means available for preferential support should be spent in the field of technology, and there is a chance of modifying this figure during the programme revision envisaged for 78–79.

The first major decision in the field of technology was the implementation of a superconducting magnet programme. In its first phase, during 76–77, it comprises 56 professional man years and a financial volume of 7.5 MUA. It aims at the development of superconducting coils for a post-JET experiment. In this field the Community favours, without weakening its own programme, a strong cooperation with other countries through the IEA. Such a cooperation was already started with the INS, an intense neutron source, which is going to be constructed in Los Alamos, USA. Other parts of the technology programme will follow, some of them would also be intended to be performed in a cooperative manner. Smaller experiments in the fields of tritium handling, blanket technology and material behaviour are already running.

In summary, the present strategy of the Community's tokamak programme is:

(1) Attack the basic questions on a broad front in order to come to reliable extrapolations.

(2) Build JET in order to get closer to reactor conditions and to check the extrapolations.

(3) Start the definition of post-JET.

(4) Develop the technologies required for the next step.

(5) Start long term planning and, in consequence, a long term programme.

(6) Increase cooperation in well defined areas in order to save time and money.

615

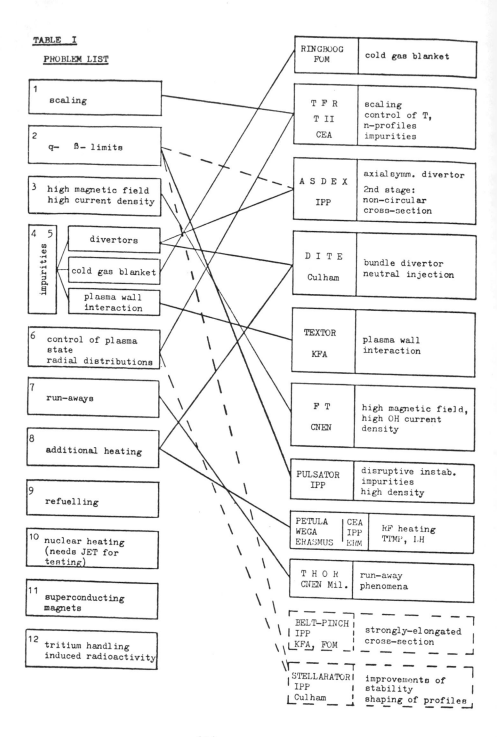

TABLE I

PROBLEM LIST

RINGBOOG FOM	cold gas blanket

| 1 scaling | | T F R T II CEA | scaling control of T, n-profiles impurities |

| 2 q- ß- limits | | A S D E X IPP | axial symm. divertor 2nd stage: non-circular cross-section |

| 3 high magnetic field high current density | | D I T E Culham | bundle divertor neutral injection |

| 4 5 impurities: divertors / cold gas blanket / plasma wall interaction | | TEXTOR KFA | plasma wall interaction |

| 6 control of plasma state radial distributions | | F T CNEN | high magnetic field, high OH current density |

| 7 run-aways | | PULSATOR IPP | disruptive instab. impurities high density |

| 8 additional heating | | PETULA CEA WEGA IPP ERASMUS EHM | RF heating TTMP, LH |

| 9 refuelling | | T H O R CNEN Mil. | run-away phenomena |

| 10 nuclear heating (needs JET for testing) | | BELT-PINCH IPP KFA, FOM | strongly-elongated cross-section |

| 11 superconducting magnets | | STELLARATOR IPP Culham | improvements of stability shaping of profiles |

| 12 tritium handling induced radioactivity | |

INTRODUCTION

In this short lecture, we will discuss only two types of reactors, the liquid metal fast breeder reactors (LMFBR) and the high temperature reactors (HTR). This does not mean that other very interesting concepts do not exist, but there are or proven light water reactors and heavy water reactors or has not reached the state of industrial development like molten-salt or gas breeder reactors.

In discussing any types of industrial development , it seems to me useful, first to indicate the reasons or motivations for this development. Then I will give a short historical review and analysis of what has been done up to now.

For HTR's a very brief status report will be presented. For LMFBR's, I will give indications of experience gained with demonstration plants and more specifically with Phenix, before listing the most important technical problems which still need more work to be fully solved. Finally, I will briefly discuss the economic status and perspectives of LMFBR's and will mention the public acceptance problem.

I. - MOTIVATIONS FOR HTR's DEVELOPMENT AND THEIR STATUS

I think that one can identify three main raisons, which were instrumental for the development of HTR.

1) One would like to have a reactor being able to compete for electricity production with already established reactors, like LWR.

2) One hopes to have a reactor which uses more efficiently natural ressources (Uranium and Thorium) than LWR.

3) One would like to have a reactor which is better suited for other
energetical uses than production of electricity: high temperature
process heat, production of hydrogen...

Status of HTR's Development

The development of HTR took some benefit of earlier development of Magnox
type reactors in United Kingdom and France, and also from the advanced
gas reactors program in United Kingdom.
The most specific achievements are the successful construction and operation of experimental reactors like Peach Bottom Reactor in United States
and similar development in West Germany.

The next step -the demonstration power plants- includes two plants of
300 MW class: . Fort St. Vrain power plant in the United States, which
after some delays in construction and startup is on the verge to achieve
nominal power output,

. The Schmehausen power plant in West Germany, which has
also experience of some delays in construction and is due for industrial
operation in two or three years from now.

As regarding commercial plants, one should first mention the very ambitious
program that General Atomic has developed in past years. This program culmi-
nated in firm orders from US-Utilities totalling some 8000 MWe of installed
capacity. Unfortunately, the delay in Fort St. Vrain plant and some increases in expected costs, obliged General Atomic to drop this program and to
withdraw from the purely commercial arrangements.

Presently, what remains is a program of development of commercial reactors
under discussion between General Atomic - US-ERDA and US-Utilities, which
may lead, in the best hypothesis, to operation of a large prototype power
plant between 1985-1990.

In parallel, there exists similar program in West Germany, which may be carried
out in co-operation with France. An agreement was signed this year between
both Governments in view to cooperate in this field. The most optimistic
target that one can forsee, is also an operation of a prototype large power

618

plant during the end of next decades.

One should mention also the strong interest of Japan for HTR and also the rather tight link existing between Europe and US-programs in HTR' field (technical exchange agreements, licensing agreements, industrial participation), which may mutually reinforce both programs.

In looking now critically on the motivations mentioned above for the development of HTR, I may make the following remarks:

1) the commercial competitivity with "proven reactors" is not established for the moment and may take some time before it will be established;

2) the saving in ressources (natural uranium) does not appear as a major incentive for the development of HTR. The calculation shows that with given demand of energy, the massive introduction of HTR in place of LWR will give only few years of delay before reaching same demand for natural uranium. One has to remember that the essential ressources are U235, so the replacement of Uranium by Thorium does not change drastically the picture of ressources.

3) The third motivation of using HTR for production of process heat, of hydrogen, seems still valid and even very important but only for the long range when one would like to use nuclear energy on a very large scale and when the electricity as a vector, will be not sufficient.

One can then conclude that there is still strong interest in developing HTR but perhaps with less sense of urgency than for other developments.

II.- MOTIVATIONS FOR LMFBR's DEVELOPMENT

One can identify two mains motivations for the development of breeder reactors and more specifically LMFBR.

1) The efficient use of natural ressources which may extend the role played by fission energy in satisfying the energy demand, from decades to thousands of years.

2) To maintain a relatively low cost of fission energy for near (next decades) and more distant future.

619

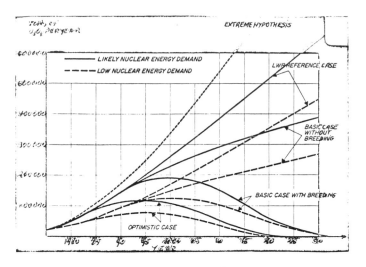

FIGURE 1 : Natural Uranium Demands

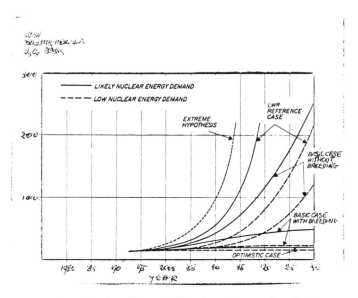

FIGURE 2 : Natural Uranium Extraction Costs

The first motivation mentioned above does not need many comments. The 50 times more efficient use of natural uranium with breeding in addition to possible use of thorium, give a pratically unlimited supply of fission energy. One has to note that it is possible to use relatively expensive uranium ories when the breeding is achieved. Professor STARR and Professor HÄFELE have shown in BNES-Journal that the potential ressources of fission energy are pratically unlimited.

Therefore, if one develop commercially valuable breeders, one has a reference system for producing energy and one can develop other perhaps more advantageous systems (fusion - solar), without undue pressures on schedule.

Now, coming to the second motivation, we can see on Figure 1, the annual requirement for natural uranium both with breeders and without. Limits for discovery and extraction of uranium have been indicated by some specialists as 150.000 tons/year, and by others as 200.000 tons/year.
. Without: likely demand curve reaches the first level in 1990; low demand curve reaches it in 2000. They reach the second limit in 1994 and 2012. Thus, discovery and extraction may become a physical problem as early as 1990, or as late as 2012. The market-place, however, will see this well in advance of either date, and probably it would react.
. With breeders: the 150.000 ton/Year limit is exceeded in only one case. The second limit is never exceeded.

In Figure 2, one can see the likely extraction cost of natural uranium with and without breeders. The most important features of this figure are the trends it indicates, rather than the absolute numbers it shows. It is recognized that there is uncertainty in the calculations, but the trends would hold at either higher or lower values. The dollars values shown are cost and not market prices. (All prices are given in constant(1975) $). If only thermal reactors are deployed, cost per pound will exceed 100$ or 200$ or even much more by 2030, depending on the assumption of nuclear energy demand and of the type of reactors used. With breeder reactors, it can be seen that costs tend to level out at moderate figures of less than 60 $/lb.

It should be noted that in times of economic stability prices may be 30-50% above extraction costs, but in the event of crisis, prices are likely to rocket upward. They could reach 10 times extraction costs, and this feature can be seen in the experience with petroleum.

621

DATE OF OPERATION	TYPE OF REACTOR	NAME OF REACTOR	POWER Thermal	Elect.
1948		Clementine	150 KW	
	Feasibility			
1951		EBR$_1$	1200 KW	200 MWe
1960		BR$_5$	5 MWt	
		DFR	60 MWt	15 MWe
	Experimental	EBR$_{II}$	60 MWt	20 MWe
		Rapsodie	40 MWt	—
1970		BOR$_{60}$	60 MWt	—
1973		BN$_{350}$	1000 MWt	150MWe ✱
	Demonstration	Phenix	590 MWt	250MWe
1976		PFR	600 MWt	255MWe

✱ + dessalination

FIGURE 3 : Breeder History

All this shows that if the breeders are not commercially available in a very near future (around 1990), the cost of natural uranium and in turn of nuclear energy, will very likely increase drastically.

III- BREEDERS HISTORY

We have summarized the development and history of breeders in Fig.: 3. On this table, one can note that between the demonstration of feasibility which took place in late 1940[5] and the demonstration of technical possibility of producing electricity in large plants (250 MWe class plants) which are now just on their way, 25 to 30 years have gone.

Without going in detailed discussion in experience gained, I would like to mention the very successful operation of experimental reactors and more precisely of the first pool type reactor in the world: EBR2, which is operating with a very high availability factor, and also the operation of the first experimental russian reactor BR5, which has demonstrated the possibility of using plutonium oxides fuel and a good compatibility of this fuel with sodium.

One other feature which is of interest, is the large agreement between different independant teams which has developed demonstration plants. All have selected sodium as coolant mixed oxide (UO2-PuO2) as fuel and stainless steel as cladding. The ranges of temperature for sodium and steam are also similar. Therefore, the experience gained in operating the demonstration plants: BN350, Phenix and PFR will be mutually beneficial.

IV.- LMFBR' FUTURE PROGRAMS

In Fig.: 4, we have represented the reactors under construction or main plans which are known to us. This table, as also Table 3, do not pretend to give a full and complete description of the situation, but rather give the main facts allowing to have a general opinion of the present status and possible development of LMFBR's.

V.- EXPERIENCE GAINED WITH DEMONSTRATION PLANTS

The operation of BN350 and PFR reactors was successful, with however, up to now, some difficulties mainly connected with steam generators.

623

Date of Operation	Type of Reactor	Name of Reactor	POWER MWt	MWe
1977 〜 1980	Experimental	KNK Fast (1)	40	
		JOYO (1)	100	
		FFTF (1)	400	
		PEC (1)	140	
		Indian (1)	40	
1980 〜 1985	Demonstration	BN 600 (1)		600
		SNR 300 (1)		300
		C.R.B.R. (2)		350
		MONJU (2)		300
1982 1984 1985 1986	Prototype or near Commercial	SUPER PHENIX (2)		1200
		C.F.R. (2)		1300
		SNR2 (2)		1200 +
		P.L.B.R. (2)		900-1200
1985	Commercial	SUPER PHENIX$_I$ (2)		2 x 1800 *
		SUPER PHENIX$_{II}$ (2)		2 x 1800 *

(1) Under construction

(2) Planned

* The exact out-put of this plant is not decided but will be comprised between 1200 and 1800 MWe.

FIGURE 4 : Breeders under Construction or Planned

One expects that PFR will produce nominal power output by the end of 1976;
BN350 to my best knowledge, has not yet reached up to now, full nominal
power, but however, has exceeded 60% of nominal power.

I will rather concentrate on the experience with Phenix, which I know better
than the experience with other reactors, but which, as I have already indicated, is mutually beneficial to other programs and may be taken as an
example of experience presently gained.

Phenix experience may be summarized in Fig.: 5. The reader can judge by
himself; looking at this very general figures, of the experience gained
with this prototype reactor, but which is in our opinion, quite favorable.

1) Description

Several times, the Phenix plant has been already described and we shall
just remind here its main characteristics:
- liquid sodium fast breeder, with power output of 250 MWe,
- fuel assemblies made of pins of plutonium and uranium oxides with 316
  stainless steel cladding,
- pool type reactor with 3 main circulating pumps and 6 intermediate heat
  exchangers immersed in a single vessel filled with sodium (see Fig. 6)
- secondary circuits with 3 independant inactive sodium loops each feeding
  a steam generator (see Fig. 7), each steam generator is constitute by
  12 modules.

2) Experience of construction and start up
   a) Schedule
Slightly more than 4 years elapsed between the start of preliminary work
related to site preparation in autumn 1968 and the filling of the circuits
with sodium at the end of 1972. As one can see on Fig. N° 8, the difference
between the initial forecast and the actual schedule, is of 5 months on a
5 years program. This testifies sufficiently that the difficulties encountered during the construction were not fondamental and the component manufacturing did not reveal very serious problems.

   b) Cost
The cumulative investment cost of Phenix in current francs from 1969 to 1973
amounts to 759 millions francs, excluding taxes, which can be broken down as
follows:

	Planned	Achieved	Difference %
Time for construction (Month) and start up	64	69	+8
Cost of construction (10^6 FF) including engineering	546	596	+9
Maximum burn up (MW days/Kg)	50	65	+ 30
Maximum power (MWe)	250	270	+8
Plant capacity factor * during the first 21 months of industrial operation (%)		67	

* One can note that in France for planning purposes the utility expects
for the two first years of operations of a proven nuclear power plant,
a plant capacity factor of 46% and for a prototype 26%.

FIGURE 5 : Phenix Experience

CUT-AWAY VIEW

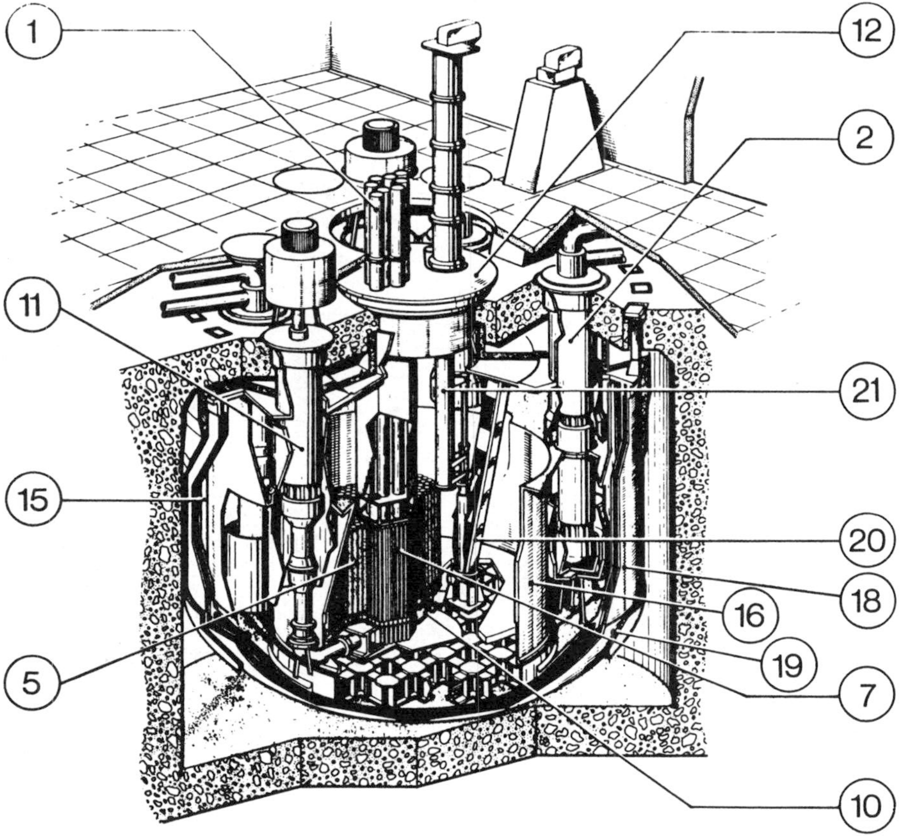

REACTOR BLOCK

1 control rod drives
2 intermediate exchanger
3 leak detector
4 upper neutron shielding
5 lateral neutron shielding
6 blanket
7 core
8 lat. shielding support
9 conical support collar
10 fuel support slab
11 primary pump
12 rotating plug
13 slab
14 roof
15 main vessel
16 primary vessel
17 core cover
18 double envelope vessel
19 primary containment
20 transfer ramp
21 transfer arm

FIGURE 6 : Phenix Reactor Bloc

627

PHENIX

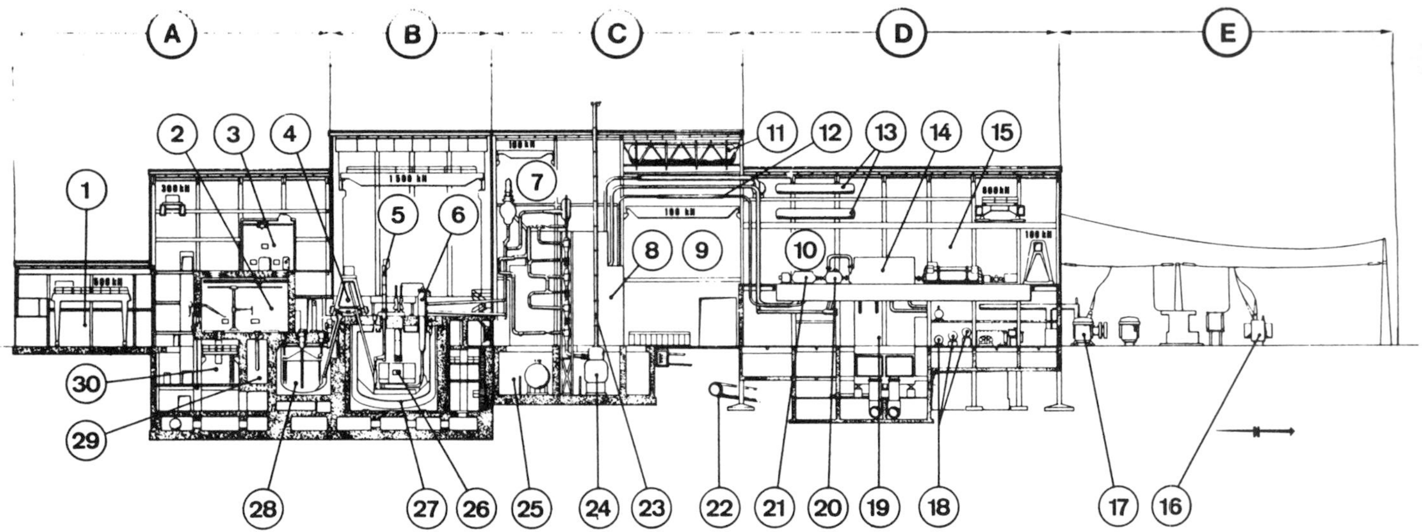

PLANT LONGITUDINAL SECTION

A fuel handling building
B reactor building
C steam generator building
D turbine hall
E switch-yard
1 stores
2 spent element cell
3 upper cell
4 loading/unloading lock
5 transfer arm
6 primary sodium pump
7 secondary sodium pump
8 steam generator
9 dismantling area for steam generator modules
10 turbine
11 buffer tanks
12 steam collectors
13 H.P. reheaters
14 L.P.
15 alternator
16 local power supply transformer
17 grid supply transformer
18 L.P. reheaters
19 condenser
20 M.P.
21 H.P.
22 condenser cooling water
23 discharge stack
24 H/Na separator
25 secondary sodium storage tank
26 core
27 reactor block
28 storage drum
29 spent element cleaning area
30 cask exit

FIGURE 7 : Phenix

628

– Engineering and construction	596 million francs
– Testing	18 million francs
– Fuel fabrication including plutonium purchases	145 million francs
	759 million francs

The engineering and construction cost estimated in 1968 amounted to 417 million francs,excluding test, fuel taxes, contingencies and provision for escalation. Of the discrepancy of 179 million francs (596-417), 129 million are due to cost escalation from 1968 to 1969-1973 period. The remainder 50 million francs are due to modifications and technical problems encountered during construction. This increase was thus limited to less than 10% of the total.

c) Operating experience

On the Figure 9, one can see the diagram of the first 18months of industrial operation. We can observe here that the fuel handling campaign takes place roughly every two months with a replacement of 1/6th of the core.

During the first year, July 74-July 75, the performances were rather high: the plant capacity factor achieved was 78%. There were few minor incidents, the main ones being 2 small sodium leaks on a secondary loop.

d) Analysis of incidents

Let us now examine the operation incidents. They represent for 1.5 year a 15,8% loss of capacity generation, very similar to the loss due to fuel handling and programmed maintenance (see Fig. 10).

The main losses, connected with the incidents, are due to leaks in water steam circuits (8%), then to the sodium leaks in secondary circuits (4,3%), the other causes being various and rather minor, but all connected with conventionnal equipment.

As for the conventionnal part of the water-steam circuit, several minor incidents occured on valves and we tried to improve the situation by suppressing a certain number of them.

One can note here that the turbo-generator used in Phenix plant is a conventional one; more of 35 similar turbo-generator systems operate on Electricité de France grids in connection with fossil fuel plants.

629

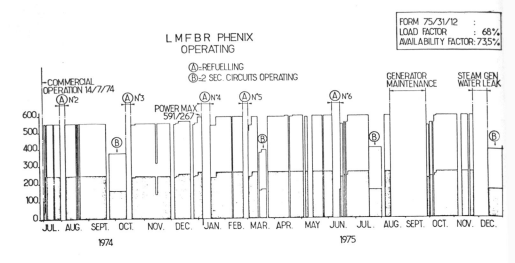

FIGURE 8 : Phenix Construction Planning

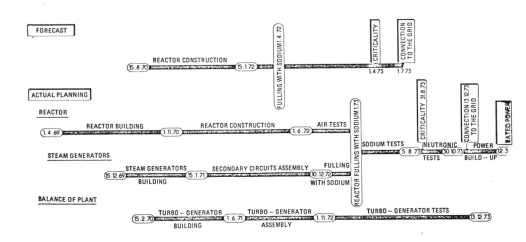

PHENIX CONSTRUCTION PLANNING

FIGURE 9 : LMFBR Phenix Operating

ANALYSIS OF UNAVAILABILITY DURING 1.5 YEAR OF INDUSTRIAL OPERATION
===

(from 1974.7/14 to 1975.1/14)

	Fraction of unavailability	Number of unscheduled shutdown
Fuel handling and plant maintenance	16.1 %	--
Control, monitoring and regulation	1.9 %	27
Water and steam leaks	8 %	9
Sodium leaks (Secondary Na loops N° 2)	4.3 %	3
Operator mistakes	.4 %	6
Miscellaneous (Electrical grid-thunder-tests)	1.2 %	1
TOTAL	31.9 %	46

FIGURE 10 : Analysis of Unavailability during 1.5 year of Industrial Operation

e) Behaviour of main components

. Fuel

The core had been designed for a burn up of 50 000 MWD/T. Now the highest
burn up obtained exceeds this figure by more than 30% and we never had
real difficulties in fuel handling.

No fuel pin rupture occured till now. We had only a small intermittent
fission gases leakage starting in June 1975, giving from time to time
small and short duration, increases, in cover gas activity which did not
create any difficulties for operation; we never had signals on delayed
neutron monitors.

One fuel assembly has been extracted from the core at every step of
10 000 MWD/T to be examined and measured so as to decide the possibility
of going to higher burn ups. Presently, the operators of Phenix are
authorized to achieve fuel burn ups up to 65.000 MWD/T.

On the other hand, several experimental fuel assemblies are under irradiation in view to improve   the fuel performances in the next plants
and first in Super Phenix.

. Pumps

performed more than 17000 hours of operation for the primary pumps,
20 000 hours for the secondary, with no failures during the industrial
operation.

. Steam generators

have operated satisfactorily: the global hydrogen diffusion rate background for the hydrogen monitoring apparatus, as foreseen, has decreased
from 500 mg/h per steam generator and stabilized at 250 mg/h. No single
one sodium or water leak in the steam generator itself was detected up
to now.

. The fuel handling system

performed 9 fuel handling campaigns without significative difficulties.
About 400 loadings, 350 unloadings and 500 transfers were performed during
these fuel handling campaigns.

General remarks

From the point of view of general results, and after 21 months of operation,
we can make the following comments:

- very good stability of the plant only a few adjustments per day are
  necessary,
- the shut down and start up procedures have been improved: for example
  for the refueling campaigns:
    . 8 hours from full power to the refueling conditions
      (shut down at 250°C)
    . 24 hours from shut down at 250°C to full power,
- the secondary circuits and steam generators allowing cooling by air with
  natural convection gives good possibilities for heat dissipation and
  disconnection from the water circuits which is very useful for their
  maintenance,
- the independance of the three secondary circuits allows easy operation
  with two loops at 2/3 of nominal power, which minimizes the loss of power
  generation due to an incident on one loop,
- as we are in a time where the environment considerations are taking every
  day a greater importance, it is interesting to note that the operation of
  Phenix confirms the advantage of the LMFBR in this field:

    . the gaseous radioactive effluents are very low (generally less than
      0.35 Ci/day). In one year, .16% of the allowed release;
    . very small liquid effluents: 8,5 Ci in 1.5 year,
    . the global thermal efficiency is 45%,
    . as for the irradiation of operation and maintenance personnel due
      to the compacity of the design, it has been kept very low; the ave-
      rage value in 1975 is 25 m.Rem that is to say 0.4% of the maximum
      admissible dose of 5 rem per year.

f) Conclusion

At the end of this rapid survey of the construction and operation experience,
we can draw the following conclusions:
- The sodium cooled, plutonium,uranium oxide fuel fast breeder confirms
  to be a good and performing concept; the pool design operates satisfactorily and shows interesting characteristics in terms of operations, maintenance, effluents release.
- For the future, the methods of computation and construction having been
  checked with experience, can now be applied to the following projects with
  a good probability of success.

We have spent quite a lot of time on the Phenix experience but we think,
it was useful to show how relatively well the different technical solutions
are demonstrated.

VI. - TECHNICAL PROBLEMS NEEDING MORE DEVELOPMENT

We would now like to mention few technical problems which still need more development to be fully solved for a commercial reactor.

1) Fuel

As it was mentioned before, we start to have statistics in Phenix burn ups up to 70 MWdays/kg, and with connected damages on structural material represented by 70 up to 80 displacements/atom. This is higher than planned value of 50 MWdays/kg, however on a commercial plant one would like to increase the burn-up to 100 MWdays/kg or even up to 140 MWdays/kg. There is a good possibility to achieve this higher burn-up and I wish to mention two routes to achieve this goal:

a) use of a different structural material than stainless steel 316 presently used. There is a good prospect for material which swells less than the presently used material.

b) use a heterogeous core chich allows to achieve higher burn-up with the same damage to the material (displacement per atom or neutron total fluence).

I wish also to mention that this higher burn-ups are certainly useful from economical point of view, but there is no magic number and when one plot the cost of fuel cycle versus burn-up, one has the hyperbolic type of courve which shows that the gain obtained from the increase of the average burn-up above 100 MWdays/kg, does not influence drastically the fuel cycle cost.

2) The other area which needs more development is the steam generators. The Phenix steam generators are once through high performance steam generators and their operation was very successful. However, they consist of relatively small modules around 16 MWth. each, and for commercial reactor it will be very likely, necessary to increase the size of the modules.

In Super Phenix, there exists one big steam generator per loop of 750 MWt This large increasing size was decided after many theoretical and experimental works were performed and one may be confident of the good operation of this large steam generators; however, the full demonstration will be obtained only through Super Phenix steam generators operation.

3) Reprocessing of spent fuel

The technical feasibility of aqueous reprocessing was demonstrated by
the operation, for more than two years, of a pilot plant in La Hague.
For Phenix'fuel, one forsees the use of dilution technics and reprocessing in the La Hague plant designed for LWR fuel. However, for the introduction of fully commercial reactors, it will be necessary to build larger
specific plant for reprocessing of fast reactors fuels which still need
some more development.

4) Doubling time

The Phenix and Super Phenix reactors were not designed to achieve short
doubling time. Typically for Super Phenix systems, one expects some 40
years of doubling time. However, the Super Phenix technology with only
a different optimization (diameter of pins, thickness of blanket...) will
permit to obtain some 20-25 years of doubling time. If one wish to decrease this doubling time further, one has many possible routes to do it:

- . develop heterogeneous cores,
- . develop carbide or nitride fuels
- . develop technics allowing short reprocessing and cooling times,
  forexample dry reprocessing.

It is reasonable to assume that any of these routes or combinations of
them will lead to doubling times as low as 10 years or even lower. One
should also note that it is not clear today, how short doubling time will
be necessary after year 2000, this will depend namely of the demand for
nuclear energy. From now to year 2000, the short doubling time of fast
reactors is not essential. Most of the plutonium will come anyway from
thermal reactors, therefore we can see that there is still time to develop technics allowing shorter doubling times.

VII. ECONOMICAL ASPECTS

A very interesting report on Breeder Reactors Economics were published last
year by T.R. STAUFFER-Harvard University-, R.S. PALMER-General Electric Cy,
and H.L. WYCKOFF-Commonwealth Edison Company-.
The results are summarized in Figure 11. It shows that for the period that
one can reasonable consider for commercial introduction of breeders,
(1985-1990), one can accept a capital cost of breeders 73% higher than the

ALLOWABLE CAPITAL COST OF BREEDER OVER LWR
===

Levelized U_3O_8 Value of Plant lifetime supply (75$) ($/lb.)	Projected year of occurrence	Allowable ratio of total plant cost [1,2] (any year and any rate of inflation) (3) (Breeder/LWR)
20	1965 to 1970	1.29
60	1985 to 1990	1.73
100	2000 to 2005	2.18

(1) The reference case cost for an LWR is $ 400/kW(e) in 1975 dollars ($600/kW(e) for a plant installed in 1982)

(2) There ratios are for a plant design capacity factor of 70%. The comparable numbers for an 80% capacity factor are: U_3O_8 - $20/lb.-1.34
U_3O_8 - $60/lb.-1.84
U_3O_8 -$100/lb.-2.09
factor of 65%: U_3O_8 - $20/lb.-1.27 - U_3O_8 - $60/lb.-1.69 - U_3O_8-$100/lb. 2.09

(3) This table is valid for any year when the value of U_3O_8 is in 75$'s. For years other than 1975, it may prove more convenient to convert the U_3O_8 values to current dollars.

FIGURE 11 : Allowable Capital Cost of Breeder Over LWR

636

capital cost of LWR and still be competitive with the last type of reactors.
Our own evaluation being perhaps a little more pessimistic about fabrication
and reprocessing cost for fuel of fast reactors, will show that for the same
period of time, one should achieve not more than the factor 1.5 in LMFBR
capital cost compared to LWR capital cost, if one wish to maintain the
competitivity.

We do not have for the moment, an exact figure of capital cost for Super
Phenix, which is a large prototype fast reactor in the most advanced planning
stage (construction should start by the end of 1976). However, my personal
estimates will be that the total construction cost of this plant, excluding
interest during construction and first core, should not exceed 800-850 $/KWU).
One expects for the commercial reactor some reduction of this cost due to
the: . building of two or more plants on the same site,
     . improving the overall design (more or less same size of components
       and higher rating)
     . effects of duplication.

It is therefore not unreasonable to expect that the capital cost of
construction of the 5th Super Phenix type plant will be reduced by some 30%
and therefore amount to some 600 $/KW (we are using here constant (1976) $.).
This has to be compared to some 400 $/KW for the LWR. Therefore, even if the
economical competitivity is not fully demonstrated today, there is a good
prospect to achieve it during the end of the next decades.

II- PUBLIC ACCEPTANCE

There is clearly a problem of public acceptance of nuclear energy and more
specifically of fast breeder reactors.
Some of the oponents make a confusion between fast breeder reactors and
an atomic-bomb. There is also a trend to identify with LMFBR the dangers
connected with the use of large quantities of plutonium and often the concern
of the possible diversion of highly enriched fuel like plutonium. This is
certainly a difficulty and may be even an increasing difficulty for peoples
who propose the commercial development of LMFBR'.
The situation varies however from country to country, forexample the positions
in US, in Western European Countries or in Eastern European Countries are not
the same.

When one try to analyse objectively the safety of breeder reactors and to compare it with the safety of other types of thermal reactors, one can find positive and negative aspects in each type of reactors, but the almost unanimous feeling of peoples working on these programs, is that both types could be built to high enough standards of safety and in this case will not present an undue hazard to the public.

As regarding the diversion of plutonium, we have to stress here that plutonium, up to year 2000, will mostly come from thermal neutron reactors and not from breeders. Therefore, if the breeders are not introduced, we will have a very difficult problem of storage of this material and perhaps a problem of prevention of diversion depending if we use or not the plutonium for recycling in thermal reactors. With the breeder reactors introduction, we can easily adjust the breeding ratio downwards and always have only the quantity of plutonium needed for running the reactors. This is probably the best way to solve the storage problem.

In addition, the fast breeder reactors have the unique possibility of using highly active and poisoned fuel. One can leave in the fuel up to some 20% of fission products. This is of course connected with a commercial development of new technics for fabrication and reprocessing fuel; technics which were experimented on EBR2 reactor. The use of this very highly active fuel is probably one of the best way to prevent diversion. In addition, the EBR2 type technics will probably be well suited for small or medium size fuel reprocessing and fabrication plants, which may therefore be situated on the same site that power plants producing some 5 to 10.000 MWe. In this way, one will minimize the transportation of highly active fuel and minimize the risks associated with the transportation.

The LMFBR plants have also some other favorable caracteristics concerning the environmental impact of these plants. I will mention here:
- much higher thermal efficiency than LWR and consequently less problems with heat rejection;
- the possibility of having practically no active effluents during operation (see Phenix operation);
- there is a potential possibility to burn some waste -long life isotopes- in fast reactors.

One can therefore conclude that there is presently a problem of public
acceptance of fast breeder reactors but if one makes the necessary efforts
of explanation of the effective reality, one should be able to convince
the public that these reactors introduce no more problems, from environmental point of view (environment being taken in its very broad sense), than
other types of reactors and very likely no more problems than other conventional power plants. One has however to note, that the public acceptance
may not be based only on objective comparison of different means of
energy production. Some of the oponents are trying to attack the essential
and less established type of fission reactors -breeders-, but ultimately
they are looking for a drastic limitation of fission energy and perhaps
even of the production of any type of energy.

CONCLUSION

We can say that the efforts devoted to the development of HTR are today not
very large but still substantial. This situation is likely justified by the
fact that HTR are a very interesting concept which has important possibilities
and potentials, but the needs for them are not really urgent.

As regarding LMFBR, the effort devoted to their development is large and it
appears that there is an urgent need for this type of reactors. The technical
and economical demonstrations of breeders are well on their ways and will likely
be completed in ten years from now.

The major problem for the moment, at least in some countries, is the public
acceptance and the more or less connected field or regulation and licensing
rules. But this problem is perhaps more a problem of fission energy as a whole
than a specific problem of breeder reactors.

LIST OF PARTICIPANTS

Arcipiani, B.	CNEN, Casaccia (Italy)
Bagatin, M.	University of Padova, Padova (Italy)
Bohdansky, J.	CCR, Ispra (Italy)
Bonnevier, B.	RIT, Stockholm (Sweden)
Booth, J.	JET, Abingdon (U.K.)
Borrass, K.	IPP, Garching (Germany)
Brandt, B.	FOM, Jutphaas (The Netherlands)
Coccorese, V.	JET, Abingdon (U.K.)
Dustmann, C.-H.	I.E.K.P., Karlsruhe (Germany)
Farfaletti-Casali, F.	CCR, Ispra (Italy)
Fenici, P.	CCR, Ispra (Italy)
Grieger, G.	CEC, Brussels (Belgium)
Grolli, M.	CNEN, Frascati (Italy)
Gruber, J.	Hahn-Meitner-Institut, Berlin (Germany)
Inoue, K.	Hitachi, Ozenji (Japan)
Jassby, D.L.	PPL, Princeton (U.S.A.)
Klippel, H.Th.	Reactor Centrum, Petten (The Netherlands)
Krauth, H.	I.E.K.P., Karlsruhe (Germany)
Lawson, J.D.	Rutherford Laboratory, Didcot (U.K.)
Manley, O.	USERDA, Washington (U.S.A.)
Norem, J.H.	ANL, Argonne (U.S.A.)
Pedretti, E.	CNEN, Casaccia (Italy)
Raeder, J.	IPP, Garching (Germany)
Reiter, F.	CCR, Ispra (Italy)
Rey, G.	CEA, Grenoble (France)
Richter, F.	KFA, Jülich (Germany)
Rocco, P.	CCR, Ispra (Italy)
Rutherford, P.H.	PPL, Princeton (U.S.A.)
Sand, F.	CNEN, Frascati (Italy)
Scherzer, B.M.U.	IPP, Garching (Germany)
Snykers, M.	SCK/CEN, Mol (Belgium)
Sørensen, H.	DAEC, Risø (Denmark)
Spadoni, M.	CNEN, Frascati (Italy)
Speth, E.	IPP, Garching (Germany)
Tone, T.	JAERI, Tokai (Japan)
Toschi, R.	CNEN, Frascati (Italy)
Wilhelm, R.	IPP, Garching (Germany)
Wolf, G.	KFA, Jülich (Germany)
Zankl, G.	IPP, Garching (Germany)